INFORMATION AGE

Second Edition

Roman Kuc

Professor of Electrical Engineering
School of Engineering & Applied Science
Yale University

CENGAGE
Learning™

Australia • Brazil • Japan • Korea • Mexico • Singapore • Spain • United Kingdom • United States

**The Digital Information Age,
Second Edition**

Roman Kuc

Publisher: Timothy Anderson

Senior Developmental Editor: Hilda Gowans

Team Assistant: Sam Roth

Senior Content Project Manager:
Jennifer Ziegler

Production Director: Sharon Smith

Team Assistant: Ashley Kaupert

Intellectual Property Analyst:
Christine Myaskovsky

Intellectual Property Project Manager:
Amber Hosea

Text and Image Researcher:
Kristiina Paul

Manufacturing Planner:
Doug Wilke

Copyeditor: Shelly Gerger Knechtl

Proofreader: Erin Buttner

Indexer: Rose Kernan

Compositor: MPS Limited

Senior Art Director: Michelle Kunkler

Internal Designer: MPS Limited

Cover Designer: Stratton Design

Cover Image: © agsandrew/Shutterstock.com

Library of Congress Control Number: 2014933745

ISBN-13: 978-1-305-07771-3

ISBN-10: 1-305-07771-7

Cengage Learning
200 First Stamford Place, Suite 400
Stamford, CT 06902
USA

Cengage Learning is a leading provider of customized learning solutions
with office locations around the globe, including Singapore, the United
Kingdom, Australia, Mexico, Brazil, and Japan. Locate your local office at:
international.cengage.com/region.

Cengage Learning products are represented in Canada by
Nelson Education Ltd.

For your course and learning solutions, visit
www.cengage.com/engineering.

Purchase any of our products at your local college store or at our
preferred online store **www.cengagebrain.com.**

Unless otherwise noted, all items © Cengage Learning.

Unless otherwise noted, all photos © Roman Kuc.

Matlab is a registered trademark of The MathWorks,
3 Apple Hill Road, Natick, MA.

For my students, ever eager to learn,
For wife Robin, a true companion, and
For Our Father, from Whom all blessings flow.

Photo Courtesey of Michael Marsland/Yale University

Roman Kuc ("*Koots*") received the BSEE from the Illinois Institute of Technology, Chicago, IL, and the PhD degree in Electrical Engineering from Columbia University, New York, NY. He started his engineering career as a Member of Technical Staff at Bell Laboratories where he designed electromechanical, analog, and digital systems, and developed efficient digital speech coding techniques. After completing his PhD he was postdoctoral research associate and adjunct professor in the Department of Electrical Engineering at Columbia University and the Radiology Department of St. Luke's Hospital, where he applied digital signal processing techniques to diagnostic ultrasound signals. He then joined the Electrical Engineering faculty at Yale University, where he enjoys teaching courses in system theory, digital signal processing, microcontroller programming, and capstone design projects. He also enjoys describing engineering problem solving to non-science majors, which attracted 780 Yale students during one memorable semester. As Director of the Intelligent Sensors Laboratory, he explores digital signal processing for extracting information from sensor data and implementing brain-based sensor systems. Motivated by biological sensing systems, such as echolocating bats, he has implemented biomimetic and neuromorphic sensing systems for robots.

Prof. Kuc has authored more than 150 technical papers and three books, *Introduction to Digital Signal Processing, Electrical Engineering in Context,* and *The Digital Information Age,* now in its second edition. He authored the biomimetic sonar chapter in *Echolocation in Bats and Dolphins,* and co-authored the Sonar Sensing chapter with Lindsay Kleeman in the *Springer Handbook of Robotics.*

Prof. Kuc is a member of the Connecticut Academy of Science and Engineering, Fellow of the Shevchenko Scientific Society, Senior Member of the IEEE, and past chairman of the Instrumentation Section of the New York Academy of Sciences. His honors include Yale Engineering's Inaugural Sheffield Distinguished Teaching Award, Yale's Order of the Golden Bulldog, and an honorary doctorate from the Glushkov Institute of Cybernetics in Ukraine.

CONTENTS

3 Combinational Logic Circuits   52

4 Sequential Logic Circuits   78

5 Converting Between Analog and Digital Signals — 92

6 Modeling Random Data and Noise — 114

We live in a digital information age and encounter information constantly in its various forms. We become sources of digital information whenever we use a smart device or a credit card. We become consumers of information when we watch a video on a Web page, receive a tweet or listen to digital audio. Hence, it is natural to inquire how this information is generated, encoded, stored, transmitted, and used. It is also necessary that a citizen of this information-pervasive society be aware of the basics and the potential uses of this technology. These concerns led to the development of this text.

The big picture is treated in this text, using broad strokes to form the relationships among these aspects of digital information. The creation of information is considered from the simple switch activations, which form the basic informational bits, to sensors that are within our smartphones and digital scanning procedures. The robustness of digital data representation is described in error detection and correction techniques. The conversion from an analog format to a digital format is described, along with the possible pitfalls. Data are expressed in binary digits, or bits, which can then be reliably stored, transmitted, and manipulated. Digital data are processed initially with basic logic gates, which then lead intuitively to logic circuits that are contained within smart devices. Information is modeled as an experiment whose outcome is unpredictable or random. The quantity of information can be measured as a numerical quantity called entropy. Destruction of information occurs when data are corrupted by random errors, whose magnitudes are at first small, and hence inconsequential, but can grow, eventually obliterating the original data. Information can be coded for error detection and correction techniques for robust storage and encrypted for security. Data are transmitted over a variety of channels, from the simplest one-way data channel illustrated with credit-card swipes, to interactive data exchange over the computer network comprising the Internet.

This book describes digital information systems, presenting them in a context that provides intuition and understanding about the particular path the information revolution has taken. We then gain an appreciation of where we are and an educated guess about where we are going. It is important to have an informed electorate that makes wise decisions in an increasingly technological society.

The goal of this book is to describe how common-place information systems work and why they work that way by illustrating how clever engineering solutions are incorporated to solve technical problems, and the tradeoffs that are encountered in making a system work. The systems considered are those encountered on a daily basis and implemented in smartphones and computers. In addition to teaching students about electrical engineering, this book invites the student to be an engineer for a term: To think quantitatively and to design useful systems.

The ideas and principles employed in digital information systems are described with minimal reliance on prerequisites in mathematics and science. All that is required is some experience in algebra. Unlike a traditional science text, which starts with theory and then illustrates it with applications, this book starts with the application and then presents the physical theory and mathematical analysis required to understand the application. The applications illustrate how ideas can be expressed mathematically and how mathematical models are used in practice.

ADDITIONS TO THE SECOND EDITION

The second edition of the *Digital Information Age* updates material in the first edition to include smart devices and the wireless networks that service them, and digital signal processing basics that explain practical data communication. The topics have been reorganized to introduce material that is covered in conventional electrical engineering programs. The new text offers a flexible text that is suitable for a variety of introductory courses for both science and non-science students.

One major addition to the second edition is the inclusion of Excel projects that allow students to do more interesting problems and probe the issues at a deeper level. A chapter that includes a primer on Excel and examples that illustrate the ideas presented in the text has been added. The use of Excel has been well-received by students because it provides an additional lasting value to the course by giving student the ability to perform data analysis and system simulations. Abstract ideas, such as those encountered in probability and digital signal processing are illustrated with numerous Excel examples, which are contained in a separate chapter that has sections that relate to their respective chapter material. Programming skills are introduced through Visual Basic for Applications (VBA) to compose simple macros. VBA extends the functionality of Excel to implement counters and to compose worksheets that simulate the operation of a complete digital communication system. The provided Excel examples can assist students to find errors in their worksheets and VBA code if their results do not match those in the example. Projects can be tailored to the skill level of particular students, from beginners who perform slight modifications and extensions of the examples, to advanced students who can investigate novel problems and simulate complete communication systems. The resulting computational capability that students achieve is helpful for further and deeper study in follow-on courses in science, engineering, and statistics.

The second edition allows the instructor to structure a course that meets particular curricular needs. For example, the initial sections of most of the chapters can serve a survey course that overviews digital devices and coding data for compression and error detection. The Excel examples illustrate the theory with simulations and the projects provide interesting extensions. The Excel projects also offer the opportunity to form courses that probe selected topics in more depth. Projects can be easily extended to allow students to attempt creative solutions and examine open-ended problems. For example, pseudo-random number generators illustrate use of probability for simulating random data and adding noise to transmitted waveforms. Digital signal processing describes a simple matched filter to process data signals corrupted with noise. Its application to transmitting data challenges students to design signals for optimal performance in the presence of noise and for serving multiple simultaneous users. The most challenging projects simulate a complete communication system to estimate its probability of error and monitor the performance as the channel degrades.

For an introductory course in an intensive electrical engineering program, the similar text *Electrical Engineering in Context* contains additional chapters covering circuits and electronics and uses Matlab® for the projects.

In addition to the print version, this textbook will also be available online through MindTap, a personalized learning program. Students who purchase the MindTap version will have access to the book's MindTap Reader and will be able to complete homework and assessment material online, through their desktop, laptop, or iPad. If your class is using a Learning Management System (such as Blackboard, Moodle, or Angel) for tracking course content, assignments, and grading, you can seamlessly access the MindTap suite of content and assessments for this course.

In MindTap, instructors can:

- Personalize the Learning Path to match the course syllabus by rearranging content, hiding sections, or appending original material to the textbook content

- Connect a Learning Management System portal to the online course and Reader
- Customize online assessments and assignments
- Track student progress and comprehension with the Progress app
- Promote student engagement through interactivity and exercises

Additionally, students can listen to the text through ReadSpeaker, take notes and highlight content for easy reference, and check their understanding of the material.

I gratefully acknowledge the assistance and encouragement of Editor Hilda Gowans of Cengage Learning, whose guidance and advice eased the author's burden. The following reviewers generously contributed their pedagogical experience to improve early versions of the manuscript. These include Pelin Aksoy, George Mason University, Steve P. Chadwick, Embry-Riddle Aeronautical University, Michael Chelian, California State University, Long Beach; Simon Y. Foo and Bruce A Harvey, FAMU-FSU, Roland Priemer, University of Illinois at Chicago, Michael Reed, University of Virginia, Christopher Schmitz, University of Illinois, and Andrew Tubesing, University of St Thomas.

Over the thirty years of teaching at Yale, I have had conversations with numerous students at various stages of their academic and professional careers: from high school students taking a college summer course, college first-years thinking about majoring in Electrical Engineering, seniors completing an Electrical Engineering program, and to recent graduates who were employed as electrical engineers. This book attempts to address the concerns the former students had: *What will I learn in an electrical engineering program?* and include the advice that the latter students offered: *What do electrical engineers do?* The material has been classroom tested with Yale and non-Yale students, ranging from bright high school seniors and freshmen intending to major in engineering to non-science majors interested in technology.

I hope that the reader has as much fun learning about the *Digital Information Age* as I have in teaching about it.

ROMAN KUC
New Haven, CT

THE DIGITAL
INFORMATION AGE

CHAPTER

1

INTRODUCTION

LEARNING OBJECTIVES

After completing this chapter, the reader should be able to:

- Understand the design process in the electrical engineering profession.
- Understand the importance of computers in smart systems.
- Appreciate the types of systems and signals that carry information.
- Know the magnitudes of relevant time and frequency scales in information systems.

1.1 INTRODUCTION

Electrical engineering is a broad profession that encompasses a wide variety of systems that deal with the movement of electrons to communicate information, control operation, and provide power. The scale of hardware varies from nanodevices to transworld power and communication networks. The associated problem-solving activity is firmly based in science and mathematics, making an electrical engineering education a desirable and sought-after preparation for many careers.

This chapter provides a general introduction to electrical engineering, and the topics discussed in this book in the following manner.

Problem descriptions—Electrical engineers formulate problems in terms of block diagrams that indicate the main component of a system and apply their tools to design a solution.

Analog and digital signals—Electrical systems communicate through waveforms that typically start and end in analog waveforms, but more often, much of the processing is performed by computers on digital values. Digital signals form the basis for communications because of their ability to form perfect copies.

Future trends—Digital systems follow Moore's law to indicate improvements that double every 18 months, which bodes a bright future for digital devices.

1.2 ELECTRICAL ENGINEERING FOR THE DIGITAL AGE

This is the age of computers, smart systems, and the communication networks that link them. Computers are becoming cheaper and more powerful, and smart devices that use computers easily connect to networks and each other. Electrical engineers play a major role is designing the devices, components, systems, and processes that result in the operations that meet specified requirements.

1.2.1 Block Diagrams

Electrical engineers describe system operation using *block diagrams*. A block diagram uses arrows to indicate the direction of signal flow and blocks to represent specific operations. It is a convenient way to summarize the role of components in a system, to show information flow, and to present the *big picture*.

In this book, *information is the quantity of data that is required to complete a task*. Information in its most basic form is transmitted by some form of energy.

- We see things because light energy is scattered from objects.

- We hear sounds because of acoustic energy caused by pressure variations produced by an instrument or generated by a loud speaker.

- We feel mechanical energy in the vibration caused by a small motor in our cell phone to indicate an incoming call.

- We exert mechanical energy to press a switch to answer a call.

- Our cell phone communicates wirelessly using antennas that transmit and receive electromagnetic energy.

- A thermometer in our computer senses thermal energy, and when it is too high, the operating frequency or voltage is reduced to prevent thermal damage.

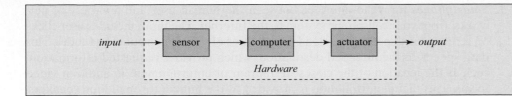

Figure 1.1

Block diagram that summarizes the operation of many smart systems.

Figure 1.1 summarizes the systems described in this book using a simple block diagram that illustrates informational flow through the basic components. Smart systems at a variety of sizes (from smartphones, to robots, to communication networks) have the following common features.

Input: Information is embodied in some form of energy that occurs naturally, such as light reflected from objects, or is produced by humans in the sounds generated by speaking or button presses. This is the human communicating with the smart system.

Hardware: Sensors convert the energy into electrical signals that are processed by computers to generate electrical signals that drive actuators.

Output: Actuators convert electrical signals into perceptual energy, such as the light generated by displays and mechanical vibrations produced by motors. This is the smart system communicating with the human.

Cell phones and Smartphones   EXAMPLE 1.1

Cell phones and smartphones are two of the most common digital devices in society. They are differentiated as follows.

Cell phones are devices that are designed primarily for wireless voice communication over a cellular system defined by antennas and transmission rules. A cell phone digitizes speech signals, encodes them for efficient transmission, and stores user data, such as the phone numbers of contacts, to simplify operations required to complete a connection.

Smartphones are cell phones that also connect to the Internet and are implemented with enhanced computers and displays to provide video features.

Both cell-phone and smartphone operations, circuits, and capabilities are described in this book.

To illustrate the application of the techniques discussed in this book, Figure 1.2 shows three systems that represent the current trends in electrical engineering.

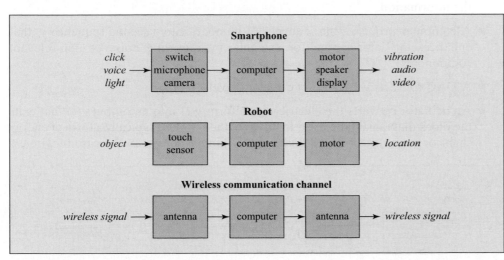

Figure 1.2

Block diagrams of three systems considered in this book.

1. *Smartphones* form the interface between the human user and a variety of networks, from your cell-phone system to the Internet. The input includes user clicks on screen icons, speech commands, or pictures of two-dimensional bar codes. This data speeds to the intended destination, which provides requested information, such as the location of the closest restaurant or entertainment in audio or video forms. This digital information is provided to the human through loud speakers, color displays, or vibrating motors.

2. *Robots* perceive their environments by acquiring sensor information (for example, a touch sensor that detects the presence of an object). This information is processed, and aided with data stored in memory—actuators are energized to produce manipulation or movement. In complex environments that produce ambiguous sensor reading, robots accept instructions through teleoperation (that is, aided by sensory information transmitted back to a base station to accomplish a desired task). Robots on Mars present particular challenges to teleoperation because their signals are very weak and the travel time delay is long.

3. *Wireless communication channels* acquire and transmit signals through antennas. Detected data are processed, stored, and encrypted to communicate data from smart device to smart device or from computers located across the globe. Various applications present different challenges, from fast download times to real-time viewing without annoying pauses.

1.2.2 Describing System Operation

How do these devices work? What are their limitations? Such questions are in the domain of electrical engineering, which is a profession of analysis, design, and problem solving. Engineers use mathematical tools and scientific principles to design devices that improve life and solve problems that occur in society. This book focuses on *electrical engineering* (EE) and shows how engineers control the movement of electrons to implement the amazing devices we encounter in our daily lives.

A typical system consists of the following parts, as illustrated in Figure 1.3, which shows a block diagram of cell-phone operation for transmitting digitized speech over a wireless channel.

- An input, typically some form of energy $e(t)$ that varies over time t.

- The sensor or transducer converts $e(t)$ into an electrical analog waveform $x(t)$.

- An ADC converts $x(t)$ into a sequence of numbers x_i, which is indexed with i. Typically, we are interested in the data x_i for $i = 0, 1, 2, \ldots, n_x - 1$ that encodes the information.

- A computer processes the sequence of numbers into a second sequence y_i that enhances the information (for example, by removing noise or non-relevant components).

- A DAC converts y_i back into an analog waveform $y(t)$.

- An actuator converts the electrical waveform $y(t)$ into an output $e(t)$ that exits the block diagram on the right. In this text, the output is typically a form of energy (light or sound) that can be perceived by human senses. In the case of wireless

Figure 1.3

Block diagram of cell-phone operation for transmitting digitized speech over a wireless channel.

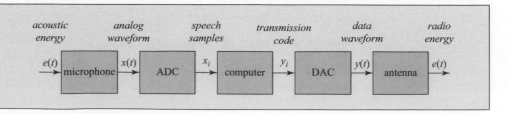

Figure 1.4

Smartphones have cameras that capture images of two-dimensional bar codes.

systems, the energy is in the form of radio waves or electromagnetic radiation that communicates with other systems that are described with their own block diagrams.

Smartphone camera   EXAMPLE 1.2

Figure 1.4 shows a smartphone camera taking a picture of a two-dimensional bar code. The camera generates an image array of pixel values. A small computer within the smartphone decodes the bar-code image into digital data and transmits the data wirelessly to the Internet to obtain additional information. Similar functions occur in intelligent robots.

The following technical questions occur in the design of a system that interprets the bar code.

- The camera image contains the entire symbol, but the symbol does not occupy the entire image. What processing steps are required for the cell-phone computer to decode the image?
- How will your camera know that it interpreted the symbol correctly and there is no error?
- The distance between your camera and the symbol increases as you move away from the symbol. Under what conditions will your camera no longer be able to decode the symbol correctly?

Teleoperation of a robot on Mars   EXAMPLE 1.3

Mobile robots are equipped with sensors to navigate around obstacles. Figure 1.5 shows a base station on Earth transmitting data to and from a camera-wielding mobile robot on Mars through a satellite relay orbiting Mars. To prevent accidents on Mars, teleoperation by an operator viewing camera images overrides autonomous operation when the situation is questionable. The signals are transmitted between antennas with a strength that is limited by the available power. The vast distance between Earth and Mars causes signals from Mars to be very weak, and signal processing is required to extract data from noise. A teleoperated robot system raises technical questions, including the following.

- What are the data-carrying capacities of the various channels?
- What transmitted signals produce the best images in a reasonable amount of time?
- With the travel delay between Earth and Mars and available power, how should the processing tasks be divided between the robot and the base station on Earth?

Figure 1.5

A base station on Earth transmits data to and from a camera-wielding mobile robot on Mars through a satellite relay orbiting Mars. The size of the arrows indicates the capacity of the various data channels to display an image from Mars.

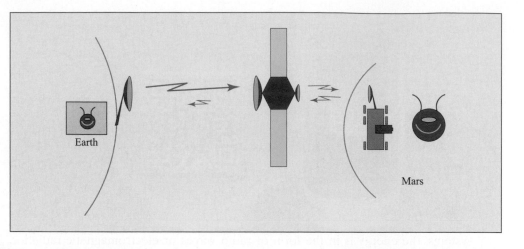

Earth

Mars

EXAMPLE 1.4

Digital cell phone system

Figure 1.6 shows the transmission activity in a cell-phone system, in which many users transmit and receive simultaneously. A computer sorts the routing information and data from different sources and directs them to their respective destinations.

Technical questions that occur in the design of a digital cell-phone system include the following.

- The telephone system operates in a particular audio frequency range, typically up to 3,500 Hz. The recording microphone produces sound waveforms having frequency components up to 10,000 Hz. A low-pass filter typically connects to the microphone output to condition the signal for digitization. The filter connects to an analog-to-digital converter (ADC) that samples the audio waveform and uses an 8-bit quantizer to store the samples into a digital memory. What values of sampling period (T_s) and quantizer resolution affect the audio quality?

- To make the system secure, the data is encrypted using a random number generator. How much encryption provides sufficient security?

- Your data is placed on a communication channel shared by other users. How can data-transmission signal design affect the number of users within a particular cell?

- How does the spacing of antennas affect system performance?

- On the receiving end, digital data are decrypted and are applied to a digital-to-analog converter (DAC) that reconstructs the original speech waveform. How does the DAC design affect the audio quality?

Figure 1.6

In a cell-phone system many users transmit and receive digital data simultaneously. A computer sorts the routing information and directs digital speech data from sources to their respective destinations.

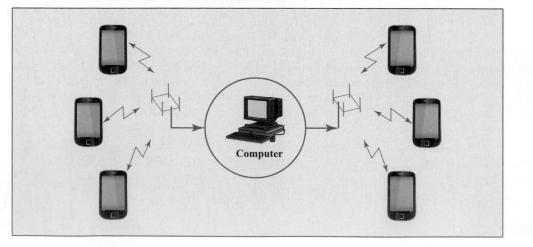

Computer

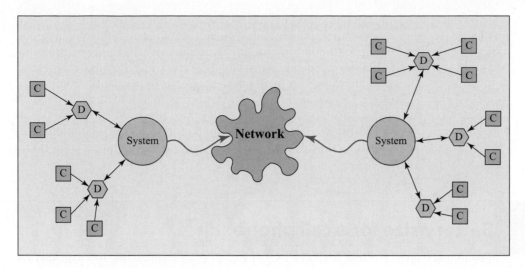

Figure 1.7

Electrical engineers design at various levels: components (C), devices (D), systems, and networks.

1.2.3 What Electrical Engineers Do

Current smart devices contain a mix of analog electronics and digital system components. Electrical engineers, although familiar with the basic principles, typically design at one of four levels of size and complexity: components, devices, systems, and networks, as illustrated in Figure 1.7. We examine each level separately here.

Components

Electrical components are the basic electronic elements that exploit physical principles to produce the electrical signals at the *front end* of devices. Components in a smartphone include the camera, microphone, and processor. A robot has sensor components to indicate where it is and actuator components to move it to where it should be.

Even components can be subdivided into more fundamental parts. Physical interactions at the atomic level convert electrical, optical, thermal, or mechanical energy into electron flow that is captured and amplified by other electronic devices, such as transistors. Such interactions occur at the micro-scale with dimensions measured in micrometers (μm) or nano-scale, measured in nanometers (nm). At these scales, engineering scientists or applied physicists devise clever techniques to detect individual ions, electrons, or light photons or to construct novel materials that may contain carbon nanotubes to have larger surface areas for a given volume, which is important for high-capacity battery fabrication. Components are designed to be manufactured inexpensively to operate under a specified temperature and other environmental range and to require a simple power supply, such as a single battery.

Smartphone camera   EXAMPLE 1.5

The camera in a smartphone uses photo-detector components to convert optical energy into electrical signals and amplifiers to generate the signals that can be stored or transmitted. Electrical engineers do not need to design a camera each time they need one in a system; rather, they simply specify an existing camera component and design its integration into a smartphone. The design challenge is to select the best camera, which usually involves a trade-off of cost, size, weight, power consumption, availability, reliability, and resolution.

The concept for a new component needs to be viable (that is, feasible and practical) and based on scientific and mathematical principles. While there are many good ideas, like time travel and perpetual motion machines, many are simply not feasible because they violate well-known scientific principles. Conversely, if a physical effect is known to occur and is stated in a formula, an engineer can implement a component to make it happen. Other ideas (like long-lasting batteries), while being feasible and important, may not be practical, because they result in prohibitively expensive or unwieldy components. *Back-of-the-envelope* (simple and approximate) calculations provide ball-park values that quickly determine their practicality for commercial portable devices.

EXAMPLE 1.6   Battery size for a cell phone

This example illustrates how to "think like an engineer" by enumerating the steps that are part of the design process.

An electrical engineer determines the practicality of a long-lasting cell-phone battery using the following steps.

STEP 1. Required power. A typical cell phone is meant to fit in the hand and uses one Watt (1 W) of power while it is operating.

STEP 2. Available power. A lithium-ion battery (LIB) has an energy density of 200 Watt-hour (W-hr) per kilogram of mass or 200 W-hr/kg.

STEP 3. Back-of-the-envelope calculation. An LIB weighing 0.1 kg (1/4 pound) provides 20 W-hr of energy. Such a LIB powers the 1 W cell phone for 20 hours, is easy to hold and carry, but requires a nightly recharge.

STEP 4. Conclusion. A battery weighing 0.5 kg (1.1 lb) provides 100 W-hr of energy and will last for about four days, but would make the cell phone five times larger and heavier. Such a cell phone would not fit comfortably in the hand. Therefore, a long-lasting battery using LIB technology is not practical.

Factoid: Scientific and mathematical principles provide the fundamentals that govern the basic operation and form the limitations of component performance.

Devices

Electrical devices are products that are designed by interconnecting components and have a particular purpose. A smartphone is a stand-alone device that performs all of the functions needed to communicate digital data. A device is typically given a model number that indicates the particular list of components and features that are included.

While engineers typically design individual components, an engineering *manager* coordinates their designs in order to meet a set of constraints. The design of a device typically balances or *trades-off* different criteria, such as cost, size, weight, speed, reliability, and accuracy, as well as marketing ideas, such as new features and ease of use. New models are introduced when the features they offer are significantly better—usually by a factor of two or more—than the current or competitor's model.

Mobile robot   EXAMPLE   1.7

A mobile robot is a device that has sensor components, including a GPS for global localization, a detection sensor for obstacle avoidance, a computer that analyzes the data to determine a path that leads to the destination, and motors that propel the robot in the desired direction.

Systems

An electrical system is an assembly of devices that allows them to operate through communication links. These links may be physically wired connections or wireless connections using radio waves. The devices connecting to a system may have different capabilities, such as the variety of smartphones that can connect to a specific system (AT&T or Verizon). A system contains an organizational management that sets the rules for operation of interconnecting devices. Even systems have constraints, such as the frequency spectrum (within which it can transmit) and environmental factors (size and location of transmission towers). System control becomes complicated, because devices can connect or disconnect at random times and locations.

Challenges for a cellular telephone system   EXAMPLE   1.8

Before 1982, the telephone network in the U.S. was a Bell System monopoly. With the support of Bell Labs, the telephone quality and reliability were very good, because all of the devices connected to the network were manufactured by the Bell System. Telephone connections were made more reliable by implementing digital switching technology to replace electromechanical switches. Enlarging an infrastructure mainly involved laying more wires and radio antennas to address the increased demand for telephone service. Planning for future expansion was easier, because there were no competitors and the incoming revenue was reliable.

In the current cellular telephone market, there are several major competitors that vie for customers by providing adequate service and offering differentiating features for a reasonable fee. The smartphones are no longer made by a cell phone company but by high-tech companies. This leads to negotiations about what features can be made available in the smartphone computer and yet be supported by the transmission capabilities of the cell phone company network. To provide the features that consumers demand, generations of improved communication networks have evolved and are currently in the fourth generation (4G). These new generations are made possible through ever-improving electronic and nanotechnology devices that operate at faster speeds, lower power, and increased complexity.

Networks

In today's digital information age, there are many systems that communicate with each other over the most complex network that has ever been implemented: the Internet. Electrical engineers face many challenges in trying to make a network function quickly, robustly, and securely under a wide variety of traffic conditions. Networks need to accommodate ever-increasing data communication needs, such as increased data throughput, or providing large quantities of data and minimum delay so that the video on your smartphone does not pause. Also, these need to be accomplished without errors and in a secure manner to guarantee privacy.

What makes a network particularly complex is that each system operates at its own pace and uses its own data format. Hence, this makes it difficult (even impossible) for a network to have a central control structure. At best, the central organization managing the network specifies guidelines in the form of a network architecture that needs to be followed to allow systems to communicate.

EXAMPLE **1.9**

Internet of Things

Not only smartphone users communicate over the Internet. Researchers are planning for the *Internet of Things* (IoT) that involves putting wirelessly connected computers into things, such as cars, traffic monitoring stations, household appliances, etc. In the IoT, any device that interacts with humans or affects their activity will have a presence on the Internet.

The most ambitious forecast talks about "6A connectivity:" Anything, Anytime, Anyone, Any place, Any service, and Any network. Clearly, advances in computer cost, power, connectivity, and flexibility are needed to realize this vision.

1.3 ANALOG AND DIGITAL SIGNALS

Signals come in two basic flavors: *analog* and *digital*. Figure 1.8 shows these signals viewed on an *oscilloscope*, which is an instrument that displays waveforms.

- *Analog signals* are waveforms produced by sensors and typically vary smoothly over time, such as sinusoidal waveforms or signals that encode music. Analog audio signals, such as speech or music, vary on the millisecond time scale. The prototypical analog waveforms are sinusoids. Sinusoidal waveforms at different frequencies are used for analyzing the frequency behavior of systems and for the spectral decomposition of more complex signals.

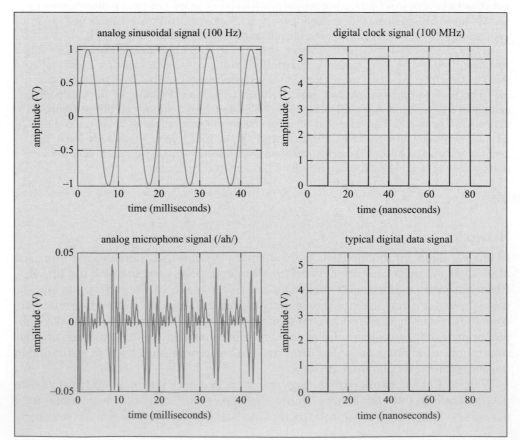

Figure 1.8

Analog and digital signals. Analog signals typically vary smoothly over time, while digital signals change abruptly between specific levels.

■ *Digital signals* occur in logic circuits and computers and are produced by *detectors* that signal the occurrence of an event. Digital signals typically take on only discrete values (usually only two), such as 0 and 5 V (Volts) that encode binary logic levels (0/1) in a computer. The constraint in digital waveforms allows computers to quickly, reliably, and flexibly process digital data.

1.3.1 Converting Between Analog and Digital Signals

The device that transforms an analog waveform, such as speech, into a digital form consisting of a sequence of numbers is called an *analog-to-digital converter* (*ADC*). Once in digital form, the digital samples can be processed, stored, or transmitted using digital technology. In the digital audio case, the acoustic signal is converted to an electrical analog waveform by a microphone. This analog signal is applied to an ADC that produces a sequence of numbers in binary (0/1) form. This data representing audio information is stored in a computer memory on your mp3 player or smartphone. These digital data are converted back into an electrical waveform using a digital-to-analog converter (DAC). This electrical signal is fed into an amplifier that drives a speaker that reproduces the original acoustic signal.

Transforming an analog waveform into a sequence of numbers must be done carefully to avoid information loss and be undistinguishable from the original. Specifically, two questions must be answered to perform analog-to-digital conversion correctly.

■ How often do we need to sample the analog signal in order to retain all the information that it contains?

■ How accurately must the resulting digital samples be represented in order to form a satisfactory reproduction of the original analog signal?

To answer the first question, we must describe a signal in terms of its frequency content (how many different frequency components are present in the analog signal). The important parameter is the highest frequency component present. To answer the second question, we investigate the process called *quantization*, which approximates a numerical value using a specified number of bits that provides a fixed precision. Transforming analog signals to digital representation comes at a price: While an analog signal can take on any value (within limits), its digital representation needs to be expressed in a finite number of bits. Figure 1.9 shows that analog-to-digital conversion results in a finite-precision approximation called *quantization*. The resulting error is viewed as

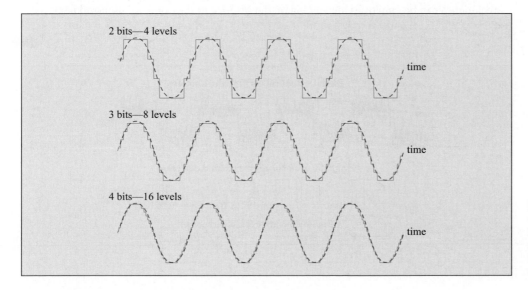

Figure 1.9

Representing an analog waveform using a finite number of bits results in a finite-precision approximation called quantization.

noise that has been added to the analog signal. A large number of bits, typically eight or more, is used to make this noise tolerable so that the performance of a particular system is not degraded by becoming digital.

1.3.2 Advantages of Digital Signals

Modern communication systems transmit data in digital form because digital data are *robust*. Many signals originate as analog waveforms, such as those produced by a microphone that converts acoustic energy produced by your utterances into electrical waveforms. While such waveforms can be transformed back into acoustic energy by driving a speaker, there are advantages in first converting analog waveforms into digital signals using analog-to-digital conversion.

Data Restoration Digital values can be restored exactly. Unlike analog signals that can take on any value within limits, digital signals are constrained to have one of two allowed values. Small deviations produced by noise can be restored to their original values. This operation is not possible for analog signals, because their waveforms can have any value between limits.

Figure 1.10 shows digital signals being restored to their original values even after the signals have been corrupted with noise. While all analog signals degrade with time, restoring digital values is possible, because digital signals are allowed to have only certain values, with the figure showing -1 V and $+1$ V. Signals corrupted with small amounts of noise typically will be close to these values. Data signals that are between 0.8 V and 1.2 V can be reset to 1.0 V.

EXAMPLE 1.10

Restoring corrupted digital signals

Consider digital waveforms that can be either -1 V or $+1$ V. The following digital signals were corrupted by additive noise during transmission to produce the detected signal values:

$$-1.1, \ 0.9, \ -0.95, \ 1.1, \ -.75, \ 0.85$$

Recognizing that these values correspond to digital levels at either -1 V or $+1$ V, the original values can be restored by setting them by using threshold $\tau = 0$ with values $< \tau \rightarrow -1$ and $\geq \tau \rightarrow +1$. Thus,

$$-1.1 \rightarrow -1, \ 0.9 \rightarrow 1, \ -0.95 \rightarrow -1, \ 1.1 \rightarrow 1, \ -0.75 \rightarrow -1, \ 0.85 \rightarrow 1$$

If the noise causes a wrong decision to be made by threshold detection, various error-correction techniques can be employed, depending on the severity of the noise, as discussed next.

Figure 1.10

Processing digital data waveforms corrupted with noise to retrieve data.

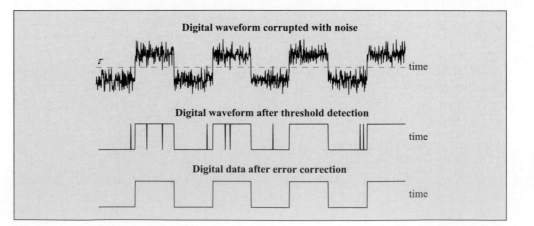

Error Detection and Correction Occasionally, the corruption is sufficiently severe to cause an erroneous restoration. Unlike analog signals, digital data also have the capability of detecting and even correcting errors. Adding some additional digital values to the original data provides error detection, and adding even more provides error correction. Of course, the penalty for these capabilities is that additional data must be transmitted, increasing both data transmission times and memory requirements. The following example illustrates this process using the simplest method of error detection and correction.

Error detection and error correction in transmitted data

EXAMPLE **1.11**

To detect transmission errors, we can transmit each digital value twice: An original 0 is transmitted as 00 and an original 1 as 11. If the detected sequence of data pairs is

$$00 \ 11 \ 01 \ 10 \ 11 \ 00$$

it is readily apparent that the third and fourth digital pairs contain errors. In the important data-entry application, a digital device can easily detect such a mismatch and instruct the user to re-enter the data.

To correct transmission errors, we can transmit each digital value three times: An original 0 is transmitted as 000 and an original 1 as 111. If the detected sequence of such data triplets is

$$000 \ 111 \ 010 \ 110 \ 111 \ 000$$

it is readily apparent that the third and fourth digital triples contain errors. A majority count process allows these errors to be corrected:

$$010 \rightarrow 000 \ \text{(original value} = 0) \qquad 110 \rightarrow 111 \ \text{(original value} = 1)$$

Factoid: The ability to restore digital signals and to achieve error correction provides a *data permanence* that was not possible with analog signals. So, while *ALL* forms of analog signals degrade with time, digital data can theoretically last forever.

Compression Digital *data* is different from digital *information*. A completely blue screen may require one million bytes of data to generate but contains very little information. Typically, the quantity of data is much larger than its information content, which is measured using *entropy* with units of bits/symbol. Often, the data size is increased intentionally to achieve *error correction*, which is the correcting of errors that may have occurred in transmission caused by noise.

Unlike analog waveforms, it is possible to exploit the structure of digital data to reduce the number of digital values that must be transmitted or stored. The following example illustrates a simple method of reducing the number of values that represent digital data.

EXAMPLE **1.12**   # Data compression using run-length coding

Consider the binary sequence that may be typical of that produced by a particular sensor:

00001111110000000001100000011111000 . . .

Notice that binary values occur in groups. Rather that transmitting each binary value separately, note that the binary values repeat and (of course) a string of 0's is always followed by a 1 and a string of 1's is always followed by a 0. These observations lead to the run-length code that provides the numbers of consecutive 0's and 1's in the data. For our sequence, the run-length code would be

4 6 9 2 6 5 3 . . .

The convention is to assume that the sequence starts with binary value 0. If a data sequence starts with a 1, the run-length code starts with a 0, indicating that there are no 0's.

1.4 WHERE IS EE GOING?

Computers, and the devices and systems that use them, are constantly becoming more powerful and less expensive with each new generation. The computational power and connectivity of computers has increased to the point that many useful and desirable tasks can be accomplished. Small and portable hand-held devices containing computers called *microprocessors* are capable of performing tasks using *apps* that were once deemed "smart" or intelligent. Once the basic sensors and actuators are available to a large number of *app* designers, a variety of popular applications termed *killer apps* will be developed.

The measure of computer power most often cited is *Moore's law*, named after Gordon Moore, a co-founder of Intel Corporation, who observed that the number of transistors on a computer chip doubles approximately every 18 months (or 1.5 years). For example, if computer circuits contained $N_o = 2{,}500$ transistors in year $t_o = 1971$, the mathematical relation that indicates that the number of transistors in year t, which is denoted as $N(t)$, is given by

$$N(t) = N_o \, 2^{\frac{t-t_o}{1.5}} = 2{,}500 \, 2^{\frac{t-1971}{1.5}} \tag{1.1}$$

Figure 1.11 shows Moore's law on a semi-logarithmic scale indicating the trend in the number of transistors in computer chips over time. The following example illustrates the technological impact.

EXAMPLE **1.13**   # Moore's law

The original small computer or microprocessor developed around 1971 had approximately 2,500 transistors. Eighteen months later, microprocessors had about 5,000 transistors. By 2013, only 42 years or twenty-eight 1.5-year cycles later, there are

$$2^{28} \approx 256{,}000{,}000$$

TIMES MORE transistors in a typical multi-core microprocessor. This computes to 640 BILLION transistors.

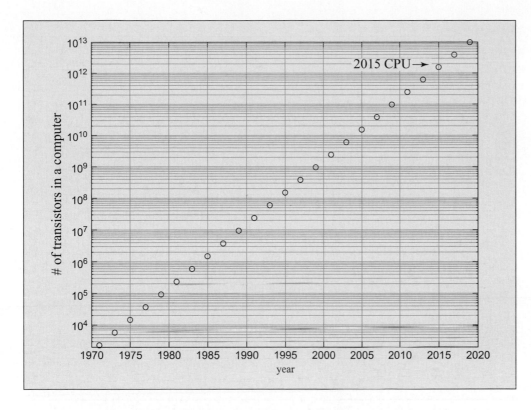

Figure 1.11

Plot of Moore's law on a semi-logarithmic scale displaying a linear trend in the number of transistors in computer chips over time. This trend also predicts other features in computing power.

Factoid: A useful approximation is $2^{10} \approx 1,000$. With $2^8 = 256$, the value $2^{28} = 2^8 \times 2^{10} \times 2^{10} \approx 256 \times 10^3 \times 10^3 = 256$ million.

Additional computer features tend to follow Moore's law.

- The speed of computer computations increases mainly through the introduction of additional computer *cores* on a single chip.

- The size of computer memories is evident in the development of *solid-state drives* (*SSDs*) that are more reliable and faster than conventional mechanical hard drives, thus reducing the nuisance of long boot-up times when your start your computer.

Other technological improvements that scale similarly to Moore's law include

- Internet data transfer speeds

- Battery power storage capacities

- Size and weight reductions of portable devices

- The number of devices connected on a network

Factoid: Computer speed and size are coupled by the laws of physics. Signals cannot travel faster than the speed of light (1 ft/ns), so smaller computers operate more quickly because they communicate signals faster over smaller distances.

Moore's law often guides companies on deciding whether it is time to upgrade to newer computers. The same trend also applies to personal computing devices, such as the smartphone and laptop: In 18 months, your smartphone will be twice as fast, twice as light, twice as intelligent, and will only need to be recharged half as often.

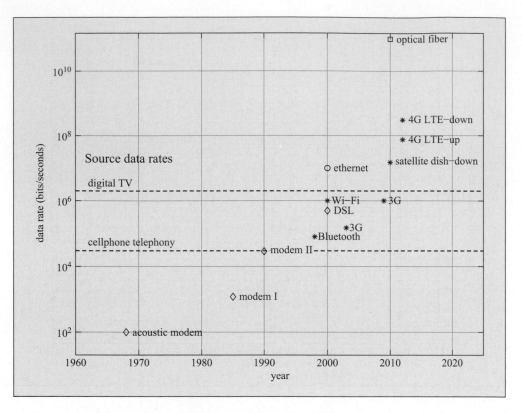

Figure 1.12

Transmission rates provided by different data channels: ◇ - standard telephone land lines, ○ - wired cable, ∗ - wireless, □ - optical fiber. Source data rates required for digital systems are shown on the left.

Data Transmission Capabilities and Requirements

Accompanying the increase in computing power is the increase in digital transmission rates provided by digital data channels, as well as the digital data rates required for popular digital services. Figure 1.12 shows how digital data communication rates evolved over time. The salient features of digital data transmission over wired channels include these properties.

> **Acoustic modem (modulator/demodulator)**—In an initial attempt to transmit data over telephone wires, different acoustic tones in the standard voice bandwidth were used to transmit 1's and 0's in both directions simultaneously, yielding a paltry 100 bps (bits/s).

> **Modem I**—Using data signals rather than acoustic tones increased the data rate to 1200 bps.

> **Modem II**—More clever coding that varied both frequencies and amplitudes increased the data rate to 28,800 bps.

> **DSL (Digital subscriber loop)**—Digital signals transmitted over tuned telephone wires increased the data rate to 0.5×10^6 bps $= 0.5$ Mbps.

> **Coaxial cable**—The increased bandwidth provided by coaxial cables increased the data transmission rate to 500×10^6 bps $= 500$ Mbps.

> **Optical fiber**—Transmitting data using light signals over optical fibers offers extremely high capabilities of up to 100×10^9 bps $= 100$ Gbps (100 Gigabits/s).

Transmitting data over wireless channels is now commonplace. One differentiation is the range of operation, over short ranges (typically less than 20 m) and over longer ranges. Short-range data transmission channels include the following.

> **Bluetooth**—Bluetooth is a wireless standard for securely exchanging data over short distances between mobile devices at a 80 kbps transmission rate. It was originally

designed to replace cables that connected computers to peripheral devices, such as printers. Smartphones and other devices equipped with Bluetooth can form *piconets*, which are networks of data-exchanging devices that are within close range of each other.

Wi-Fi—Wi-Fi is a wireless standard that was designed to replace wired ethernet connections to the Internet by forming a wireless local area network (WLAN). Higher transmission powers permit larger transmission rates and longer ranges.

Digital communication includes longer-range data channels.

Cellular service—Cellular services that communicate with your smart device, have undergone improvements classified as *generations*, such as 3G and the ten times more powerful 4G LTE (*Long Term Evolution*). The first generation, 1G, transmitted analog signals, and digital data began with 2G. Cellular systems operate with a dense network of transmitting antennas that maintain connectivity with your mobile smart device. The more powerful antenna network transmits data *down* to your smart phone at rates up to 300 Mbps, while your less powerful smart device (for battery life and safety reasons) transmits data back *up* to the network at 75 Mbps. A proposed 5G system will have increased data transmission rates by steering a narrow beam using antenna arrays at both ends.

Satellite dish—Geostationary satellites located 24,000 miles above the earth can beam down data to satellite dish antennas at rates up to 15 Mbps. The 0.25 second delay due to the speed of light (data signal speed) to bounce the data off the satellite is not significant, especially for users in remote locations who would otherwise have no access at all.

Using these data transmission channels are sources of digital information that send content to your smart device. Some information, such as Web pages and texting, are not time sensitive and display the information when it is received. Delays vary according to the congestion of data on the various networks that connect the source to your smart device. In contrast, other sources must transmit data at the rate it is produced for the system to operate successfully. These include the following.

Digital speech—When you speak into your smartphone, your speech waveform is converted into digital data at a rate of 30 kbps. For your speech to be transmitted without interruption, the cell-phone system must be able to maintain this 30 kbps average transmission rate *continuously*. Actually, your speech digital data is stored in short, sub-second durations that are transmitted quickly as separate data packets, which are recombined at the receiving end with no perceptual distortion. Modern data channels easily transmit at this data rate and use the excess channel capacity to service additional users.

Digital TV—Similarly, each standard digital TV channel transmits 8 Mbps to display the program you are watching. Note that Wi-Fi and 3G will not support even a single channel. You can watch programs on your laptop on a Wi-Fi network because the smaller data rate produces either a lower-quality video or a smaller-sized video with both relaxing the data transmission requirement.

1.5 OVERVIEW OF CHAPTERS

This book attempts to tell a story that provides a coherent framework for understanding digital systems. The following list presents a road map through the chapters.

Chapter 2 describes sensors and actuators that are present in smart devices and robots. Sensors produce informational waveforms and indicate events that form the human-to-computer interface, while actuators produce sensory excitations that form the computer-to-human interface.

Chapter 3 describes the basic combinational logic operations and their corresponding elemental logic gates. Elementary gates are interconnected to form logic circuits that implement a desired truth table to perform a useful task, such as controlling digital displays and doing arithmetic operations.

Chapter 4 introduces sequential logic circuits that perform memory and counting operations that occur within smart devices.

Chapter 5 describes analog-to-digital converters (ADCs) that transform analog waveforms into digital data for processing by computer and digital-to-analog converters (DACs) that transform digital data back to analog waveforms.

Chapter 6 introduces basic ideas in probability that this text uses to model random symbols produced by informational sources and noise that corrupts transmitted signals. This chapter also describes processing signals that contain both deterministic and random components for data detection.

Chapter 7 considers the special case of data signals with values known by the transmitter and receiver and how to use this knowledge to extract the data values. A common and important problem is the detection of data signals that have been corrupted by random noise, as in typical cell-phone transmission signals.

Chapter 8 designs signals that obey the orthogonality condition to serve multiple simultaneous users or to allow one user to transmit multiple bits during a single transmission interval. Signals are described that are orthogonal in time, in frequency, and in code, where the latter is used in cellular networks.

Chapter 9 describes the fundamental ideas in *source coding* that perform data compression to contain the same information in a smaller quantity of data and encryption to make data secure. The amount of information produced by a source or contained within a data file is measured by computing its entropy. A simple method to encrypt data illustrates the steps involved in the process.

Chapter 10 describes the fundamental ideas motivating *channel coding* methods that implement error detection and correction. The capacity of a channel to transmit data is computed from the bandwidth and signal-to-noise ratio of the transmitted signals.

Chapter 11 describes how networks, such as the Internet, cope with data that occur at random times and that have various sizes by employing data packets with additional information. Solving problems that occur when data packets interfere, called *collisions*, are described in both wired and wireless transmission networks.

Chapter 12 describes the design of machine-readable codes that facilitate the transfer of data from commonly used devices, including credit cards and bar codes.

Chapter 13 provides a primer in Excel and presents examples of programs that illustrate the calculations performed in the previous chapters.

1.6 Further Reading

Most books include references for further investigations of particular topics. One disadvantage of this traditional approach is that references quickly become outdated as new technology becomes available and hence are more of historical value. A second disadvantage is that references are often difficult and expensive to obtain, especially textbooks and papers that are provided by professional organizations.

Modern telecommunication technology offers an alternative: Internet search engines provide a means of accessing current information by searching for key words and terms. Such searches provide results almost immediately and at minimal cost. The only caveat to this on-line approach is that one often is deluged with *hits* from each query that include vendor sites and useless—or even worse—wrong information. The task is to separate the wheat from the chaff. This book identifies useful search terms by using *italic font* in the text. For example, consider learning more about Moore's law. Searching for *Moore's law* using Google or another search engine produces over one million hits. One useful resource is *Wikipedia*, which is an evolving on-line encyclopedia that is edited and updated continuously by people knowledgable in their particular field. Such search results and the links they provide allow you to learn the correct jargon or terminology, to find the latest features, or to probe as deeply as your needs require.

1.7 Summary

This chapter introduced the basic ideas that are explored in smart devices and communication systems described in this book. While analog signals are often the originating and terminal waveforms in a system, current *smart* systems convert these analog waveforms into digital data to take advantage of the flexibility and robustness inherent in the communication and manipulation of digital data. Electrical engineers having various specialties design systems that perform the tasks needed to make such systems operate successfully and reliably, including:

- Transmitting data from one point to another (Communications)
- Storing data in memories that are optimized for size and power consumption (Microelectronics)
- Manipulating data to accomplish a desired computational task (Computer engineering)
- Extracting waveform informational features for reliable operation (Signal processing)
- Operating a system in a stable fashion in the presence of unexpected disturbances and device variations (Control systems)
- Implementing systems to assist humans by combining different forms of sensory data and stored information (Robotics)
- Encoding data for reliable communications more quickly and over farther distances (Information theory)

1.8 Problems

1.1 Illuminated mouse. You often power your laptop with the battery while you are traveling. You need to buy a new mouse but want to maximize the battery life. Explain why buying the illuminated mouse is not a wise choice.

1.2 Threshold detection. Digital signals that occur within your computer are designed to be either 0 V or 5 V. Additive noise produced the detected values:

$$-0.1, \ 3.9, \ 0.9, \ 5.1, \ 0.7, \ 4.85$$

What threshold value would you use to restore the values? Explain why. Restore these detected values to their designed values.

1.3 Error correction. Threshold detection converted signal values 0 V and 5 V into binary logic values 0 and 1 respectively. For transmission over a noisy channel, each binary value is transmitted five times. A threshold detector produces the binary sequence:

$$00100 \ 11001 \ 01000 \ 10110 \ 10001$$

a. Assuming at most two errors occur per 5-bit code word, estimate the probability of error in the channel as the number of errors in the sequence divided by the number of data bits in the code words.

b. What rule would you apply to try to correct the errors?

c. Write your corrected binary sequence.

1.4 Prediction with Moore's law. Using the current year's performance as the base, how much more powerful will your computer be in six years?

1.5 Prediction with Moore's law. How long will you need to wait for your next computer to be 100 times more powerful than your current computer?

1.6 Simultaneous users on a 4G LTE network. How many digital speech signals can a 100 Mbps 4G LTE service simultaneously?

1.7 Simultaneous TV channels on an optical fiber. Assuming an HDTV program requires a data rate of 15 Mbps, how many channels can an optical fiber provide simultaneously?

1.9 Excel Projects

The following Excel projects illustrate the following features.

- Specifying a function using relative addressing and displaying it in a scatter plot.

- Varying the parameters in a function using absolute addressing.

- Changing a chart scale from linear to logarithmic for clearer data display.

Using a Narrative box (described in Example 13.4), include observations, conclusions, and answers to questions posed in the projects.

1.1 Specifying input values and plotting a linear function. Using Example 13.7 as a guide, plot a linear *funds depletion* curve. Assume you start the term, defining time = 0 with $500 for expenses. You spend $50 per week, producing a slope of −$50/week, making the curve intersect $0 at week 10. You need the funds to last at least 12 weeks. Modify the slope value so that the funds are exhausted between weeks 12 and 13. What is the resulting slope value on your chart?

1.2 Moore's Law. Extend Example 13.9 to plot Moore's law from 1971 to 2020 in three-year increments and compare linear and logarithmic plots of the y values.

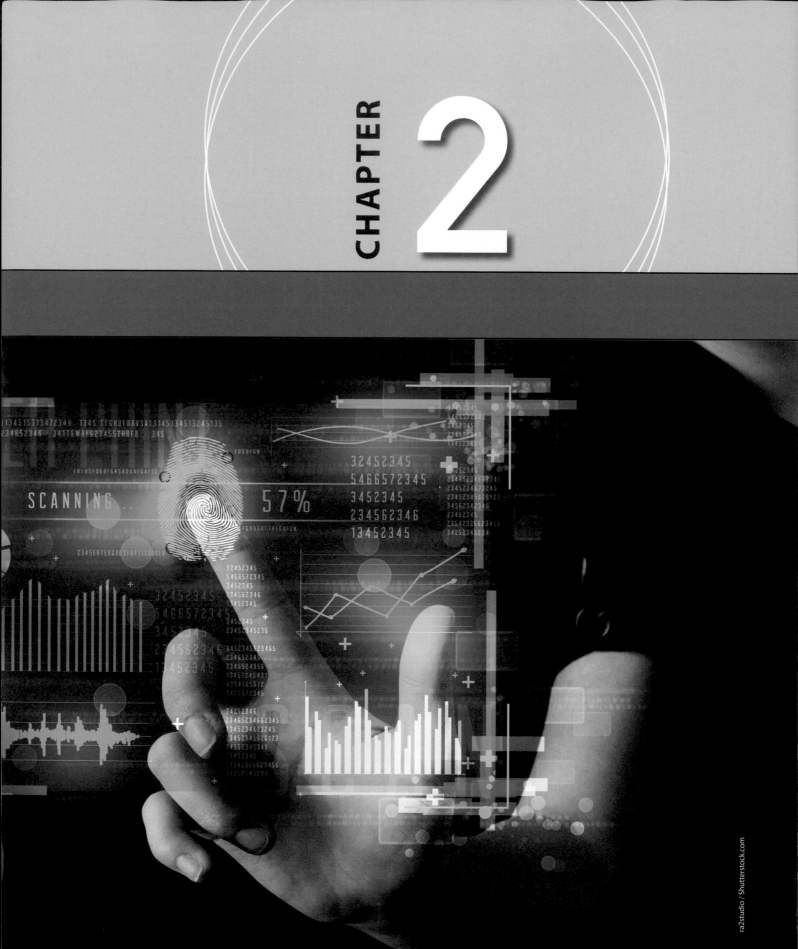

CHAPTER 2

SENSORS AND ACTUATORS

LEARNING OBJECTIVES

After completing this chapter, the reader should be able to:

■ Understand the importance of sensors and actuators in smart devices and robots.

■ Differentiate sensors that produce waveforms from detectors that produce events.

■ Appreciate human-made sensing system as analogs of biological systems.

2.1 INTRODUCTION

Sensors and actuators are devices that let the smartphone or tablet interact with the human user, permit a robot to navigate an environment, and allow devices to communicate over wireless networks. Sensors let the smartphone know what you want to do, such as when you operate the touch screen. Actuators let the smartphone indicate to you that it wants your attention through vibrations or sounds. Sensors and actuators perform dual roles. Whereas sensors convert energy that includes mechanical, optical, or electrical forms into electrical signals, actuators convert electrical signals into energy, which is something you can see, hear, or feel. Robots offer analogs to biological sensing and actuation through sensors that perceive the environment and actuators that accomplish locomotion and grasping tasks.

This chapter provides a general introduction to sensors and actuators in the following manner.

Sensors—In many smart systems, sensors form the front end by sensing the environment, monitoring system operation, and interpreting human intensions.

Mechanical sensors and actuators—Switches in the form of push buttons, keyboards, and tactile displays are a basic communication channel with smart devices, which often respond through vibrational cues.

Acoustic sensors and actuators—Acoustic sensors detect human speech and actuators play audio and sounds to form a richer communication interface between humans and computers.

Optical sensors and actuators—Optical sensors interpret the environment through images, and actuators display video to form the richest communication interface between humans and computers.

Proprioception—Accelerometers and GPS systems localize smart devices to enhance their capabilities to assist humans in a variety of tasks.

Active sensing—By probing the environment with optical or acoustic energy, active sensors are used to automate many tasks and guide robots and cars to prevent collisions.

2.2 ANALOG AND DIGITAL SENSORS

A sensor is a device that converts physical energy into an electrical signal that information processing systems typically convert into digital form for their interpretation. Sensors form the *front end* of information systems because they indicate what you want the device to do. This chapter describes sensors that provide information in a numerical or digital form.

We classify a sensor by the type of signal it produces, as Figure 2.1 illustrates.

Analog sensors produce *waveforms* (or signals) that vary continuously in time. One example of an analog sensor is a mercury thermometer whose column height rises to a height proportional to the temperature.

Digital sensors produce discrete values. For example, a digital thermometer produces a number that indicates the temperature to the nearest 0.1° Fahrenheit (*e.g.*, 98.7°F). A digital sensor is typically composed of an analog sensor that converts energy into a waveform that is followed by a device that converts the analog value into a digital approximation.

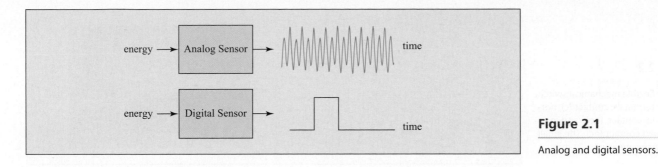

Figure 2.1

Analog and digital sensors.

Analog and digital sensors each have their own advantages. Although an analog sensor seems more accurate than a digital sensor, a digital sensor provides sufficient accuracy for most tasks. For example, an analog thermometer that reads 104.5472867329°F is not really more useful than the digital thermometer that reads 104.5°. Both indicate a fever. However, a digital thermometer that produces readings in 0.5° increments has the potential for error detection: If a nervous parent using a digital thermometer writes down the temperature of a sick child as 104.3°F, health care workers would detect the error because 104.3° is not a valid reading.

Digital Events

The simplest digital sensors are *binary sensors* that produce *binary values* that take on only one of two possible values. Typically, for reliable operation, these two values are the extreme voltages in the systems: zero volts (0 V or *ground*) and the maximum value V_{max} (corresponding to the battery voltage). To normalize variable system voltages, *logical* levels of either 0 or 1 are used. A detector that changes logic levels in response to some action indicates the occurrence of an *event*. Events considered in this book include:

- A switch closure produced by pressing a key pad or tactile display.

- The occurrence of a data signal on a communication channel.

- A significant change of an accelerometer indicating that a cell phone display is rotated to a new orientation.

The following sections describe *sensors* according to the energy form that actuates them and *actuators* that convert electrical signals into that energy.

2.3 MECHANICAL SENSORS AND ACTUATORS

2.3.1 Switches

The simplest sensor is a mechanical switch that produces a two-level or binary (0 or 1) output. For example, the open switch can represent a 0, and the closed switch can represent a 1. This type of sensor is well-suited to digital logic and computer systems that store and process data in binary form. We encounter these sensors in many applications.

The computer can implement a sense of touch with a *mechanical switch*. Figure 2.2 shows the most flexible mechanical switch, which has three terminals: *contact* (*c*), *normally open* (*no*) *contact*, and *normally closed* (*nc*) *contact*. The no contacts complete an electrical circuit when the switch is activated, while the nc contacts open an electrical circuit. A computer interprets the closed or open pair of contacts as a binary logical value.

The most common switch on an electrical device, such as a smartphone, computer, or robot, is the *on/off* switch. Figure 2.3 shows a switch that allows current I to flow to the device by connecting the battery.

Figure 2.2

The most flexible mechanical switch has three terminals: contact (c), normally open contact (no), and normally closed contact (nc). The small black square measures 1 mm × 1 mm.

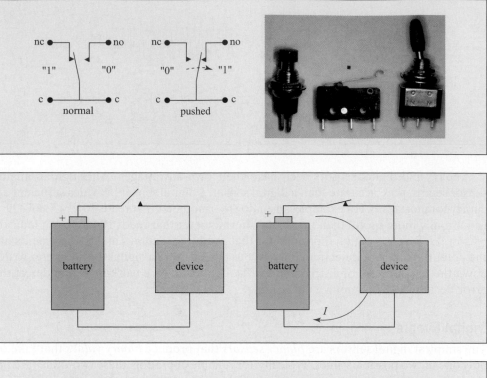

Figure 2.3

An electrical switch connects a battery to an electrical device to cause current *I* to flow.

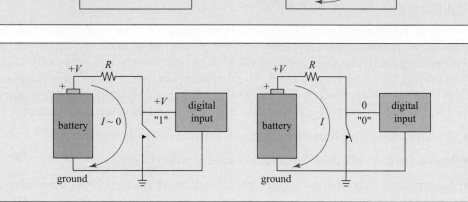

Figure 2.4

A switch closure provides a binary (0/1) indication to a digital device.

Figure 2.4 shows a switch closure that provides a binary (0/1) indication to a digital device. Resistor *R* is an electrical component shown in the connection from the battery to limit the current *I* that can flow. When the switch is open, $I \approx 0$ and the voltage at the device digital input is the battery voltage $+V$. The digital device typically interprets the $+V$ voltage as as a logical 1. There is negligible current that flows into the digital device.

When the switch is closed, the digital input is connected to *ground*, which is the negative side of the battery corresponding to 0 V. The digital device then interprets a 0 V as a logical 0.

EXAMPLE 2.1 | Robot bumper

A pair of switches can provide additional information, such as direction. Figure 2.5 shows an application of a pair of switches for mobile robot navigation and obstacle avoidance. A pair of bumper contact switches mounted on the front of the robot—one to each side—are normally in the open state. Each switch connects to a digital input on the robot controller, and both inputs are normally at logic 1. When the robot bumps into an obstacle, the switch on that side closes, converting that input to a logic 0. With two switches and their corresponding inputs, two logic value indicate to a robot when the obstacle is to the left or right. The robot is programmed to stop, reverse a little, and turn to avoid the object in its path.

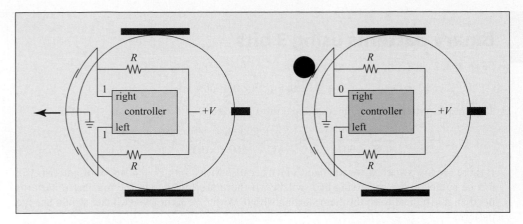

Figure 2.5

A pair of bumper contact switches indicate to a mobile robot when an obstacle is to the left or right. Right figure shows contact with object on the right side of the robot.

2.3.2 Switch Arrays

Each key of a computer keyboard is actually a switch. When you type, each key you press closes a switch that corresponds to a particular character. You could connect each switch to a pair of wires and a sensing circuit, but this would be wasteful. To reduce the number of wires, a keyboard is structured as a two-dimensional array of switches formed by the rows and columns of keys (typically 5 rows and 13 columns). Figure 2.6 shows a partial keyboard matrix.

For computer consumption, it is convenient to employ binary addresses to identify switches. If b bits are used to specify the address, 2^b unique addresses can be specified, although not all 2^b addresses need to be used. The conventional method for assigning addresses to switches uses the binary counting sequence. Consider a binary pattern with the left-most bit being the *most significant bit (MSB)* and the right-most bit the *least significant bit (LSB)*:

$$\overbrace{0/1}^{MSB} \quad 0/1 \quad 0/1 \quad \cdots \quad 0/1 \quad \overbrace{0/1}^{LSB}$$

The counting value of a b-bit pattern, V, equals

$$\overbrace{V = 2^{b-1} \times \overbrace{0/1}^{MSB} + 2^{b-2} \times 0/1 + \cdots + 2^1 \times 0/1 + 2^0 \times \overbrace{0/1}^{LSB}}^{b \text{ bits}} \tag{2.1}$$

$V = 0$ when all bits values equal 0, and $V = 2^b - 1$ when all equal 1.

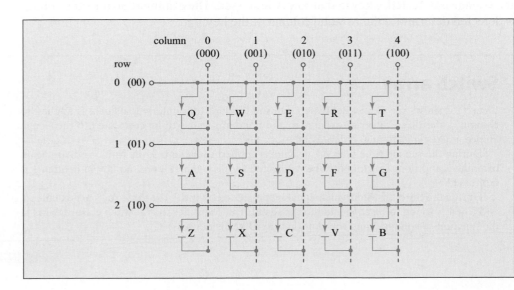

Figure 2.6

A keyboard switch array consists of a two-dimensional array of switches formed by the rows and columns of keys. Binary addresses are shown in parentheses.

EXAMPLE **2.2**

Binary patterns using 3 bits

For $b = 3$, $2^3 = 8$, and

$$V = 2^2 \times 0/1 + 2^1 \times 0/1 + 2^0 \times 0/1$$

The unique patterns and their counting equivalents are

$$\underset{000}{\overset{=0}{\frown}},\ \underset{001}{\overset{=1}{\frown}},\ \underset{010}{\overset{=2}{\frown}},\ \underset{011}{\overset{=3}{\frown}},\ \underset{100}{\overset{=4}{\frown}},\ \underset{101}{\overset{=5}{\frown}},\ \underset{110}{\overset{=6}{\frown}},\ \underset{111}{\overset{=7}{\frown}}$$

If there are five switches, we still need 3 bits, because with 2 bits, $2^2 = 4$, and four patterns will not be sufficient for specifying five switches. If there are fewer than the maximum patterns needed, it is typical to count them starting with 0. With $b = 3$, the five switches would use the patterns

$$\underset{000}{\overset{=0}{\frown}},\ \underset{001}{\overset{=1}{\frown}},\ \underset{010}{\overset{=2}{\frown}},\ \underset{011}{\overset{=3}{\frown}},\ \underset{100}{\overset{=4}{\frown}}$$

EXAMPLE **2.3**

Telephone key pad

A standard telephone key pad has 12 buttons arranged as a 4-row-by-3-column switch matrix. The four rows have 2-bit row addresses 00, 01, 10, and 11. The three columns with addresses require 2-bit column addresses, normally using the first three of the four possibilities. Hence, a 4-bit row/column address is required to determine the switch that was pressed. The code 0110 is interpreted as

$$\overset{\text{row}}{\overset{\frown}{01}}\quad \overset{\text{column}}{\overset{\frown}{10}}$$

to indicate the switch in row 1 and column 2 was pressed.

A microcomputer chip in the keyboard interprets this row-to-column connection and transmits the corresponding code to the computer. A voltage is applied to each column—one at a time—for a short time interval, typically 1 millisecond (1 ms). During this interval, each row is examined—one at a time—for the presence of the voltage. If the voltage is detected, a key in that row is depressed. The examination time the voltage is detected determines the row and column of the key.

EXAMPLE **2.4**

Switch array

Figure 2.7 shows the D key depressed. During the first time interval, a signal is applied to column 0, the three rows are examined, and none are found to be connected to the signal. Hence, no key in column 0 is depressed.

During the second time interval, a signal is applied to column 1, the three rows are again examined, and none are found to be connected to the signal. Hence, no key in column 1 is depressed.

During the third time interval, a signal is applied to column 2, the three rows are examined, and row 1 is found to be connected to the signal. This identifies that a switch closure exists at the junction of column 2 and row 1 or that the D key is depressed.

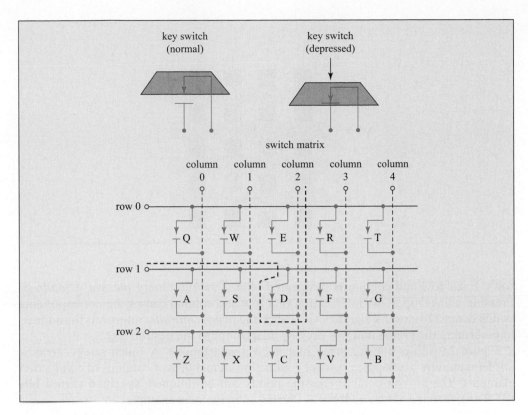

Figure 2.7

Pressing a key closes the switch that makes the connection shown by the dashed line.

2.3.3 Touch Screens

Switches can be arranged as a two-dimensional (2D) array for graphical user interface (GUI, pronounced *gooey*) applications. The switch in a touch screen display consists of a pair of contacts separated by small compressible non-conducting spacers, as illustrated in Figure 2.8. Fingertip pressure on the screen surface compresses the insulators, and the two conductors make contact.

Other touch screens work through *capacitive coupling* where only finger proximity to the screen is sufficient to register a reading. However, these are prone to false readings caused by water droplets on the screen.

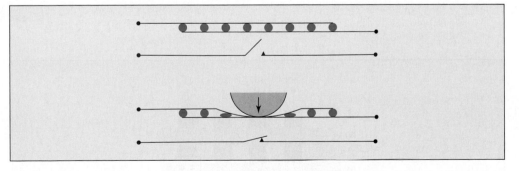

Figure 2.8

The switch in a touch screen display consists of a pair of contacts separated by the small, compressible non-conducting spacers shown as small circles.

> *Factoid:* Such pressure sensors are used in robot manipulators to pick up both delicate and heavy objects.

The 2D switch array is constructed by overlapping rows of conducting elements over columns of conduction elements. Figure 2.9 shows four columns overlapped by four

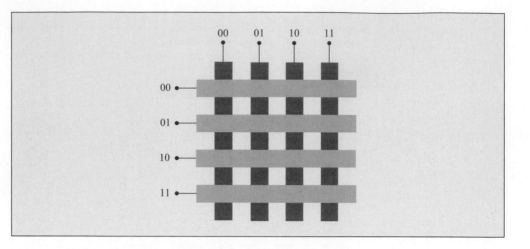

Figure 2.9

A touch screen array is implemented by overlapping rows of conducting elements over columns of conduction elements.

rows. Each row and column is designated with a unique binary pattern or *address*. Pressing a fingertip at the intersection of a row and column causes the corresponding switch contact to close. A digital device called the *switch controller* interprets the address to determine the switch that was pressed to accomplish the desired task.

Figure 2.9 shows a 4 × 4 array that is on a smartphone. A touch screen array is implemented by overlapping rows of conducting elements over columns of conduction elements. The 16 row-column crossing points can be uniquely specified with 4 bits ($2^4 = 16$), forming a 4-bit address, as illustrated in the next example.

EXAMPLE 2.5   Smartphone touch screens

A smartphone home screen display shows icons with one centered over each crossover point. Figure 2.10 shows a friends icon (smiley face) and a music icon. When we depress the friends icon, the controller decodes the row-column address as 01-11, and the music icon corresponds to the row-column address as 10-01.

To differentiate this icon set from that on another screen, the controller may append a screen address to the row-column address. If the smartphone has eight screens, it assigns a 3-bit address with the home screen address as 000. Then the friends icon has the screen-row-column address of 000-10-11.

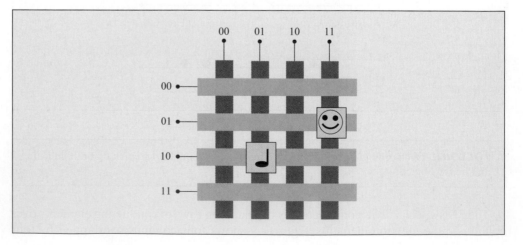

Figure 2.10

Pressing the music icon produces the row-column bit pattern 10-01, and pressing the friends icon produces 01-11.

High-Resolution Touch Screens

The resolution of a touch screen R_s is measured in terms of the number of switches per unit length to accommodate separate row and column calculations. A switch located every 1 mm results in a linear switch resolution equal to $R_s = 1$ switch/mm. As technology improves, the resolution increases. A typical fingertip closes multiple switches on a high-resolution display. To determine the fingertip position, the smartphone computer senses all the switches that are closed (using the same addressing method described previously) and computes the row and column switch addresses closest to the midpoint locations. To simplify the calculations, the fingertip that closes row switches from r_{min} to r_{max} is approximated the row midpoint r_{mid} value

$$r_{mid} = \text{floor}\left(\frac{r_{min} + r_{max}}{2}\right) \tag{2.2}$$

where the floor operation ignores any remainder, which is the normal result that occurs with integer division. A similar result occurs for the column midpoint c_{mid}.

High-resolution touch screen   EXAMPLE 2.6

This example illustrates a high-resolution touch screen by considering one-dimensional (row) switch array having a resolution $R_s = 1$ switch/mm. The array is 6.3 cm long, contains 64 switches, and is encoded using a 6-bit address starting at leftmost 0 (= 000000) to rightmost 63(= 111111).

A fingertip width measuring approximately 1 cm depresses 10 switches simultaneously. If the finger closes switches $r_{min} = 40$ to $r_{max} = 49$, then the midpoint is computed as

$$r_{mid} = \text{floor}\left(\frac{r_{min} + r_{max}}{2}\right) = \text{floor}\left(\frac{40 + 49}{2}\right) = 44$$

Gestures, such as finger swipes, are determined by sampling the midpoint locations very often and computing the velocity of the drag motion from the changes in the midpoint locations during the sample time. If T_s is the sample period, R_s is the linear switch resolution in switches/mm and δx_m is the minimum change in midpoint position during this sample period that forms a detectable change, the minimum speed of a finger swipe is

$$v_{min} = \frac{\delta x_m}{R_s T_s} \tag{2.3}$$

This threshold speed would ignore slow, unintended finger movements.

Finger swipes   EXAMPLE 2.7

As a finger moves across the array described in Example 2.6, the computer determines the finger midpoint positions every $T_s = 0.05$ seconds (s). A swipe is sensed when a midpoint position changes by more than four switch locations between position determinations, making $x_m = 4$ switches. With $R_s = 1$ switch/mm, the minimum speed of a finger swipe is

$$v_{min} = \frac{\delta x_m}{R_s T_s} = \frac{4 \text{ switches}}{1 \text{ switch/mm} \times 0.05 \text{ s}} = 80 \text{ mm/s}$$

Multiple finger presses are computed in the same manner but require a more powerful processor. Luckily, the power of newer processors increases as the processor cost remains almost constant, allowing additional features to be included in new models for approximately the same price as the previous model.

EXAMPLE 2.8 Multiple finger gestures

A computer monitors a linear (row) array with $R_s = 1$ switch/mm every $T_s = 0.05$ s. Multiple finger presses exhibit themselves as *regions* of switch closures with one region for each finger. The smartphone computer determines the row and column midpoints of each region, as shown in Example 2.6, and the velocity of each finger, as shown in Example 2.7. Two fingers press the switch array with finger $F1$ closing switches 15 to 23 and the smaller finger $F2$ closing 40 to 44. The computer determines midpoint positions using integer (rounding down) arithmetic

$$P1 = \text{floor}\left(\frac{23 + 16}{2}\right) = 19$$

and

$$P2 = \text{floor}\left(\frac{40 + 44}{2}\right) = 42$$

These finger positions produce finger separation at

$$S = P2 - P1 = 42 - 19 = 23 \quad (= 23\,\text{mm})$$

At time T_S later, $F1$ closes switches 11 through 20, producing midpoint $P1' = 15$, and $F2$ closes 48 through 52, producing $P2' = 50$. The new finger separation is

$$S' = P2' - P1' = 50 - 15 = 35 \quad (= 35\,\text{mm})$$

Comparing S' to S indicates an widening gesture. If $T_S = 0.05$ s, then the finger separation speed is

$$v_F = \frac{S' - S}{T_S} = \frac{35\,\text{mm} - 23\,\text{mm}}{0.05\,\text{s}} = 240\,\text{mm/s}$$

2.3.4 Vibrating Motor

A common actuator in a smartphone is a vibrating motor that provides a tactile sensation to indicate an incoming call. Such a motor is shown in Figure 2.11. Connecting a motor to a power supply with a switch causes the motor to turn. An eccentric (off-center) weight connected to the motor shaft causes a smartphone to vibrate to provide a silent

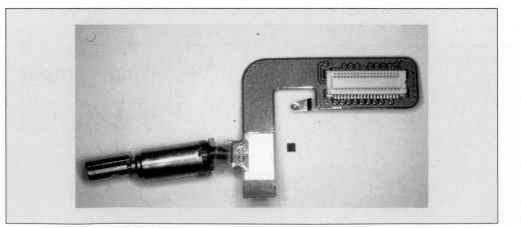

Figure 2.11

Vibrating motor in a cell phone. The black square is 1 mm × 1 mm in size.

indication of an incoming call or an alarm condition. Clever system designers implement other functions with a vibrating motor by varying its speed or on/off pattern.

2.4 ACOUSTIC SENSORS AND ACTUATORS

Acoustic energy carries the sounds we hear, including speech generated by humans, music generated by instruments, and alarms generated by loud speakers.

2.4.1 Microphone

Audio signals are pressure variations in air that contain information, although we perceive the world mostly through visual images with our eyes. Microphones are sensors that convert pressure variations into electrical signals, while a speaker and beeper convert electrical waveforms into sounds. Figure 2.12 shows a microphone, speaker, and beeper in a cell phone. The microphone waveform is typically converted to digital data with an *analog-to-digital converter (ADC)*, which is described in Chapter 7.

Microphone waveforms   EXAMPLE 2.9

Figure 2.13 shows three displays of the waveform of the utterance /ah/ (as in f/ah/ther) produced by laptop computer microphone. The ADC in the laptop used a sampling rate $f_s = 8$ kHz, sampling period $T_s = 125\,\mu$s, and encoded the sample values using 8 bits/sample. The number of different levels encoded with 8 bits is $2^8 = 256$.

The three displayed waveforms show:

1. Two seconds of microphone output that includes the silent period before the speech, the speech waveform, and the silence after the speech. This "silent period" indicates the background noise that is present.

2. The speech waveform is extracted from the 2-second microphone output by applying a threshold to find the beginning and end points of the speech.

3. A 500-sample section of the extracted waveform displays details in the speech waveform. The time window duration corresponds to 62.5 ms. Note that the waveform appears to be approximately periodic with the pitch period equal to 60 samples (7.5 ms).

Figure 2.12

Microphone, speaker, and beeper (from left to right) in a cell phone. The black square is 1 mm × 1 mm in size.

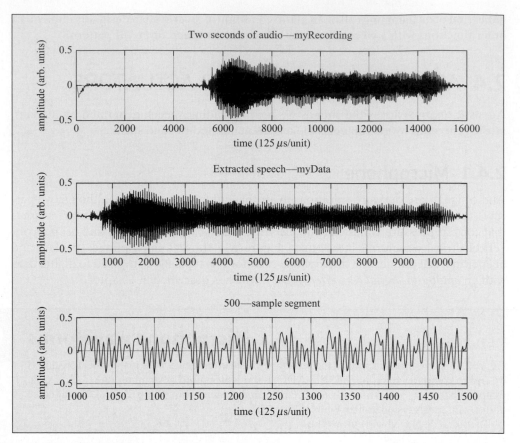

Figure 2.13

Speech waveform of the utterance /ah/ (as in f/ah/ther) produced by laptop computer microphone and sampled using sampling rate $f_s = 8$ kHz and encoding each sample with 8 bits (256 levels).

EXAMPLE 2.10 High-quality digital audio

High-quality audio is represented with digital values each having 16 bits. The number of steps representing the analog level is then

$$N_{steps} = 2^{16} = 65,536$$

If your digital player is powered with 3 V supplied by two AAA batteries, the step size Δ is

$$\Delta = \frac{3 \text{ V}}{65,536} = 4.58 \times 10^{-5} \text{ V} = 45.8 \, \mu\text{V}$$

2.4.2 Loud Speaker

A common actuator in audio systems is the familiar loud speaker. The one in your smartphone allows you to hear a telephone conversation or listen to music that is either stored in memory or transmitted over a wireless connection. A pair of speakers in your headset emits stereo sound that is slightly different to your left and right ears to provide directionality to the sound sources.

A speaker works by transforming the electric current into a coil of wire into a magnetic field that varies with current amplitude. This field interacts with a permanent magnet that is attached to a diaphragm. The magnet moves, vibrating the speaker diaphragm, which varies the adjacent air pressure that your ears sense as sound.

The audio signals in your cell phone start in digital form either stored in a digital memory or transmitted over a digital data channel. A *digital-to-analog converter*

(DAC) converts these digital samples into analog waveforms that produce the speaker current.

A second acoustic actuator found in a smartphone is the *beeper* that signals the arrival of an incoming call. The beeper uses a *piezoelectric transducer* containing a disc that vibrates like a drum surface when driven by an electrical voltage at a particular frequency (about 1 kHz) to produce the tone you hear. This device works only at this *resonant* frequency, making it very efficient (that is, producing a loud sound with very little electrical power).

2.5 OPTICAL SENSORS AND ACTUATORS

Optical energy carries the images we see, including those illuminated by sunlight or by an artificial source of light (such as a flashlight). The camera is the optical sensor in a smartphone, while LED lamps and color displays are the actuators.

2.5.1 Camera

Figure 2.14 shows a typical smartphone camera. Cameras are composed of picture elements or *pixels* that determine how accurately they record a scene. The resulting gray-scale (black-and-white) image can be considered to be a two-dimensional array where each pixel has an x and y location, and the value at that location is the pixel intensity. An intensity value equal to zero displays as black and a maximum value as white. The maximum value depends on the number of levels that are stored in the camera. Typical maximum values include 255 (8-bit camera) and 1023 (10-bit camera).

A color camera stores the intensity in a three-dimensional array with two dimensions representing (x, y) position and the third dimension having three levels, corresponding to the energy in red, green, and blue light, which conventionally is referred to as *RGB*.

Cameras are typically specified in terms of number of pixels. For example, a 1 megapixel camera would have one million pixels that may be structured as one thousand rows and one thousand columns, representing a square image. Rectangular image formats are more common, having more columns than rows with the number of pixels being equal to the number of rows times the number of columns.

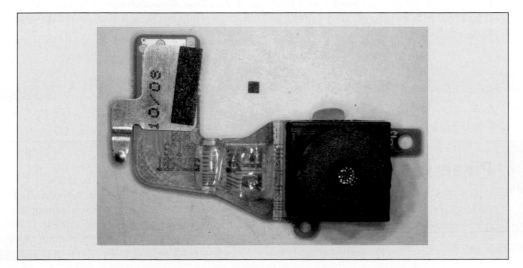

Figure 2.14

The camera circuit in a smartphone. The black square measures 1 mm × 1 mm.

EXAMPLE **2.11**

Smartphone cameras

Most current smartphone camera sensors have between 5 and 13 megapixels and easily rival stand-alone digital cameras. While some camera sensors are considering >40 megapixel sensors, other sensor systems are exploring additional features by including additional cameras—beyond the front and back facing cameras used for video/audio communications. These additional cameras could gather environmental data, perform face tracking, and produce other useful effects.

2.5.2 Display

Optical actuation occurs by displaying binary digits stored in memory as an image on a digital display, which range in size from a smartphone to monitor measuring 27 inches or more. In any case, light sensors are replaced by tiny light emitters, such as *light emitting diodes (LEDs)*, that produce red, blue, and green (RGB) light in each picture element (pixel) in the display. Controlling the intensity of each R-G-B LED pixel triplet in a display produces the colors that we see. Newer organic LEDs (OLEDs) are smaller, brighter, and more efficient than standard LEDs. Liquid crystal displays (LCDs) that are commonly used in inexpensive electronics, do not produce light; instead, they block light from passing through them. Hence, LCDs are the most energy efficient but have limited applications. Display sizes vary from small (smartphones) to very large (highway billboards or *jumbotrons* in sports venues).

EXAMPLE **2.12**

Number of LEDs in a display

Consider a 15-inch laptop monitor that has a 2D array of 1440 (horizontal) by 900 (vertical) pixels with each pixel containing a set of R, G, and B LEDs. Hence, there are

$$1440 \times 900 \times 3 = 3{,}888{,}000 \text{ LEDs}$$

Such large numbers are usually expressed using scientific notation ($x.xx \times 10^y$). In this case,

$$3{,}888{,}000 \text{ LEDs} = 3.89 \times 10^6 \text{ LEDs}$$

Each LED is typically controlled with 256 levels (8 bits, $2^8 = 256$) to produce a corresponding brightness with 0 corresponding to fully off and 255 to fully on. Hence, each image on your monitor is encoded with

$$3{,}888{,}000 \text{ LEDs} \times 8 \text{ bits/LED} = 31{,}104{,}000 \text{ bits} = 3.11 \times 10^7 \text{ bits}$$

EXAMPLE **2.13**

Pixels in an image

Figure 2.15 shows a gray-scale digital camera image that is zoomed in to show the individual pixels. Pixel sizes in high-resolution cameras are so small that individual pixels cannot be perceived with the eye. When viewed from close up, (magnified) pixels become evident.

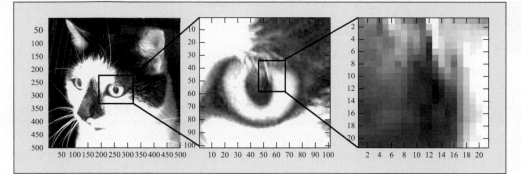

Figure 2.15

Digital camera images are composed of square pixels that have constant values and exhibit sharp transitions at the edges.

Digital image processing applications

EXAMPLE 2.14

With digital images represented by a matrix of numbers, digital signal-processing techniques allow interesting and clever features to be implemented with fast computers, allowing the results to be produced quickly. Two examples come to mind.

- Camera images can be processed to locate faces. Face recognition techniques rely on the expected positions of two eyes, a nose, and a mouth. These features can be quickly detected in the image data.

- Modern cameras watch you and recognize your face and eyes when you are watching a video. This allows smartphones to pause the video when you are not looking at the screen.

2.6 PROPRIOCEPTION

Proprioceptor sensors indicate the current positional state of the system—either a smartphone or robot. Just like nerves in your arm tell you where your arm is located (such as when you reach for a glass), smart devices have similar self-location sensors. The two that we consider here are accelerometers and global positioning system (GPS).

2.6.1 Accelerometers

An *accelerometer* is a sensor that measures acceleration, which typically comes in one of two forms.

1. Acceleration is a change in velocity either in its magnitude (known commonly as speed) or in its direction (as when your car makes a turn). The accelerometer in your smartphone can also count your footsteps, because each step causes abrupt accelerations.

2. Accelerometers are also sensitive to the force of gravity. This *always-downward* component of acceleration tells your smartphone what its orientation is when you are viewing the display.

> *Factoid:* Researchers are investigating accelerometer waveforms produced by walking to identify a person's physical condition and identity.

EXAMPLE 2.15 — Digital accelerometer

A digital accelerometer produces an indication (similar to a switch closure) of when the acceleration exceeds a threshold value. A digital accelerometer detects sudden changes in motion that are greater than those encountered in normal usage.

- If your laptop were to fall from your desk to the floor, a modern laptop would contain a digital accelerometer that senses the sudden motion and puts the hard drive into a collision-safe mode (very quickly) before your laptop hits the ground.

- A digital accelerometer in a car activates an air bag in an accident. As a car comes to rest during a front-end collision, it undergoes a very large negative acceleration. The accelerometer has a built-in threshold, so relatively small accelerations caused by speeding up, braking, driving over potholes, and bumping another car while parking are not sufficient to activate the air bag.

EXAMPLE 2.16 — Smartphone accelerometer

An accelerometer in your smartphone also measures g, which is the gravitational acceleration. Figure 2.16 shows how a pair of accelerometers detect the g value to determine how to orient the display on your smartphone.

2.6.2 GPS

Global positioning systems (GPS) use time-stamped information transmitted from a set of satellites at known locations to compute their positions. While only time differences from additional satellites are sufficient for localization, consider here the simpler case in which the travel times from satellites to the GPS themselves are known. For example,

Figure 2.16

A pair of accelerometers measure the gravitational acceleration g to determine how to orient the display on your smartphone.

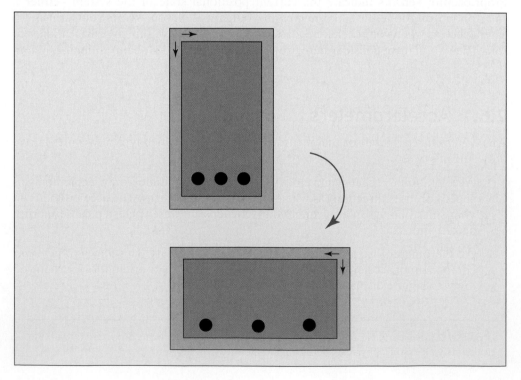

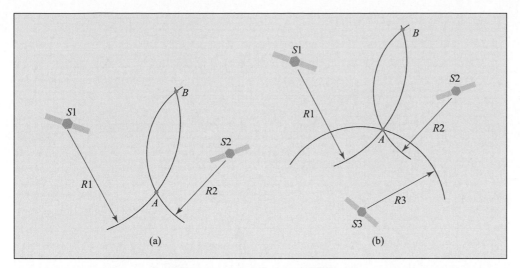

Figure 2.17

GPS localization. (a) Two satellites provide ambiguous localization in two dimensions. (b) Three satellites are needed for unambiguous localization in two dimensions.

Figure 2.17 shows two satellites $S1$ and $S2$ transmitting the current time signal simultaneously. The GPS measures the travel times from the delays in the received signals. The signal delay from $S1$ determines its range $R1$ and that from $S2$ determines $R2$. With two satellites, there are two locations on a planar surface A and B that correspond to these ranges. By adding another satellite $S3$, the GPS can differentiate A from B to determine a unique location in the plane. A fourth satellite provides data to locate the GPS in three dimensions.

GPS calculations　　EXAMPLE 2.17

To illustrate GPS localization, consider the simple two-dimensional case shown in Figure 2.18. Geometry indicates that we can simplify the problem by defining an axis that passes through $S1$ and $S2$, and centering the coordinate system on $S1$, $S2$ is at $(d, 0)$, and

$$x_G^2 + y_G^2 = R1^2$$

where $R1$ is the distance from $S1$ to the GPS derived from the travel time of the signal from $S1$ to GPS. Consider knowing the distance $R2$ derived from the signal travel time from $S2$ to GPS. Then that GPS lies somewhere on a circle with the $S2$ located at its center. Analytically,

$$(x_G - d)^2 + y_G^2 = R2^2$$

Equating both to y_G^2

$$y_G^2 = R1^2 - x_G^2 = R2^2 - (x_G - d)^2$$

Expanding the right equation, we get

$$R1^2 - x_G^2 = R2^2 - x_G^2 + 2dx_G - d^2$$

yielding the unique solution for x_G, as

$$x_G = \frac{R1^2 - R2^2 + d^2}{2d}$$

Inserting this value into the $R1$ equation yields the two solutions for y_G as

$$y_G = \pm\sqrt{R1^2 - x_G^2}$$

Solving similar equations for $R1$ or $R2$ and $R3$ leads to a unique location for the GPS.

This analysis assumes that the range values are exact and correct. Figure 2.18b shows the case when there are errors in the $R1$ and $R2$ values. While particular $R1$ and $R2$ values do yield solutions, errors cause the true GPS location to lie within a region whose dimensions increase with the error magnitudes. Signal processing techniques are applied to the satellite signals to reduce this region size to one that is useful in practice.

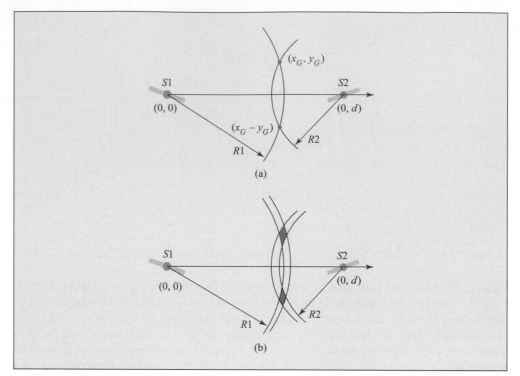

Figure 2.18

GPS localization in two dimensions. (a) Exact range values provide two locations. (b) Range errors reduce the location resolution from a point to a region.

GPS-equipped smart devices provide position and velocity data for the family car as well as autonomous mobile robots. This information is superimposed on maps stored within the smart device memory to indicate road locations and their speed limits. If you add your trip destination, the smart device uses the distance traveled along a desired path (either shortest, quickest, or toll-free) and their speed limits to estimate your arrival time. The time automatically updates as you encounter traffic delays and detours.

EXAMPLE 2.18

GPS on smartphones

GPS-equipped smartphones connected through wireless networks to the Internet are providing increasingly information-rich utilities that include the following.

Personal travel guide: As long as your smartphone can access the satellites, your location is known, and the device plans a route to your destination that includes public transportation.

Points of interest: If you are traveling and searching for a restaurant, a smartphone provides the locations of nearby dining facilities as well as their menus.

Traffic monitoring: If every vehicle on the road was equipped with a smart device that communicates with a computer, traffic flow could be monitored and alternate routes suggested to relieve congestion.

Golf caddy: When on a golf course, a smartphone tells you the distance to the center of the green, keeps track of your hitting distances, and (after learning your golfing ability) suggests which club to use on the next shot.

Sorting pictures: When you take a picture, the smartphone also records its GPS location. Then, if you wanted to view pictures taken when you visited your grandparents, you could sort them according to the GPS location of their home.

GPS for autonomous vehicles: While a GPS provides location information, it does not provide the location of other nearby (non-GPS equipped) vehicles or obstacles—at least currently. For this information, we need active ranging sensors for obstacle detection and object localization, which are discussed in the next section.

2.7 ACTIVE SENSORS

The sensors discussed previously are *passive* in that they detect energy produced by another source. *Active* sensors produce their own energy that interacts with the environment and is then sensed. Such sensors detect the presence of objects, such as people counters at store entrances, or bar-code readers. Other sensors determine the range of an object, such as infrared ranging systems found in auto-focus cameras and sonar systems commonly used in robotics.

2.7.1 Optical Detection

An active optical detector consists of two components: a transmitter and a receiver. Figure 2.19 shows an infrared (IR) light-emitting diode (LED) in a clear enclosure and an IR detector in a dark enclosure. The detector enclosure is dark, because visible light does not pass through it, but it is transparent to IR light. When current is passed through an LED, it generates optical power.

Figure 2.20 shows these two components positioned so that the transmitter directs the IR light beam directly into the receiver. When no object interrupts the beam, the receiver detects the light energy and produces an output signal interpreted by the system that no object is present. It may be helpful to think of the receiver operation in terms of a switch: The receiver acts as an open switch when a beam is detected. When an object interrupts the beam, the receiver produces a signal interpreted by the system that an object is present. In this case, the receiver acts as a closed switch to indicate the presence of the object.

The transmitter and receiver are separated by distances from centimeters for counting small objects (such as pills entering a bottle) to meters for counting large objects (such

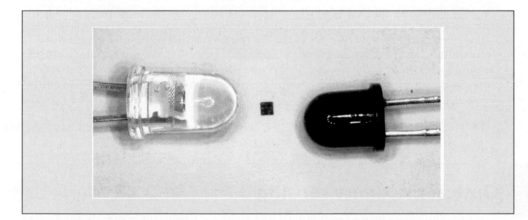

Figure 2.19

Infrared (IR) light-emitting diode (LED on left) and IR phototransistor (PT). The square measures a 1 mm × 1 mm. The LED is located in the remote control and the PT in the TV.

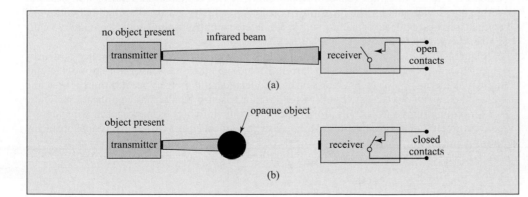

Figure 2.20

IR beam interrupt sensor. (a) When no object is present, the receiver detects the IR beam, producing an open pair of contacts. (b) When an object interrupts the IR beam, the receiver causes the contacts to close.

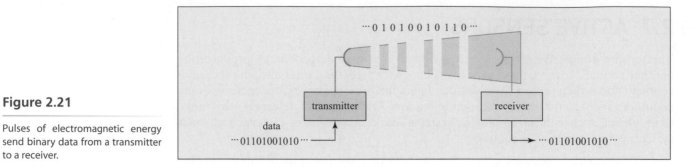

Figure 2.21

Pulses of electromagnetic energy send binary data from a transmitter to a receiver.

as people entering a store or boxes moving down a conveyer in a manufacturing plant). Such systems are used in automatic garage doors to detect the presence of cars or people beneath the door and in home burglar alarms to detect the presence of a burglar.

EXAMPLE 2.19 | Binary data transmission

The basic idea of the beam interrupt sensor also applies to sensors that detect radio waves used in digital TV and cellular telephone systems. A similar transmitter/receiver system communicates data many hundreds of kilometers (radio links) or thousands of kilometers (satellite links). Instead of an object blocking the optical beam, the transmitter modulates (turns it on and off) the beam to transmit binary data. For example, the beam is switched on to transmit a 1 and switched off to transmit a 0. A remote receiver senses this modulated beam to detect the data, as shown in Figure 2.21.

One desirable feature of a digital data transmission system is that the actual beam intensity is not important as long as it causes the receiver to reliably detect the signal. For example, a sensor output value that is a thousand times greater than that needed to exceed a threshold produces the same interpretation by the system than the value that is only twice the threshold. This feature is important because weather and other environmental conditions produce fluctuations in the beam intensity as it travels between the transmitter and receiver. These fluctuations do not affect the reliable transmission of data as long as the detected signal exceeds the threshold at all times. Chapter 7 describes digital signal processing that helps to extract data values from sensor signals, and Chapter 10 describes methods to correct errors when they do occur.

EXAMPLE 2.20 | Optical proximity sensing

An optical proximity sensor transmits and receives optical energy over very short distances. Unlike the beam interrupt sensor, the proximity sensor contains the emitter and detector within one device, making the proximity sensor compact. The sensors that read bar codes and compact disks are optical proximity sensors that detect non-reflecting objects that are typically transparent or black in color.

Figure 2.22 shows the operation of a proximity sensor that detects the presence of nearby objects from the light they reflect back to the detector. When the sensor does not detect any light, it acts as an open switch. Upon sensing a reflected light intensity greater than a threshold value, the detector acts as a closed switch. Proximity sensors have many applications. These sensors count items, such as pills dropping from a chute into bottles, packages moving along a conveyor belt, or bars in a bar code. Robots use proximity sensors to detect physical obstacles, thereby preventing collisions.

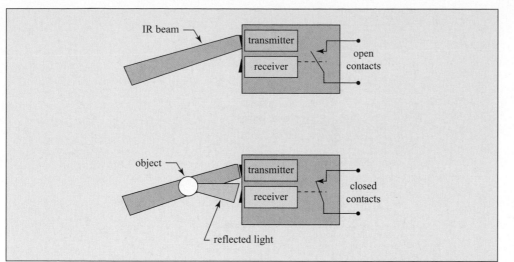

Figure 2.22

An infrared (IR) proximity sensor provides an switch closure when a reflecting object is present.

A wide variety of detectors use optical energy to function. Some employ the red neon-helium laser, such as the UPC symbol scanners in supermarkets.

Bar code scanner   **EXAMPLE 2.21**

Figure 2.23 shows a laser scanning a black-and-white bar code. When the laser spot shines on a white region, light bounces back to the detector, and when it shines on a black region, the light is absorbed and no light returns to the detector. This figure shows that a black bar produces a binary 1 (a high level in the waveform) and a white space produces a binary 0 (a low level).

2.7.2 Infrared Ranging

Not all optical energy is in the visible range. Many sensors use infrared (IR) light, because being in the low (below red) end of the light spectrum, it can be reliably detected under various lighting conditions.

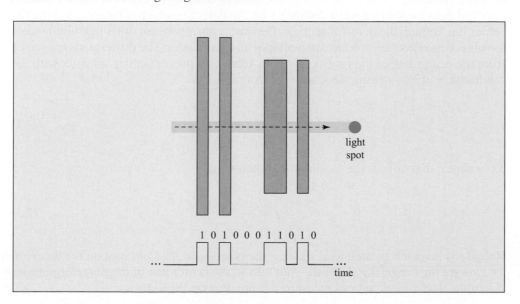

Figure 2.23

Laser scanning a bar code.

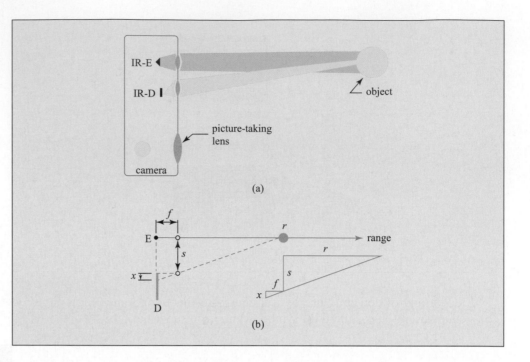

Figure 2.24

Infrared (IR) range sensor. (a) The emitter (E) produces a narrow IR beam that reflects from a object at range r. The reflected light produces a small spot on the detector (D) at a location that determines the range. (b) A pinhole model used for the calculation of range. The pair of similar triangles leads to the formula for range of $r = sf/x$.

An autofocus camera uses a variation of the proximity sensor to determine the range of an object for setting the focus. Figure 2.24a shows an auto-focus camera that includes an infrared (IR) emitter and an IR detector, which are similar to the proximity sensor. The emitter sends a thin beam of infrared light using a lens. When the beam encounters a reflecting object, the light scatters. The detector lens collects some of the scattered light. The detector determines the object's range from the location of the light spot on the detector.

To illustrate the operation more clearly, the two lenses are modeled as pinholes, as shown in Figure 2.24. These pinholes enable us to trace the path of the light emitted from the transmitter and reflected back to the detector. Lenses concentrate and collect the light rays, making the system more sensitive. The distance from the receiver pinhole to the focal plane is f, and the separation between the transmitter and receiver pinholes is s. The values of f and s are fixed by the camera design. An object at range r reflects a spot of light back to a point a distance x on the detector, which is measured along the center line though the detector pinhole. Because a narrow beam of IR light illuminates the object, its reflection produces a small spot on the detector. The distance x is measured from the center line of the receiver pinhole, forming a pair of similar triangles with the relationship of r to s being the same as f to x:

$$\frac{r}{s} = \frac{f}{x} \tag{2.4}$$

The range r that adjusts the camera focus then equals

$$r = \frac{sf}{x} \tag{2.5}$$

Range r is inversely related to x, which is the position of the light spot on the detector. The longer the range, the closer the spot falls to the center line of the detector pinhole. A simple electrical circuit can measure this position on the detector.

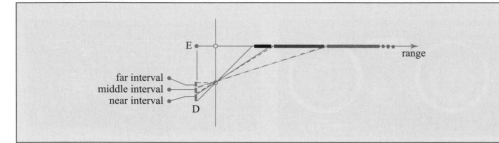

Figure 2.25

Digital IR ranging system. Multiple detector elements determine the range interval of an object rather than the exact range value. The detector array (D) sets the camera's focus without any calculations.

EXAMPLE 2.22 · IR ranging

Consider a camera with an IR emitter and detector spaced 2 cm apart ($s = 2$ cm) and the detector positioned 0.5 cm behind the lens ($f = 0.5$ cm). An object casts a spot image 0.1 mm from the center line of the detector lens ($x = 0.1$ mm). The range of the object equals

$$r = \frac{sf}{x} = \frac{0.02 \text{ m} \times 0.005 \text{ m}}{0.0001 \text{ m}} = \frac{10^{-4} \text{ m}^2}{10^{-4} \text{ m}} = 1 \text{ m}$$

We can also digitally determine the distance between the camera and object without doing any calculations. To create a simple digital IR ranging system, the IR range sensor's single detector is divided into an array of smaller detector elements, as shown in Figure 2.25. Each detector element corresponds to a particular range interval. When a detector element senses the light spot reflected from the object, it sends a signal that adjusts the lens focus to the appropriate range interval. Note that this digital sensor uses a *qualitative method*, which requires no calculations, to determine the range interval. The detector element that senses the reflected light sets the focus.

EXAMPLE 2.23 · Digital IR range sensor

Consider a camera that has an IR emitter and detector spaced 2 cm apart ($s = 2$ cm) and an array of detector elements positioned 0.5 cm behind the lens ($f = 0.5$ cm). Each detector element has a width of 0.1 mm. The light reflected from an object produces a spot on the second detector element from the detector pinhole centerline.

The second detector element spans $x = 0.1$ mm to $x = 0.2$ mm. From these two limits we can find the range interval that extends from the near range limit r_N (corresponding to $x_N = 0.2$ mm) to the far range limit r_F (corresponding to $x_F = 0.1$ mm). When we convert all measurements to meters, these two limits equal

$$r_N = \frac{sf}{x_N} = \frac{0.02 \text{ m} \times 0.005 \text{ m}}{0.0002 \text{ m}} = \frac{10^{-4} \text{ m}^2}{2 \times 10^{-4} \text{ m}} = 0.5 \text{ m}$$

$$r_F = \frac{sf}{x_F} = \frac{0.02 \text{ m} \times 0.005 \text{ m}}{0.0001 \text{ m}} = \frac{10^{-4} \text{ m}^2}{10^{-4} \text{ m}} = 1 \text{ m}$$

Any object 0.5 m to 1 m in range casts a light spot onto the second detector element. When the second detector element senses a spot anywhere along its width, it focuses the camera at an intermediate range, such as 0.7 m.

Figure 2.26

Sonar ranging system that operates in air at 40 kHz. (a) Transmitting (T) and receiving (R) transducers. (b) Integrated circuit chips and components. The square is 1 mm × 1 mm.

2.7.3 Acoustic Ranging

Sonar is a common sensor for mobile robots because it is inexpensive, works in dark environments, and provides the location of obstacles. Figure 2.26 shows a popular inexpensive sonar. A sonar transmitter emits an acoustic pulse having a frequency around 40 kHz, as shown in Figure 2.27a. The acoustic energy is contained within a beam, which is similar in shape to the beam produced by a flashlight. The sound pulse travels at the speed of sound at $c = 343$ m/s in air. A reflection is produced when the acoustic pulse encounters an object. If the reflection is detected by the sonar receiver, usually located near the transmitter, the echo is detected. Most sonars detect echo arrival times with the first echo determining the range to the nearest obstacle.

Figure 2.27a shows the sonar echo waveform applied to a threshold τ. Thresholding eliminates artifacts caused by noise. The time at which the echo amplitude first exceeds the threshold defines the echo *time-of-flight* (TOF), which is the time the acoustic pulse travels the round-trip distance between the sonar and closest object. The range to the closest object is computed from the sound speed c and the TOF using

$$r = \frac{c \times \text{TOF}}{2} \qquad (2.6)$$

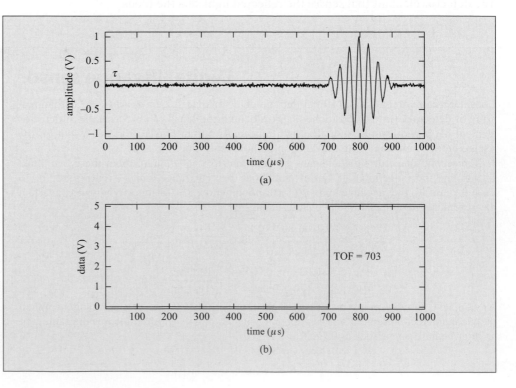

Figure 2.27

Waveform produced by echo from object at 0.12 m range. (a) Echo waveform. (b) Digital TOF indication. Acoustic pulse was transmitted at time $t = 0$, and the echo exceeds the threshold τ at time $t = 703\,\mu s$.

Sonar ranging EXAMPLE 2.24

A sonar transmits an acoustic pulse, thus defining time $t = 0$. Figure 2.27 shows the echo waveform exceeds threshold τ at time TOF $= 703\,\mu$s. The closest object range detected by sonar is

$$r = \frac{c \times \text{TOF}}{2} = \frac{(343\,\text{m/s}) \times (703 \times 10^{-6}\,\text{s})}{2} = 0.121\,\text{m}$$

Radar Ranging

Sonar ranging works well for short ranges, less than 5 m. Obstacle detection for a car traveling on a highway requires detection for ranges that are ten times farther. Radar (radio detection and ranging) systems provide this extra range and operate using time-of-flight similar to sonar—except that the speed is that of light ($c = 3 \times 10^8$ m/s, or 1 foot per nanosecond). Fast electronics determine the range to the nearest 0.1 m, which is sufficiently accurate for long-distance detection.

Radar detection for automobiles EXAMPLE 2.25

A radar transmits a pulse, thus defining time $t = 0$. The echo waveform exceeds threshold τ at time TOF $= 30$ ns. The closest object range detected by radar is

$$r = \frac{c \times \text{TOF}}{2} = \frac{(3 \times 10^8\,\text{m/s}) \times (30 \times 10^{-9}\,\text{s})}{2} = 4.5\,\text{m}$$

2.8 Summary

This chapter describes the sensor inputs to and actuator outputs from smart digital systems. These are motivated by tasks that these systems serve to perform. The sensor and actuator modalities follow the human senses of touch, hearing, and vision. The following chapters indicate the processing performed by computer and communication systems to acquire and process these sensor output data to form the signals that drive the actuators.

2.9 Problems

2.1 Music icon address. What screen-row-column address would the controller assign to the music icon shown in Figure 2.10 if the icon is located on the third screen of 16 possible screens?

2.2 Calculator switch array. A scientific calculator has 50 keys for digits and logarithmic and trigonometric functions arranged in five rows and ten columns. Specify a binary row-column address code to indicate which key was pressed.

2.3 Forming a touch screen switch array. A touch screen array has a count of rows and columns that sums to 10.

What is the structure of the array that accommodates the maximum number of keys?

2.4 Finger swipe along a switch array. Extending Example 2.7, a linear switch array is 10 cm long and has a resolution $R_s = 2$ switches/mm. A swipe motion is detected if the mid-point location changes by more than 8 switches. If the sampling period $T_s = 0.1$ s, what is the minimum finger swipe speed along the linear array that indicates a swipe motion?

2.5 Multiple finger gesture. Extending Example 2.8, a linear switch array is 10 cm long and has a resolution $R_s = 2$

switches/mm and a sampling period of $T_s = 0.1$ s. If $P1 = 20$, $P2 = 40$, $P1' = 22$, and $P2' = 36$ is sensed as a gesture, what is the finger swipe speed? Is it widening or spreading?

2.6 **Number of bits in a large color LED display.** A large color billboard is a two-dimensional array of $2^{10} \times 2^{10}$ pixels with each pixel containing a red, green, and blue LEDs. (Single LED packages contain separate R, G, and B LEDs inside.) Assuming that each LED is controlled to shine at one of 256 levels, how many bits are needed to specify a color image on the billboard? How many different colors can each 3-LED pixel display?

2.7 **Number of possible images in a large color LED display.** A large color billboard is a two-dimensional array of $2^{10} \times 2^{10}$ pixels with each pixel containing red, green, and blue LEDs. Assuming that each LED is controlled to shine at one of 256 levels, what is the number of different images that can be displayed? Express answer as a power of 10.

2.8 **Bit rate to generate a full-screen movie.** A video game displays images on your laptop monitor having a resolution of 1680×1050 pixels. Each pixel contains a red, green, and blue LED, and each LED is controlled to shine at one of 256 levels. The game produces a new image on the screen 60 times per second. How many bits per second are being sent to your monitor while you are playing your game? Give answer in scientific notation $(x.xx \times 10^y)$.

2.9 **Smartphone location from two range measurements.** This problem considers the location information using the range values measured by two antennas. Let antennas $A1$ and $A2$ be located 5 km apart. Determine the two possible locations for the smartphone relative to antenna $A1$ when the smartphone range from $A1$ is 3 km and from $A2$ is 3.5 km.

2.10 **Smartphone location region caused by range errors.** Sketch and determine the four points defining the region that contains your smartphone when the range measured from antenna $A1$ is (3 ± 0.1) km and that from $A2$ is (3.5 ± 0.1) km.

2.11 **Pulse time for a bar code scan.** In Example 2.21, if a laser spot moves across the bar code at 10 m/s and the width of the thinnest bar is 1 mm, what is the duration of the shortest pulse produced by the scanner? Give answer in μs (10^{-6} s).

2.12 **IR range sensor.** In an IR autofocus camera, the emitter and detector are separated by 1 cm and positioned 1 cm behind the lenses, which are modeled as pinholes. The light reflected from an object produces a spot 1 mm from the centerline of the detector pinhole. What is the range of the object from the camera in meters (m)?

2.13 **Digital IR range sensor.** In a digital IR autofocus camera, the emitter and detector are 1 cm apart, and the detector array is 1 cm behind the lens. An IR detector element has near and far limits $x_F = 0.01$ mm and $x_N = 0.02$ mm that senses light reflected from an object located from r_N to r_F in range. Determine the values of r_N and r_F in m.

2.14 **Digital IR range sensor dimensions.** In a digital IR autofocus camera, the emitter and detector are 1 cm apart and the detector array is 1 cm behind the lens. What are the detector element's near and far limits (x_F and x_N) that senses light reflected from an object located 1 m to 4 m away? Give answer in millimeters (mm).

2.15 **Sonar ranging—range to TOF.** A sonar system operates in air up to a maximum range of 4 m. What is the maximum TOF? Give answer in ms (10^{-3} s)?

2.16 **Sonar ranging—TOF to range.** A sonar system observes a TOF = 10 ms. What is the object range in meters (m)?

2.17 **Sonar ranging resolution.** A sonar system experiences a jitter in the echo arrival time because of dynamic temperature variations in air, which limits the TOF resolution to ΔTOF $= \pm 50$ μs. What is the corresponding sonar range resolution Δr in mm?

2.18 **Radar ranging—range to TOF.** A radar system operates up to a maximum range of 100 m. What is the maximum TOF?

2.19 **Radar ranging—TOF to range.** A radar system observes a TOF = 0.1 μs. What is the object range in meters (m)?

2.20 **Radar ranging resolution.** A radar system is specified to have a range resolution of ± 0.1 m. What is the corresponding resolution in the radar TOF?

2.10 Excel Projects

The following Excel projects illustrate the following features:

- Inserting shapes into the worksheet.
- Composing a VBA Macro to implement a counter.
- Assigning a Macro to a shape.
- Implementing a touch screen display.

Using a Narrative box (described in Example 13.4), include observations, conclusions, and answers to questions posed in the projects.

2.1 **Implementing a counter.** Modify Example 13.10 to place the counter in cell A2. Replace the rectangles with your favorite shape.

2.2 **Implementing a 2 × 4 switch array.** Extend Example 13.11 to implement a switch array having two rows and four columns. Provide the binary addresses of the row and column location with each bit in a separate cell.

2.3 **Specifying your favorite cell color.** Example 13.12 shows the VBA Macro that increases and decreases the red component of a cell color. Complete the worksheet by writing the VBA code that also specifies the green and blue components.

2.4 **Color display specifications.** Compose a worksheet that specifies a color RGB display with each LED having 256 levels having values n_{row} (number of rows), n_{col} (number of columns), and f_{frame} (frame rate with units frames/second) and computes n_{pixels} (number of pixels), n_{LED} (number of LEDs), n_B (number of bytes per frame), n_b (number of bits per frame), and $\mathcal{D}$ (data rate in bits/second).

COMBINATIONAL LOGIC CIRCUITS

LEARNING OBJECTIVES

After completing this chapter, the reader should be able to:

- Relate voltage levels to logic variable values.
- Form truth tables, logic circuits, and logic equations for simple logic circuits.
- Understand the process of designing and analyzing a combinational logic circuit.
- Design logic circuit alternatives that use fewer logic gates.
- Implement digital logic circuits employed in smart devices, robots, and communication systems.

3.1 INTRODUCTION

In this chapter, digital *signals* are considered to be *logic symbols* that are manipulated with logic circuits. Elementary logic gates are interconnected to implement digital logic circuits that perform useful tasks. These include controlling a seven-segment display and performing arithmetic operations.

The main topics in this chapter are covered in the following sequence.

Digital logic gates—The binary voltage levels in digital circuits lend themselves for analysis using AND, OR, and NOT logic operations that form the basic building blocks of digital logic circuits.

Digital logic circuits—Digital logic circuits implement a truth table that defines the desired output values when considering all possible input combinations of input logic variables.

Logic equations—Logic circuits are implemented on a digital computer using logic equations that define the desired logic operations.

Efficient digital logic circuits—Digital logic circuits can be implemented with a fewer number of gates by manipulating logic equations and by applying simplifying observations.

3.2 LOGIC VARIABLES AND LOGIC EQUATIONS

Digital signals represent binary data (0/1) by having only two discrete values, such as 0 and 5 V. Such digital signals were the earliest electrical signals, typically produced by switch closures in early telegraphs. When a switch is *open*, no electrical current was transmitted. When the switched *closed*, current began to flow from the transmitter to the receiver.

A *logic variable* is an abstract descriptor that converts digital signals expressed in volts to logic levels that equal either 0 or 1 and represents mutually exclusive alternatives such as true/false, yes/no, or closed/open. We denote logic variables with capital letters, such as A, B, and Y. Logic variables have their own mathematical operations that are described using *Boolean algebra*.

Combinational (also called *combinatorial*) *logic* produces an output logic variable that is related to only the current values of input logic variables. In other words, the input logic variables *combine* to form the output logic variable value. The relationship between the input values and the output logic value that they produce is specified in three ways.

1. A *logic equation* uses Boolean algebra. While there are various ways of expressing logic operators, this book uses the conventions that can be typed on computer keyboards. The operators and their symbols are given in Figure 3.1. There are

Figure 3.1

Logic operations performed by the elementary NOT, AND, and OR gates.

Logic Operation	Symbol	Example
NOT	–	$Y = \bar{A}$
AND	·	$Y = A \cdot B$
OR	+	$Y = A + B$

row #	A	B	C	Y
0	0	0	0	0
1	0	0	1	0
2	0	1	0	0
3	0	1	1	0
4	1	0	0	0
5	1	0	1	0
6	1	1	0	0
7	1	1	1	1

Figure 3.2

Truth table with input logic variables A, B, and C and output logic variable Y.

many ways to write the logic operators, and we chose these for their simplicity and their relation to the arithmetic operations of addition ($+$) and multiplication ($\cdot$). The problem with this notation is that the symbols, especially the $-$ cannot be typed in the instructions used in programming languages.

2. A *truth table* consists of two sections, as shown in Figure 3.2. The input section lists all of the possible logic conditions of the input variables, denoted by A, B, and C. The output section indicates the output value Y that is produced for each set of inputs, as required by the desired task.

3. A *logic circuit* is the interconnection of elemental gates. The elemental gates include the AND, OR, and NOT gates described later in this chapter. All other combinational logic functions can be implemented using these three gates.

While the basic binary nature of data signals remain unchanged in their analysis, the devices that produce these signals have changed dramatically with improvements in technology. Initial implementations of digital circuits used mechanical switches that controlled current flow. Big and manually operated switches were replaced with electrically operated switches called relays. These relatively slow mechanical relays were replaced by faster electrical switches, initially implemented with vacuum tubes. Large, hot, and energy inefficient, vacuum tubes were replaced by semiconductor transistors. Transistors have continued to shrink in size and energy usage, currently with billions embedded within integrated circuit chips. Digital electronics that implement these gates in modern computers consist of integrated circuits using energy-efficient field effect transistors (FETs).

3.3 ELEMENTARY LOGIC GATES

To illustrate elementary logic equations, truth tables, and logic circuits, this section describes the most basic combinational logic gates: the AND, OR, and NOT gates. All other combinational logic functions can be implemented with circuits using these three elementary gates.

3.3.1 NOT Gate

The simplest logic gate is the NOT gate, which produces a 1 at its output only if a 0 is present at the input, thus forming the logic complement. The NOT gate symbol, logic equation, and truth table are shown in Figure 3.3. By convention, inputs enter the circuit elements from the left, and the outputs are on the right. Because there are only two values for our binary logic variables, it is often said that one value is *not* the other—hence, the name NOT gate.

Figure 3.3

NOT gate symbol, logic equation, and truth table. The input is *A* and the output is $Y = \overline{A}$. The output of the NOT gate is the logical opposite or *complement* of the input.

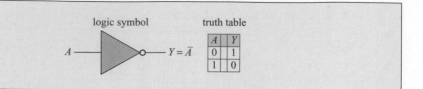

The logic equation *Y equals NOT A* is written as

$$Y = \overline{A} \tag{3.1}$$

with the overscore indicating the logical complement. In words, *Y equals not A*. The NOT operation is *negation* (or *logical complement*): $Y = 1$ when $A = 0$, and $Y = 0$ when $A = 1$, which is clearly shown in its simple truth table. The input section is labeled *A* with its two possible values 0 and 1. The output section, labeled *Y*, shows the value of the NOT gate output when the input is the value of *A* in the same row.

3.3.2 AND Gate

The AND logic element (AND gate) produces a logic 1 at its output only when logic 1's are applied to all of its inputs. Figure 3.4 shows a two-input AND gate logic symbol, logic equation, and truth table.

The logic equation for two input logic variables *A* and *B*, producing output logic variable *Y*, is written

$$Y = A \cdot B \tag{3.2}$$

In words, *Y equals A and B*.

The truth-table input section lists all of the possible values of inputs *A* and *B* using a binary counting order: 00 (=0), 01 (=1), 10 (=2), and 11 (=3). This counting order is the convention for structuring the input section of any truth table. This convention allows the input section to be written down first, before the values of the output logic variable are determined. We note that $Y = 1$ only when $A = 1$ *and* $B = 1$. If either $A = 0$ or $B = 0$, or if both equal 0, then $A = 0$. The logical AND is identical to algebraic multiplication.

Figure 3.5 illustrates how two switches can be connected to form an AND circuit. A 1 designates a closed switch and a 0 designates an open switch. The output $Y = 1$, which is

Figure 3.4

AND gate symbol, logic equation, and truth table. Logic variables *A* and *B* are inputs and $Y = A \cdot B$ is the output. *Y* equals 1 only when both *A and B* equal 1.

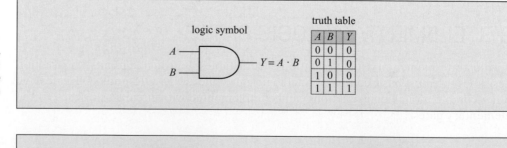

Figure 3.5

An AND gate formed with a pair of switches. (a) The output *Y* is 1, which is a closed switch only when both switches *A* and *B* are closed (=1). (b) The output is 0, which is an open switch when either switch or both switches are open (=0).

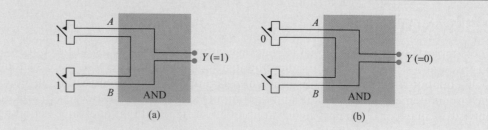

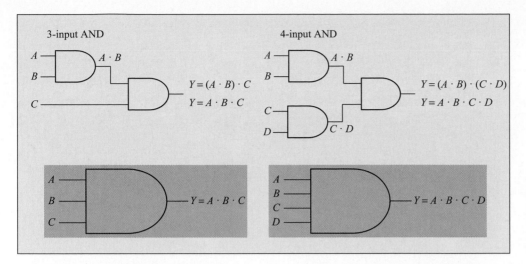

Figure 3.6

Standard 2-input AND gates combine to form 3-input and 4-input AND gates. Shaded boxes shows simplified symbols.

a closed switch that occurs only when both switches A and B are closed, corresponding to $A = 1$ and $B = 1$. The output $Y = 0$ is an open switch that occurs when either switch or both switches are open ($=0$).

Many applications require more than two inputs to an AND gate. Figure 3.6 shows how to combine basic 2-input AND gates to include additional inputs. Three inputs A, B, and C applied to an AND gate would be expressed with the logic equation

$$Y = A \cdot B \cdot C \tag{3.3}$$

Figure 3.6 shows parentheses added to illustrate how the outputs of the initial AND gates are combined at the output of the last AND gate. Although not necessary in this simple case, parentheses are used in more complex logic equations. The simplified logic symbols are shown below the interconnected 2-input AND gates. *The output of a multiple-input AND gate equals 1 only when all the inputs are 1's.*

3.3.3 OR Gate

An OR gate produces an output logic 1 when any of the inputs is a logic 1. An OR gate having two inputs is shown in Figure 3.7. $Y = 1$ when $A = 1$ OR $B = 1$. If both $A = 0$ and $B = 0$, then $Y = 0$. If both $A = 1$ and $B = 1$, then $Y = 1$—not 2 as would the case in normal addition, because logic variables take on only 0 and 1 values. The logic equation Y *equals A OR B* is written

$$Y = A + B \tag{3.4}$$

The OR gate can be extended to accommodate additional logic variable inputs, as shown in Figure 3.8. As in the 2-input OR gate, the output of the multiple-input OR gate is a 1 when any of the inputs is a 1.

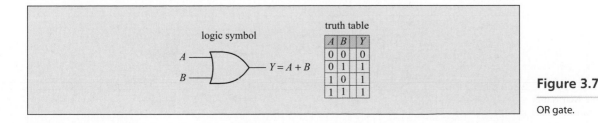

Figure 3.7

OR gate.

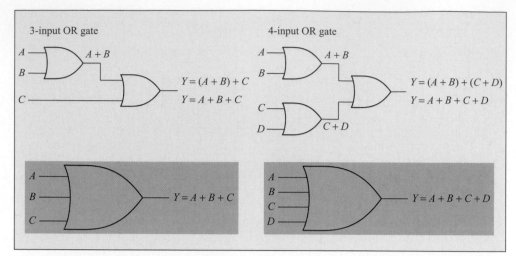

Figure 3.8

Two-input OR gates combine to form 3- and 4-input OR gates. Shaded boxes shows simplified symbols.

3.4 BUILDING-BLOCK GATES

The NOT, AND, and OR gates are the basic gates from which all other gates and combinational logic circuits are implemented. However, these are not the most flexible in terms of implementing a logic circuit using the smallest number of chips. This section describes the logic gates that are commonly used to actually construct logic circuits either because they are the easiest to implement with digital electronics or because they can result in the smallest chip count.

3.4.1 NAND Gate

The NAND (NOT-AND) gate is the most common gate in digital logic implementations because of its flexibility and simple structure. Figure 3.9 shows a NAND gate and a digital integrated circuit package that contains four NAND gates.

The logic equation for a NAND gate is

$$Y = \overline{A \cdot B}$$

The flexibility of the NAND gate is shown in Figure 3.10 in that it can form a NOT gate when the two inputs are connected together. The resulting NOT gate can be connected to the output of a NAND gate to form an AND gate.

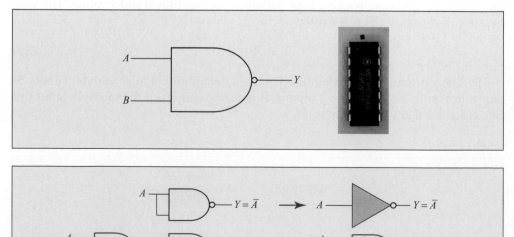

Figure 3.9

NAND gate and digital integrated circuit that contains four NAND gates.

Figure 3.10

A NAND gate can form a NOT gate, and two NAND gates can form an AND gate.

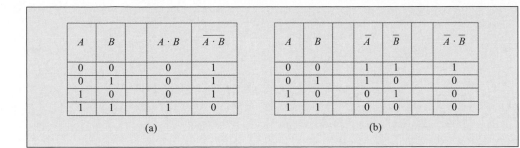

Figure 3.11

Truth tables for Example 3.1.

While it is tempting to think $\overline{A \cdot B} = \overline{A} \cdot \overline{B}$, it is wrong. The best way to remove doubt is to compare truth tables. Figure 3.11a shows the truth table of $A \cdot B$ and its NOT value $\overline{A \cdot B}$. In contrast, Figure 3.11b shows the truth table for $\overline{A} \cdot \overline{B}$ that gives values of inputs $\overline{A}$ and $\overline{B}$ for forming the output value. Comparing the values in the output columns for the same A and B values demonstrates that

$$\overline{A \cdot B} \neq \overline{A} \cdot \overline{B}$$

3.4.2 NOR Gate

Figure 3.12 shows a two-input NOR (NOT-OR) gate. The NOR gate is described by the logic equation

$$Y = \overline{A + B}$$

The truth table of $A + B$ and its NOT value $\overline{A + B}$ are given in Figure 3.13.

The NOR truth table shows only a single 1 in the output section. Hence, we can write the logic equation of the NOR gate, $\overline{A + B}$, and equate it to the condition of the inputs that produce the logic 1 at the input, $\overline{A} \cdot \overline{B}$, which results in

$$\overline{A + B} = \overline{A} \cdot \overline{B}$$

This logic equation is one of *DeMorgan's laws*, which describes the equivalents of logical expressions. These laws are described in computer engineering courses to provide flexibility in implementing efficient logic circuits. If we consider the right side of the equation, we see the AND of two complemented logic variables. Figure 3.14 shows how the NOR gate is implemented from two NOT gates and one AND gate.

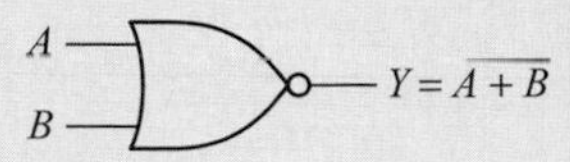

A ———⟩○— $Y = \overline{A + B}$
B ———

Figure 3.12

Two-input NOR gate.

Figure 3.13

NOR truth table that starts with the OR truth table.

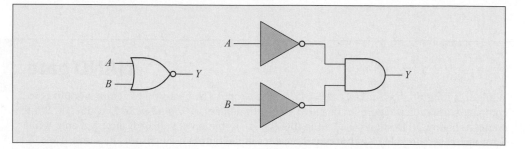

A	B		A + B	$\overline{A + B}$
0	0		0	1
0	1		1	0
1	0		1	0
1	1		1	0

Figure 3.14

Equivalent logic circuits. NOR gate can also be implemented using two NOT gates and one AND gate.

3.4.3 ExOR Gate

The exclusive-OR gate (ExOR) is a useful gate for comparing two binary numbers for encrypting data and for arithmetic operations. The ExOR gate produces an output 1 only when one—but not both—of the inputs is 1. Let A and B be the inputs and Y be the output. The ExOR logic operation is denoted by the symbol $\oplus$ and is expressed as

$$Y = A \oplus B \tag{3.5}$$

Figure 3.15 shows the ExOR logic symbol and its truth table. The following examples show the ExOR gate used in comparison and in encryption tasks.

EXAMPLE 3.3

Comparing two binary numbers

Consider two 4-bit binary numbers given by $A_3 A_2 A_1 A_1$ and $B_3 B_2 B_1 B_0$. We want a logic circuit that compares these two numbers and produces a logic output NE with $NE = 1$ only when the two numbers are not equal (that is, there is a difference in at least one bit position in the two numbers).

Figure 3.16 shows an ExOR gate at each bit position A_i and B_i for $0 \le i \le 3$. If $A_i = B_i$, both are either 0 or 1, and the output of the ExOR gate is 0. If $A_i \ne B_i$, the output of their ExOR gate is 1. If the two 4-bit binary numbers are not equal, they must differ in at least one of the bit positions, causing that ExOR output to equal 1. The OR gate combines all of the ExOR gate outputs, and hence it will be a 1 if any of the ExOR outputs is a 1. Hence, $NE = 1$ indicates that the two numbers are not equal.

Figure 3.15

Exclusive OR (ExOR) gate produces an output that is a logical 1 when only one of its two inputs is a logical 1.

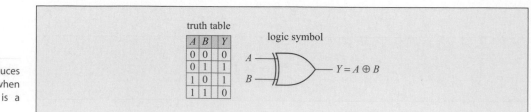

truth table

A	B	Y
0	0	0
0	1	1
1	0	1
1	1	0

logic symbol

$Y = A \oplus B$

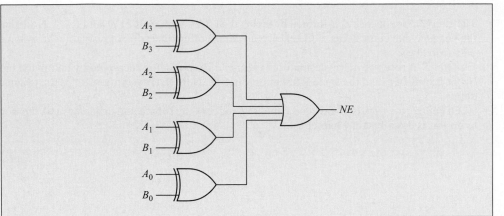

Figure 3.16

ExOR comparison circuit. $NE = 1$ when $A_3 A_2 A_1 A_1 \neq B_3 B_2 B_1 B_0$.

Performing encryption with ExOR gate

EXAMPLE 3.4

The easiest way to implement an encryption scheme is with ExOR gates. For encrypting, we apply the binary data and a random binary sequence—one bit position at a time—as inputs to an ExOR gate. The ExOR gate produces the encrypted data sequence. Applying the encrypted data and the same random binary sequence—again one bit position at a time—to an ExOR gate yields the original data sequence.

To demonstrate this let D_i for $i = 0, 1, 2, \ldots, n_x - 1$ denote the original binary data produced by a source and RBS_i for $i = 0, 1, 2, \ldots, n_x - 1$, which is a random binary sequence having the same number of bits as in the original data. Then, using this bit-wise ExOR operation, the encrypted message, which is denoted by $\mathcal{E}_i$ for $i = 0, 1, 2, \ldots, n_x - 1$ is formed by

$$\mathcal{E}_i = D_i \oplus \text{RBS}_i \text{ for } i = 0, 1, 2, \ldots, n_x - 1$$

The encrypted message $\mathcal{E}_i$ is transmitted to the destination, where the receiving party either knows or can generate the same RBS_i used in the encryption. To retrieve the original data D_i, the ExOR operation is performed on $\mathcal{E}_i$ with RBS_i, or

$$D_i = \mathcal{E}_i \oplus \text{RBS}_i \text{ for } i = 0, 1, 2, \ldots, n_x - 1$$

thus reproducing the original data at the destination.

Consider encrypting hexadecimal numbers (0 through F that represent 4-bit sequences) generated by a source that produces the D_i for $i = 0, 1, 2, \ldots, n_x - 1$ ($n_x = 28$) shown in Figure 3.17a. The randomly generated RBS_i contains the same number of bits as in D_i.

5	7	E	4	0	2	1	(Original data)
0101	0111	1110	0100	0000	0010	0001	(D_i)
$\oplus$	$\oplus$	$\oplus$	$\oplus$	$\oplus$	$\oplus$	$\oplus$	$\oplus$
0010	1010	1110	0100	1011	0010	1101	(RBS_i)
0111	1101	0000	0000	1011	0000	1100	($\mathcal{E}_i$)
7	D	0	0	0	0	C	(Encrypted data)

(a)

7	D	0	0	0	0	C	(Encrypted data)
0111	1101	0000	0000	1011	0000	1100	($\mathcal{E}_i$)
$\oplus$	$\oplus$	$\oplus$	$\oplus$	$\oplus$	$\oplus$	$\oplus$	$\oplus$
0010	1010	1110	0100	1011	0010	1101	(RBS_i)
0101	0111	1110	0100	0000	0010	0001	(D_i)
5	7	E	4	0	2	1	(Original data)

(b)

Figure 3.17

Tables showing (a) data encryption and (b) decryption in Example 3.4.

The encrypted sequence $\mathcal{E}_i$ is formed by taking the bit-by-bit ExOR of D_i and RBS_i. Note that the hexadecimal equivalents of the encrypted data (bottom line) do not resemble the original data (top line).

After $\mathcal{E}_i$ is received at the destination, Figure 3.17b shows it is decrypted by taking the bit-by-bit ExOR of $\mathcal{E}_i$ and the same RBS_i to produce D_i. Note that decryption yields the original data.

Of course, the trick is to have the same RBS_i at both the source and destination. A method to do this is described in Chapter 9.

▶ 3.5 FROM LOGIC GATES TO LOGIC CIRCUITS

Using the three basic gates (AND, OR, and NOT), we show that any combinational logic function can be implemented as a logic circuit. This construction of a logic circuit proceeds from the truth table, which explicitly lists all the possible sets of input logic variables and their associated output values. The truth table input section is formed using the conventional binary counting sequence, while the output section meets the requirements of the task. Several tasks are considered to illustrate the process, including digital arithmetic operations and controlling a seven-segment display.

The design of a logic circuit includes the following steps.

STEP 1. Construct a truth table of all the possible input logic combinations and their desired output logic values.

STEP 2. Identify the input logic patterns that produce logic 1 outputs.

STEP 3. Implement a logic circuit that recognizes these input patterns.

3.5.1 Constructing a Truth Table

If a logic circuit has n input logic variables, there are 2^n possible patterns of 0's and 1's. These patterns are listed in numerical order with the first row starting with zero expressed in binary form using one bit per input logic variable. In this way, the input patterns for all truth tables are the same and depend only on n, which is the number of input logic variables.

EXAMPLE 3.5

Truth table input section

The elementary AND and OR logic gates described previously had two inputs A and B $(n = 2)$, and the $2^n = 2^2 = 4$ possible binary input patterns are given by

$$00 \quad 01 \quad 10 \quad \text{and } 11$$

Each pattern represents one row in the truth table input section. Figure 3.18 shows truth tables with $n = 2$ and $n = 3$ input logic variables. A logic circuit having four inputs $(n = 4)$, which are denoted A, B, C, and D, has $2^4 = 16$ possible input binary patterns ranging from 0000 (0_{10}) to 1111 (15_{10}). The corresponding truth table then has 16 rows.

Generalizing, n input logic variables produce 2^n rows in the truth table.

inputs		output		inputs			output
A	B	Y		A	B	C	Y
0	0			0	0	0	
0	1			0	0	1	
1	0			0	1	0	
1	1			0	1	1	
				1	0	0	
				1	0	1	
				1	0	1	
				1	1	0	
				1	1	1	

Figure 3.18

Truth tables describing two and three input logic variables.

3.5.2 Binary Pattern Recognition

The output section of a truth table is specified to satisfy a particular task. We consider specifying the logic outputs for various tasks later in this chapter. The next step in implementing a logic circuit involves recognizing each particular *binary pattern* of input variables that generates an output equal to 1. For example, a 4-bit pattern can be represented as

$$b_3 \, b_2 \, b_1 \, b_0$$

The standard method of recognizing a specific bit pattern is by using an AND gate. Because the output of an AND gate is a 1 only when all of its inputs are equal to 1, we need to transform the specific pattern we are trying to recognize into one having all 1's. We do this by converting each 0 in the pattern into a 1 by using a NOT gate. The logic variables that are equal to 1 in the desired pattern are applied directly to an AND gate, while those that equal 0 are first applied to a NOT gate and then to the AND gate. Then when the desired bit pattern occurs, the AND gate input will have all inputs equal to 1 and produce a logic 1 output. This procedure is illustrated in the following example.

Logic circuit to recognize the 4-bit pattern $b_3 b_2 b_1 b_0 = 1100$ **EXAMPLE 3.6**

We want to design a logic circuit that produces the output P_{1100}, which equals 1 only when $b_3 = 1$ **and** $b_2 = 1$ **and** $b_1 = 0$ **and** $b_0 = 0$. The logic equation that corresponds to this pattern recognition is given by

$$P_{1100} = b_3 \cdot b_2 \cdot \overline{b_1} \cdot \overline{b_0}$$

To implement the logic circuit that has $b_3 b_2 b_1 b_0$ as inputs and P_{1100} as the output, we connect logic variables b_3 and b_2 directly to an AND gate, because their values in the desired pattern equal 1. The logic variables b_1 and b_0 equal 0 in the desired pattern and hence are first applied to NOT gates to convert them to 1, as shown in Figure 3.19. The output of the AND gate equals 1 only when the desired input pattern occurs.

Figure 3.19

Logic circuit to recognize the 4-bit pattern $b_3b_2b_1b_0 = 1100$. NOT gates convert the 0's in the desired pattern to 1's. $P_{1100} = 1$ only when $b_3b_2b_1b_0 = 1100$.

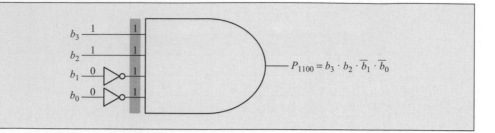

3.6 LOGIC CIRCUIT IMPLEMENTATION USING SUM OF PRODUCTS

Logic circuits typically need to recognize a set of input patterns with each pattern being an AND product of logic variables. When more than one pattern needs to be recognized, each pattern must be recognized separately by its own AND gate. The output of each AND gate is connected to one of the inputs of an OR gate to produce their sum. Hence, the OR gate output indicates when one of the desired patterns is present at an AND gate input. Such an implementation is called a *sum-of-products* logic circuit.

EXAMPLE 3.7

Recognizing binary pattern corresponding to 7_{10} or 11_{10}

Four bits $b_3b_2b_1b_0$ can represent integers between 0 and 15. The two numbers to be recognized, 7_{10} and 11_{10}, have the 4-bit representations given by

$$7_{10} = 0111_2 \quad \text{and} \quad 11_{10} = 1011_2$$

A separate logic circuit is implemented to recognize each pattern. The outputs of these two logic circuits are connected to a 2-input OR gate. The circuit that recognizes 7_{10} has output P_7. From the binary pattern representing 7, the logic equation for this circuit is

$$P_7 = \overline{b_3} \cdot b_2 \cdot b_1 \cdot b_0$$

The circuit that recognizes 11_{10} has output P_{11}. From the binary pattern representing 11, the logic equation for this circuit is

$$P_{11} = b_3 \cdot \overline{b_2} \cdot b_1 \cdot b_0$$

The output of the OR gate, indicated by $P_{7\ \text{or}\ 11}$, equals 1 only when one of the desired bit patterns occurs. The corresponding logic equation is written as

$$P_{7\ \text{or}\ 11} = P_7 + P_{11}$$

Figure 3.20 shows this logic circuit. As is evident from the figure, logic circuit diagrams can get complicated rather quickly with some signal lines crossing over others. To indicate the crossing lines that are actually electrically connected, a solid circle is drawn at their intersection. No solid circle indicates that there is no connection. Sometimes, a curved bridge in one of the crossing wires indicates that there is no electrical connection. In simple logic circuits it is often obvious when crossing wires electrically connect, so no additional notation is necessary.

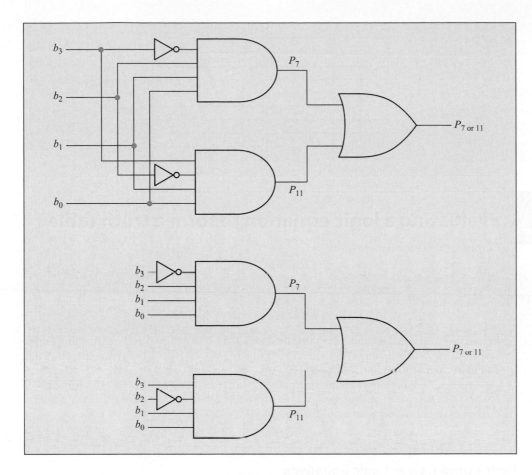

Figure 3.20

Equivalent drawings of a logic circuit to recognize binary patterns that represent either 7_{10} or 11_{10}.

3.7 TRUTH TABLE FROM LOGIC EQUATION AND LOGIC CIRCUIT

The previous topics described how to implement a logic circuit from a truth table. For analyzing a given logic equation or logic circuit, the inverse operation needs to be done. This section describes determining the truth-table output section either from a given logic equation or from a given logic circuit.

Forming a Truth Table from a Logic Equation

Logic equations can become very confusing. The evaluation of the logic operators in a logic equation follows the convention used for the operations in a complicated algebraic expression, which I recall as *My Dear Aunt Sally*, or first perform multiplications, then divisions, then additions, and finally the subtractions. Considering a complemented logic value as an individual logic variable, a logic equation is evaluated by first performing the AND (product) followed by the OR (addition).

For our purpose of implementing a truth table, parentheses can make logic equations simpler to understand by expressing them as the sum (OR) of product (AND) terms. Use parentheses to enclose AND-ed logic variables into a single logic expression. This will leave a set of terms that are combined with a logic OR operation. Evaluating the products containing the fewest logic variables first, where each product term in parentheses leads to one or more rows in the truth table being set = 1. The remaining rows that do not evaluate to 1 become 0. The following examples illustrate this approach.

A	B	C	Y	Reason
0	0	0	0	Remaining term = 0
0	0	1	1	$(\overline{A} \cdot \overline{B} \cdot C) = 1$
0	1	0	1	$(B \cdot \overline{C}) = 1$
0	1	1	0	Remaining term = 0
1	0	0	1	$A = 1$
1	0	1	1	$A = 1$
1	1	0	1	$A = 1$
1	1	1	1	$A = 1$

Figure 3.21

Evaluating logic equation to form a truth table.

EXAMPLE 3.8

Evaluating a logic equation to form a truth table

Consider the logic equation given by

$$Y = A + B \cdot \overline{C} + \overline{A} \cdot \overline{B} \cdot C$$

Use parentheses to enclose the product terms to express the logic equation as the sum of product terms:

$$Y = (A) + (B \cdot \overline{C}) + (\overline{A} \cdot \overline{B} \cdot C)$$

Evaluating the products containing the fewest logic variables first, we set the truth table output rows to 1 by first evaluating (A), then $(B \cdot \overline{C})$, and finally $(\overline{A} \cdot \overline{B} \cdot C)$. The remaining rows that do not evaluate to 1 become 0.

Figure 3.21 shows the resulting truth table. Note that the row $ABC = 110$ satisfies both $A = 1$ and $(B \cdot \overline{C}) = 1$. Evaluation of $A = 1$ produced four 1's and eliminated the need for further evaluations for those input patterns.

Distributive Law in Logic Equations

To assist in expressing a logic equation as the sum of product terms, logic equations can be simplified by using a *distributive law* applied to the AND operation:

$$A \cdot (B + C) = A \cdot B + A \cdot C = (A \cdot B) + (A \cdot C) \tag{3.6}$$

The distributive law also holds in standard arithmetic multiplication:

$$2 \times (3 + 4) = 2 \times 3 + 2 \times 4 \tag{3.7}$$

Note that the distributive law does not hold for the OR operation. For example,

$$A + (B \cdot C) \neq A + B \cdot A + C \tag{3.8}$$

To show this, consider why the distributive law does not hold for standard arithmetic addition:

$$2 + (3 \times 4) \neq (2 + 3) \times (2 + 4) \tag{3.9}$$

EXAMPLE 3.9

Distributive law in logic equation to form a truth table

Consider the logic equation given by

$$Y = A \cdot (B + \overline{C})$$

Use the distributive law and parentheses to enclose the product terms to express the logic equation as the sum of product terms:

$$Y = A \cdot B + A \cdot \overline{C} = (A \cdot B) + (A \cdot \overline{C})$$

Figure 3.22 shows the resulting truth table. Note that the row $ABC = 110$ satisfies both $(A \cdot B) = 1$ and $(A \cdot \overline{C}) = 1$.

A	B	C	Y	Reason
0	0	0	0	Remaining term = 0
0	0	1	0	Remaining term = 0
0	1	0	0	Remaining term = 0
0	1	1	0	Remaining term = 0
1	0	0	1	$(A \cdot \overline{C}) = 1$
1	0	1	0	Remaining term = 0
1	1	0	1	$(A \cdot B) = 1$
1	1	1	1	$(A \cdot B) = 1$

Figure 3.22

Distributive law applied to logic equation to form a truth table.

Forming a Truth Table from Logic Circuit

Any combinational logic circuit is constructed by interconnecting elemental logic gates. The simplest way of forming the truth table from a logic circuit is to translate the logic circuit into its equivalent logic equation and using the method in the previous section. The following example illustrates this method.

Logic circuit to truth table   EXAMPLE   3.10

Consider the logic circuit shown in Figure 3.23. The logic equation is given by

$$Y = \overbrace{(A + \overline{B})}^{\text{OR gate}} \cdot C$$

Use the distributive law and parentheses to enclose the product terms to express the logic equation as the sum of product terms:

$$Y = A \cdot C + \overline{B} \cdot C = (A \cdot C) + (\overline{B} \cdot C)$$

Figure 3.23 shows the resulting truth table. Note that the row $ABC = 101$ satisfies both $(A \cdot C) = 1$ and $(\overline{B} \cdot C) = 1$. The table shows the reason to be $(A \cdot C) = 1$ because it was the first term considered, and once an output becomes 1, no other input terms need to be considered.

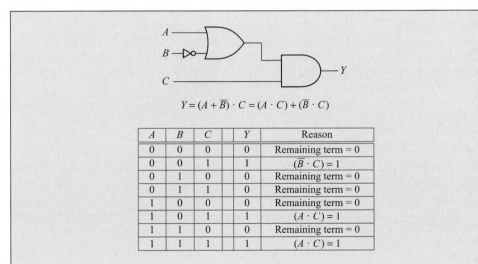

$$Y = (A + \overline{B}) \cdot C = (A \cdot C) + (\overline{B} \cdot C)$$

A	B	C	Y	Reason
0	0	0	0	Remaining term = 0
0	0	1	1	$(\overline{B} \cdot C) = 1$
0	1	0	0	Remaining term = 0
0	1	1	0	Remaining term = 0
1	0	0	0	Remaining term = 0
1	0	1	1	$(A \cdot C) = 1$
1	1	0	0	Remaining term = 0
1	1	1	1	$(A \cdot C) = 1$

Figure 3.23

Logic circuit converted to logic equation for determining the truth table.

▶ 3.8 DESIGNING EFFICIENT LOGIC CIRCUITS

When implementing a logic circuit, the number of gates should be kept as small as possible to make the logic circuit simple and inexpensive. This section considers two techniques that may reduce the number of gates in a logic circuit.

1. Compare original and complemented output logic variables to find the one that has the smaller number of 1's.

2. Appropriately assign "don't care" values ($\times$) to logic 0 or 1, that is, it doesn't matter if the logic value is 0 or 1, because that input pattern is not observed in practice.

To implement an efficient logic circuit that produces the output Y, the design procedure starts by comparing the number of 1's in the truth table output section with the number of 0's to find the smaller count. If the number of times Y equals 1 is smaller, we proceed with the implementation of the logic circuit using the procedure just described.

If the number of times Y equals 0 is smaller than the number of times it equals 1, we consider implementing the NOT of the output variable or $\overline{Y}$. We do this by first complementing the 0's in the output section to 1's and the 1's to 0's (the NOT operation) to form the values of $\overline{Y}$. Having done this, the number of 1's is smaller in this *complementary* output of the truth table. We proceed with the logic circuit implementation of $\overline{Y}$ with the procedure previously described. After the design is finished, the output of this *complementary* logic circuit, $\overline{Y}$, is connected to a NOT gate that produces the original Y values specified in the truth table.

The next example illustrates that implementing a complementary logic circuit can lead to a much simpler design.

EXAMPLE 3.11 — Implementing a truth table

Consider the truth table shown in Figure 3.24. The truth table has two inputs A and B and the output Y. The complement of Y, or $\overline{Y}$, is also shown as the complementary output of the truth table.

We implement two logic circuits using the two methods previously described.

In the first design, we find the values of A and B that make $Y = 1$ and implement an AND gate to recognize each pattern. There are three such patterns, and the outputs of the corresponding AND gates are connected to the OR gate to implement the logic circuit.

In the second design, it was noted that when there were fewer cases when $Y = 0$ in the truth table, so a logic circuit to implement $\overline{Y}$ was designed. Because $\overline{Y} = 1$ for only one set of A and B values, this design is very simple. To implement the original Y function, $\overline{Y}$ was applied to a NOT gate. Note that a simpler logic circuit results. This second logic circuit is the NAND (Not-AND) gate.

This example illustrates that an engineering solution to a problem is not unique (that is, several different systems can be built to perform the same function). Each particular implementation has its own advantages and disadvantages. Designing a system is really an art that often results in an elegant and clever implementation.

de Morgan's Laws

Considering alternative means of implementing a logic circuits leads to two interesting equations used in *propositional logic* and simplifying logic circuits called *de Morgan's laws*, which are given by

$$\overline{A \cdot B} = \overline{A} + \overline{B} \tag{3.10}$$

$$\overline{A + B} = \overline{A} \cdot \overline{B}$$

The following example illustrates the proof of the first law.

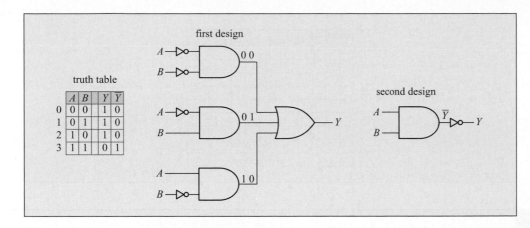

Figure 3.24

Two logic circuits having different structures implemented from the truth table shown at left.

Proving one of de Morgan's laws | EXAMPLE 3.12

This example proves de Morgan's first law

$$\overline{A \cdot B} = \overline{A} + \overline{B}$$

by using truth tables.

First, consider the logic equation given by

$$Y = \overline{A} + \overline{B}$$

In words, Y equals not A or not B. Hence, $A = 0$ or $B = 0$ makes $Y = 1$, otherwise $Y = 0$, generating the truth table shown in Figure 3.25. With $Y = \overline{A} + \overline{B}$ having three 1's and one 0, a more efficient logic circuit would result by implementing $\overline{Y}$, which has one 1 and three 0's, and negating the output of that logic circuit with a single NOT gate.

The logic equation for $\overline{Y}$ from the truth table gives

$$\overline{Y} = A \cdot B$$

Negating this logic equation to obtain Y gives

$$Y = (\overline{\overline{Y}}) = \overline{A \cdot B}$$

Equating the initial logic equation with this last result, we arrive at de Morgan's first law:

$$\overline{A \cdot B} = \overline{A} + \overline{B}$$

Assigning Don't Care Values

Some truth tables have input sequences that will never occur in practice. In such cases, the output value for these *impossible* sequences is assigned an × symbol, which indicates a "don't care" condition. The output value can be assigned either a 0 or 1 value—whichever is more convenient for implementing an efficient logic circuit.

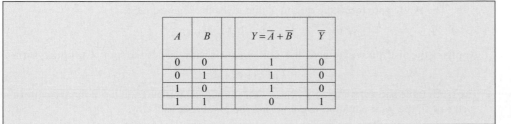

A	B		$Y = \overline{A} + \overline{B}$	$\overline{Y}$
0	0		1	0
0	1		1	0
1	0		1	0
1	1		0	1

Figure 3.25

Truth table for Example 3.12.

Figure 3.26

Truth table with the complement of the output shown and containing don't-care ("X") values.

A	B	C	Y	$\overline{Y}$
0	0	0	1	0
0	0	1	1	0
0	1	0	1	0
0	1	1	1	0
1	0	0	1	0
1	0	1	1	0
1	0	1	1	0
1	1	0	X	X
1	1	1	0	1

EXAMPLE 3.13

Efficient logic circuit implementation

Consider the truth table shown in Figure 3.26. We consider two logic circuits that implement a truth table, the first with $\times = 0$ and the second with $\times = 1$.

Clearly, the number of 1's in $\overline{Y}$ is smaller than in Y—no matter what the value of $\times$. Hence, we implement $\overline{Y}$ and use a NOT gate to obtain Y.

For the $\times$ value, let us consider each value separately to determine the number of 1's.

For $\times = 1$, $\overline{Y}$ has a single input pattern that makes it 1. Thus,

$$\overline{Y} = A \cdot B \cdot C$$

and the logic equation for the truth table is

$$Y = \overline{A \cdot B \cdot C}$$

For $\times = 0$, $\overline{Y}$ has two input patterns that makes it 1: $A = 1$ and $B = 1$, while C can equal either 0 or 1. So,

$$\overline{Y} = (A \cdot B \cdot C) + (A \cdot B \cdot \overline{C})$$

Hence, the value of C does not matter, and the logic equation can be simplified to be

$$\overline{Y} = A \cdot B$$

Since $\overline{(\overline{Y})} = Y$, the logic equation for the truth table is

$$Y = \overline{A \cdot B}$$

Hence, making $\times = 0$ produces a more efficient design—one that uses fewer gates.

EXAMPLE 3.14

Logic circuit analysis

Consider the logic circuit shown in Figure 3.27.

The logic equation for the logic circuit is found by proceeding from left-to-right forming the output of each gate.

- The NOT gate forms $\overline{B}$.
- The OR gate forms $A + \overline{B}$.
- The AND gate forms the output Y that gives the logic equation

$$Y = (A + \overline{B}) \cdot C$$

Applying the distributive law gives the logic equation with the desired sum-of-products form as

$$Y = A \cdot C + \overline{B} \cdot C$$

The truth table shown in Figure 3.27 was determined by applying the input binary pattern in each row to the logic equation to determine the Y value in that row.

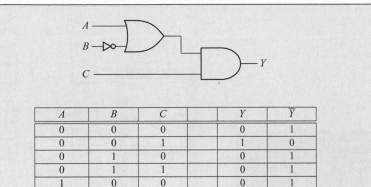

A	B	C		Y	$\overline{Y}$
0	0	0		0	1
0	0	1		1	0
0	1	0		0	1
0	1	1		0	1
1	0	0		0	1
1	0	1		1	0
1	1	0		0	1
1	1	1		1	0

Figure 3.27

Logic circuit and truth table for Example 3.14.

3.9 USEFUL LOGIC CIRCUITS

This section describes logic circuits that implement some interesting tasks that are often performed by smart devices.

3.9.1 Implementing an ExOR Gate

To implement the ExOR truth table in Figure 3.15, we use an AND gate to recognize each of the two patterns of A and B for which $Y = 1$. This occurs when $A \cdot \overline{B} = 1$ or when $\overline{A} \cdot B = 1$. This is expressed in logic equation form as

$$Y = (A \cdot \overline{B}) + (\overline{A} \cdot B)$$

To implement the ExOR gate, the outputs of the two AND gates are applied to an OR gate, as shown in Figure 3.28.

3.9.2 Seven-Segment Display Driver

A common digital display used in clocks, scales, and smart sensors is the 7-segment display shown in Figure 3.29. The 7-segment display device has seven inputs that by convention are labeled a, b, c, through g with each connected to an LED segment. Making an input 1 lights the corresponding segment. Let us implement a logic circuit that displays digits.

It is not unusual to have a digital logic circuit that has more than one output. In this case, the initial design starts with a separate logic circuit design for each output. Then if it is noticed that some gates are common to both logic circuits, they can be used for both.

Decimal digits are typically represented in a 4-bit binary code called *binary-coded-decimal (BCD)*. To form digits on the display, we need to implement a logic circuit that converts the 4-bit BCD pattern input into the appropriate seven logic circuit outputs with each one controlling its respective segment. The truth table relating the input BCD code to the seven outputs is shown in Figure 3.30. Note that, of the possible 16 patterns

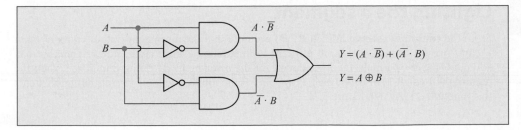

Figure 3.28

Implementing an ExOR gate with AND, OR, and NOT gates.

Figure 3.29

(a) A 7-segment display. (b) Logic variables *a* through *g* identify the segments. When the logic variable of a particular segment equals 1, the segment is lit. (c) The logic variable values that produce the ten digits are shown.

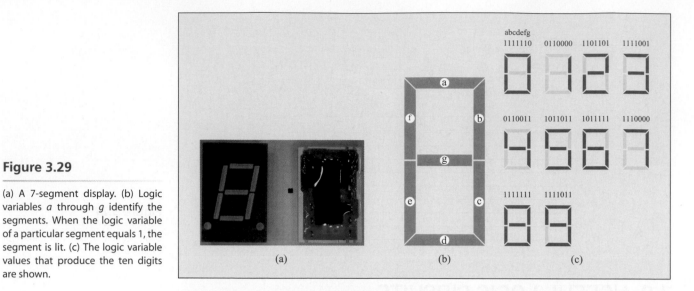

(a)   (b)   (c)

Figure 3.30

Truth table to convert binary-coded-decimal (BCD) codes to a 7-segment display.

row	digit	inputs				outputs						
		b_3	b_2	b_1	b_0	a	b	c	d	e	f	g
0	0	0	0	0	0	1	1	1	1	1	1	0
1	1	0	0	0	1	0	1	1	0	0	0	0
2	2	0	0	1	0	1	1	0	1	1	0	1
3	3	0	0	1	1	1	1	1	1	0	0	1
4	4	0	1	0	0	0	1	1	1	0	1	1
5	5	0	1	0	1	1	0	1	1	0	1	1
6	6	0	1	1	0	1	0	1	1	1	1	1
7	7	0	1	1	1	1	1	1	0	0	0	0
8	8	1	0	0	0	1	1	1	1	1	1	1
9	9	1	0	0	1	1	1	1	0	0	1	1
10	.	1	0	1	0	X	X	X	X	X	X	X
11	.	1	0	1	1	X	X	X	X	X	X	X
12	.	1	1	0	0	X	X	X	X	X	X	X
13	.	1	1	0	1	X	X	X	X	X	X	X
14	.	1	1	1	0	X	X	X	X	X	X	X
15	.	1	1	1	1	X	X	X	X	X	X	X

that 4 bits can produce, only 10 are used to display digits. The remaining binary patterns, 1010 to 1111, produce "don't care" values in the truth table output.

To design a logic circuit, we start with one of the segments (*a* through *g*) and determine the input patterns (BCD codes) that cause that output to equal 1, thus lighting that segment. Each such BCD code could be recognized using a 4-input AND gate pattern recognition circuit. The outputs of these AND gates connect to an OR gate that becomes a logic 1 whenever any one of these BCD codes appears. The OR gate output connects to the segment LED. Each segment can be treated in this manner to implement the complete BCD to 7-segment decoder.

EXAMPLE 3.15   Lighting the *a* segment

From the truth table shown in Figure 3.30, we note that the a segment output contains eight 1's and two 0's. Hence, it is more efficient to implement a logic circuit for $\bar{a}$. Then, we can add a NOT gate to obtain the logic circuit for the *a* segment. By doing this, $\bar{a} = 1$ when the pattern 0001 (digit 1) or 0100 (digit 4) appears at the input. The logic circuit that implements this function is shown in Figure 3.31.

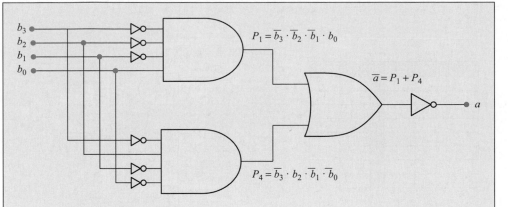

Figure 3.31

Logic circuit that lights segment a. Segment a is not lit only when either the digit 1 or 4 occurs. $P_1 = 1$ when a $1_{10} = b_3 b_2 b_1 b_0 = 0001_2$ occurs and $P_4 = 1$ when a $4_{10} = 0100_2$ occurs.

3.9.3 Binary Adder

One common task performed by a computer is to perform arithmetic operations, such as addition. We use the rules for doing addition on decimal numbers to determine how to perform the same operations on binary numbers. The references at the end of the chapter describe how the other arithmetic operations are performed.

When we add two groups of digits, we start with the least significant digits (the ones on the far right) and add these to compute the partial sum. If the sum S is less than the base of the number system (here $B = 10$), the sum is entered, and the next significant digits are added. If the sum is greater than the base, the base is subtracted, the result is entered, and a carry digit C is added to the the sum of the next significant digits. The procedure is continued until the partial sum and the carry are both zero-valued.

The same procedure is applied to binary numbers, but the process is simpler because the numbers consist of only 0's and 1's. Let the first 3-bit binary sequence be denoted by $A_2 A_1 A_0$ and the second by $B_2 B_1 B_0$. The sum is denoted by $S_3 S_2 S_1 S_0$. Note that the sum can have one more bit than the two inputs because of the possibility of a carry. The carry bits are denoted by $C_2 C_1 C_0$.

Binary addition EXAMPLE 3.16

Consider adding $A_2 A_1 A_0 = 110_2$ to $B_2 B_1 B_0 = 111_2$.

Starting with the least significant bit (LSB) and expressing the result as the binary pattern $C\,S$ (because it directly expresses the value of a binary number), we get

$$A_0 + B_0 = 0 + 1 = 1_{10} = 1_2 \rightarrow C_0 = 0, S_0 = 1$$

The sum of the next significant bits plus the previous carry bit is

$$A_1 + B_1 + C_0 = 1 + 1 + 0 = 2_{10} = 10_2 \rightarrow C_1 = 1, S_1 = 0$$

The sum of the next significant bits plus the previous carry bit is

$$A_2 + B_2 + C_1 = 1 + 1 + 1 = 3_{10} = 11_2 \rightarrow C_2 = 1, S_2 = 1$$

C_2 is not zero, so we need to continue one more step. If the most significant bit (MSB) position has already been evaluated, the next carry value being $= 1$ indicates an overflow condition. The final sum equals the carry bit from the previous stage, as

$$S_3 = C_2$$

Hence, the sum is

$$S_3 S_2 S_1 S_0 = 1101_2$$

As a verification, adding $110_2 = 6_{10}$ and $111_2 = 7_{10}$ produces $13_{10} = 1101_2$.

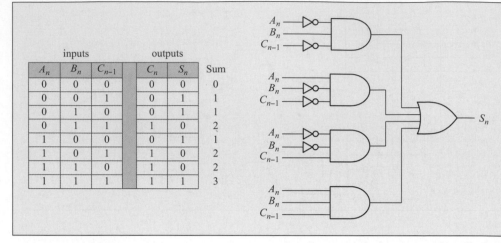

Figure 3.32

Truth table for a binary adder. Direct implementation of the sum bit S_n in a binary adder is shown. A similar logic circuit can be implemented to produce the carry bit C_n.

inputs			outputs		
A_n	B_n	C_{n-1}	C_n	S_n	Sum
0	0	0	0	0	0
0	0	1	0	1	1
0	1	0	0	1	1
0	1	1	1	0	2
1	0	0	0	1	1
1	0	1	1	0	2
1	1	0	1	0	2
1	1	1	1	1	3

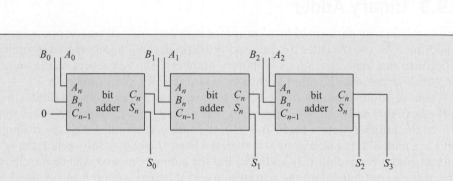

Figure 3.33

Repeating the same logic circuit, known as a *full adder*, can add two 3-bit numbers to form a 4-bit result.

Using this procedure, we can implement a logic circuit that adds two multi-bit binary numbers. In the description of binary addition just given, the main operation was the addition of the three bits formed by the two input bits, A_n and B_n, and the carry bit C_{n-1} from the previous sum to produce a sum bit S_n and carry bit C_n. The truth table that accomplishes this operation is shown in Figure 3.32. The output columns were formed with C_n followed by S_n to express binary numbers with the least-significant bit (LSB) being rightmost.

The logic circuit has three inputs and two outputs. Each output is implemented separately using the standard procedure. The logic circuit for the S_n output of the binary adder is shown in Figure 3.32.

A binary adder logic circuit is implemented by combining individual bit-adder sections. The circuit that adds two 3-bit numbers is shown in Figure 3.33. Note that the carry input to the first stage is a 0. The carry result from the last stage is the most significant bit in the sum.

3.9.4 Binary Multiplication

The direct implementation of digital addition can be extended easily to digital multiplication. The approach is the same as in the direct implementation of addition: For each combination of binary input patterns, we manually determine and specify the resulting output binary pattern.

Consider the multiplication of two 2-bit numbers $A_1 A_0$ and $B_1 B_0$ to produce product P:

$$P = A \times B$$

(3.11)

inputs		product		inputs				outputs			
A	B	P		A_1	A_0	B_1	B_0	P_3	P_2	P_1	P_0
0	0	0		0	0	0	0	0	0	0	0
0	1	0		0	0	0	1	0	0	0	0
0	2	0		0	0	1	0	0	0	0	0
0	3	0		0	0	1	1	0	0	0	0
1	0	0		0	1	0	0	0	0	1	0
1	1	1		0	1	0	1	0	0	0	1
1	2	2		0	1	1	0	0	0	1	0
1	3	3		0	1	1	1	0	0	1	1
2	0	0		1	0	0	0	0	0	0	0
2	1	2		1	0	0	1	0	0	1	0
2	2	4		1	0	1	0	0	1	0	0
2	3	6		1	0	1	1	0	1	1	0
3	0	0		1	1	0	0	0	0	0	0
3	1	3		1	1	0	1	0	0	1	1
3	2	6		1	1	1	0	0	1	1	0
3	3	9		1	1	1	1	1	0	0	1

Figure 3.34

Truth tables for direct implementation of a 2-bit multiplier. Decimal values are given in the left table and their binary equivalents in the right table.

Because A and B vary between 0 and 3, the product P varies between 0 and 9. Hence, 4 bits are required to represent the product $P = P_3 P_2 P_1 P_0$. Figure 3.34 shows the decimal table and binary truth table that describes the truth table.

3.9.5 Binary Division

The direct implementation of digital division requires the following two considerations.

1. The division implemented in a digital logic circuit is an *integer division* (that is, the result is an integer that is obtained by ignoring the remainder). For example, the integer division result of 3 divided by 2 is not 1.5, it is 1.

2. Division by 0 is not allowed and should generate an error—if and when it occurs. This is the purpose of a *divide-by-zero* error flag bit, which is a bit that is set to 1 when such a case occurs.

These considerations for integer division leads to the formulation of the truth table shown in Figure 3.35. Each combination of binary input patterns determines the resulting output binary pattern.

For example, consider the integer division of two 2-bit numbers $A_1 A_0$ and $B_1 B_0$ with the quotient being

$$Q = \frac{A}{B} \tag{3.12}$$

Because the A value of a 2-bit number varies between 0 and 3 and the non-zero denominator B varies between 1 and 3, the quotient Q varies between 0 (when $A = 0$) and 3 (when $A = 3$ and $B = 1$). Hence, 2 bits are required for the quotient $Q_1 Q_0$.

When $B = 0$, an error condition occurs. Use logic variable E to indicate an error with $E = 1$ indicating that a division by zero occurs. When $E = 1$, the values of $Q_1 Q_0$ do not matter and become a "don't care" condition, which can simplify a logic circuit design. Figure 3.35 shows the decimal and binary truth tables.

inputs		outputs			inputs				outputs		
A	B	Q	E		A_1	A_0	B_1	B_0	Q_1	Q_0	E
0	0	X	1		0	0	0	0	X	X	1
0	1	0	0		0	0	0	1	0	0	0
0	2	0	0		0	0	1	0	0	0	0
0	3	0	0		0	0	1	1	0	0	0
1	0	X	1		0	1	0	0	X	X	1
1	1	1	0		0	1	0	1	0	1	0
1	2	0	0		0	1	1	0	0	0	0
1	3	0	0		0	1	1	1	0	0	0
2	0	X	1		1	0	0	0	X	X	1
2	1	2	0		1	0	0	1	1	0	0
2	2	1	0		1	0	1	0	0	1	0
2	3	0	0		1	0	1	1	0	0	0
3	0	X	1		1	1	0	0	X	X	1
3	1	3	0		1	1	0	1	1	1	0
3	2	1	0		1	1	1	0	0	1	0
3	3	1	0		1	1	1	1	0	1	0

Figure 3.35

Truth tables for direct implementation of 2-bit division. Along with the quotient Q, and error condition $E = 1$ indicates division by zero. The "X" indicates a don't-care condition, and can take on either a 0 or 1 value.

3.10 Summary

This chapter described combinational logic circuits that operate on binary-valued logic variables to accomplish a desired task. A combinational logic circuit produces an output logic variable that is related only to the current values of the input logic variables. Combinational logic includes the AND, OR, and NOT gates. Using these three gates, any logic function can be implemented as a logic circuit. The behavior of a combinational logic circuit was described with a truth table. The construction of a digital logic circuit from the truth table was illustrated with several applications.

3.11 Problems

3.1 **Implementing an OR gate using AND and NOT.** Implement an OR gate using only AND and NOT gates. (Hint: Start with the truth table that implements $\overline{Y}$.)

3.2 **Using only NAND gates.** This problem demonstrates that only a NAND gate is needed to implement the basic gates and, hence, any combinatorial logic circuit. Using only two-input NAND gates, implement a NOT, an OR, and an AND gate.

3.3 **From logic equation to truth table.** Generate the truth table that corresponds to the logic equation given by

$$Y = A \cdot \overline{B} + B \cdot \overline{C} + C \cdot \overline{A}$$

Implement a logic circuit that has a small number of gates.

3.4 **Distributive law.** Use the distributive law to generate the truth table that corresponds to the logic equation given by

$$Y = A \cdot (\overline{B} + C)$$

Implement one logic circuit directly from the logic equation. Implement a second logic circuit from the truth table using the sum-of-products approach. Advanced courses in logic design teach how to implement logic circuits having the minimum number of gates.

3.5 **Logic circuit analysis.** Generate the truth table that corresponds to the logic circuit shown in Figure 3.36. Draw a different logic circuit that implements this truth table.

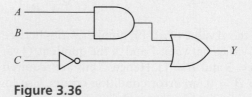

Figure 3.36

Logic circuit for Problem 3.5.

3.6 **Proving de Morgan's second law.** Starting with the truth table that implements $\overline{A + B}$, implement an alternative logic circuit that proves de Morgan's second law given in Eq. (3.10).

3.7 7-Segment display: Lighting the *b* segment. Implement the logic circuit that recognizes the BCD codes that light the *b* segment in a 7-segment display.

3.8 Digital adder: Carry bit logic circuit. Extending Figure 3.32, implement the logic circuit that produces the carry bit C_n from the inputs A_n, B_n, and C_{n-1}.

3.9 Direct implementation of binary multiplication. Using the truth table in Figure 3.34, implement the four logic circuits that produce each term in the product $P_3 P_2 P_1 P_0$.

3.10 Direct implementation of binary division. Using the truth table in Figure 3.35, design the logic circuit to produce the quotient $Q_1 Q_0$ and divide-by-zero error E.

3.12 Excel Projects

The following Excel projects illustrate the following features:

- Examine Excel logic functions.

- Perform decimal-to-binary conversion, and use it to form the input section of a truth table.

- Implement a 7-segment display using conditional formatting.

Using a Narrative box (described in Example 13.4), include observations, conclusions, and answers to questions posed in the projects.

3.1 3-Input AND gate. Extend Example 13.13 to implement a 3-input AND gate.

3.2 3-Input OR gate. Extend Example 13.14 to implement a 3-input OR gate.

3.3 Decimal to binary conversion. Extend Example 13.16 to convert a decimal number into its 4-bit representation, with each bit in a separate column.

3.4 Verifying logic equations. Using Example 13.17 as a guide, compose a worksheet that implements the truth table for the following logic equations.

1. $Y = (A \cdot B) + (\overline{A} \cdot B)$
2. $Y = A + \overline{B}) \cdot C$
3. $Y = A + B \cdot \overline{C} + \overline{A} \cdot \overline{B} \cdot C$

3.5 Implementing a 7-segment display. Using Example 13.18 as a guide, compose a worksheet that forms a 7-segment display based on the binary value of the digit, rather than the digit value.

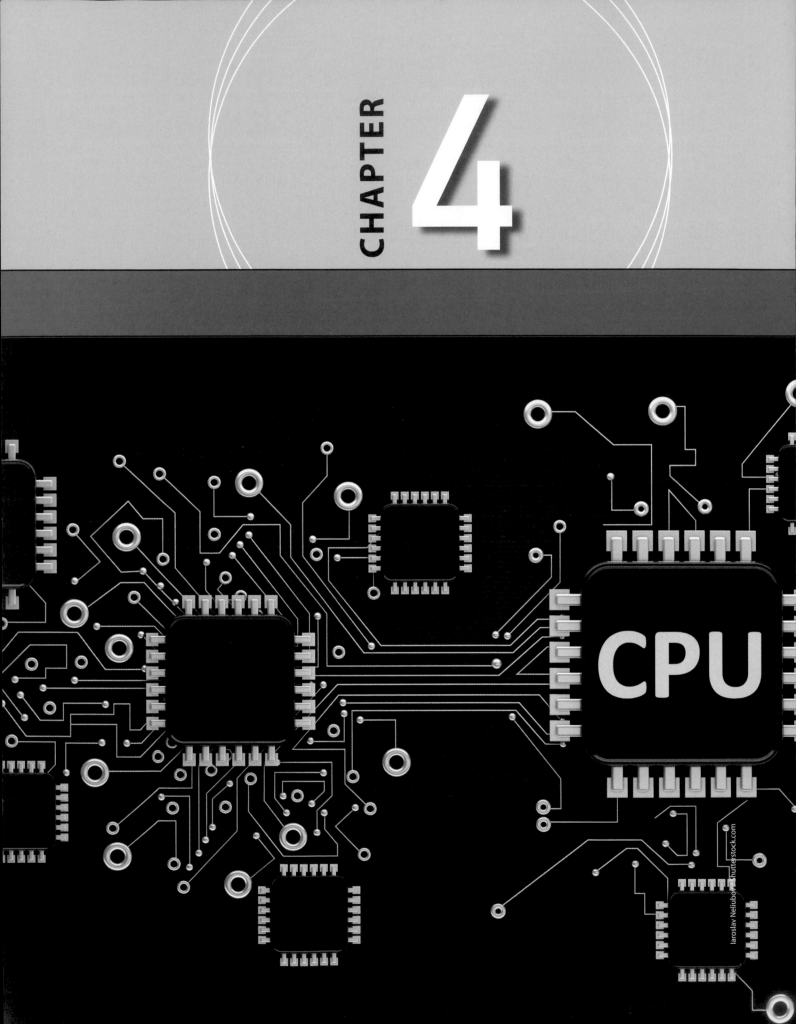

CHAPTER

4

CPU

Iaroslav Neliubov/Shutterstock.com

SEQUENTIAL LOGIC CIRCUITS

LEARNING OBJECTIVES

After completing this chapter, the reader should be able to:

- Draw timing diagrams and construct state transition tables that illustrate the operation of flip-flops.

- Understand the process of designing and analyzing sequential logic circuits that implement digital memories and counters.

4.1 INTRODUCTION

While a combinational logic circuit produces an output value that is a function of only the current input logic variables, the output of a sequential logic circuit also depends on its *state*, which is determined by the past input values. This chapter describes two elementary sequential circuits that are important components in smart devices: the *set-reset flip-flop* (SR-FF) is the basic memory element in many digital circuits and the *toggle flip-flop* (T-FF) is the basic counter element. The behavior of a flip-flop is described with a *timing diagram* and a state transition table that indicate the flip-flop output value in response to the changes in the inputs.

The main topics in this chapter are covered in the following sequence.

Set-Reset flip-flop—The SR-FF is the basic memory unit in digital logic circuits. Its implementation uses *positive feedback* that is illustrated with a pair of interconnected NOR gates.

Toggle flip-flop—The T-FF is the basic counting element in digital logic circuits.

Computers—Hardware implementation of logic circuits is replaced by computer instructions to simplify changes and upgrades. Computers are described in terms of their computational power and power consumption.

4.2 SEQUENTIAL CIRCUITS

A combinational logic produces an output that is a function of only the input logic variables. The sequential logic circuit output is a function of the inputs and also depends on its current *state*, which is a type of memory that causes the sequential circuit to produce different output values—even to the same input values.

A sequential circuit is similar to a human's *state of mind* that may elicit very different reactions to the same greeting. For example, if you studied all night, your response to a cheerful *"Good Morning"* may be quite different than if you had a full night's sleep.

This chapter describes an important class of sequential logic circuits called *flip-flops*. Flip-flops get their name from the behavior of their outputs flipping from one state to the other. While there are many different types of flip-flops used in practical logic circuits, this chapter describes the two shown in Figure 4.1 that are fundamental: the set-reset flip-flop (SR-FF), which is the basic memory unit in the computer, and the toggle flip-flop (T-FF), which is the computer's counting and pulse-generation element.

Figure 4.1

Flip-flops are an important class of sequential logic circuits. Set-reset flip-flops (RS-FFs) form memory circuits. Toggle flip-flops (T-FFs) form counting and pulse-generating circuits.

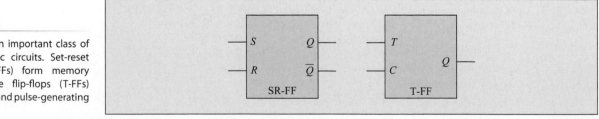

4.3 SET-RESET FLIP-FLOP

The set-reset flip-flop (SR-FF) is the basic *memory* element in digital devices. The SR-FF can be implemented using two OR and two NOT gates (or two NOR gates). What is novel and *non-combinational* is the presence of *feedback*, which connects an output back to the input. This feedback *reinforces* the asserted input values to implement the memory function.

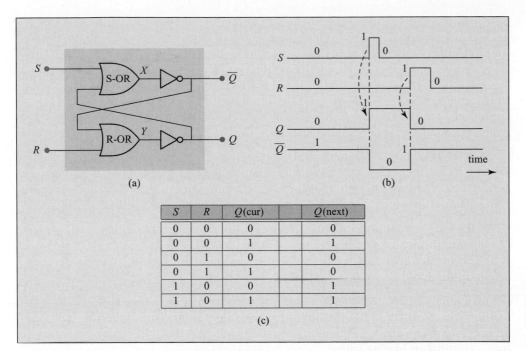

Figure 4.2

The SR-FF: (a) implementation using OR and NOT gates, (b) timing diagram, and (c) state transition table.

Figure 4.2a shows the SR-FF implemented using OR and NOT gates. It has two inputs denoted S for *set* and R for *reset*, and two outputs conventionally denoted Q and its complement $\overline{Q}$. Having access to a logic variable and its complement often results in a simpler digital circuit because it may eliminate the need for a NOT gate. The SR-FF provides both for free.

While the truth table plays a major role in the design of combinational logic circuits, the *timing diagram* illustrates how the sequential circuit output changes in response to changes in the input. Figure 4.2b shows the timing diagram for the SR-FF. It is typical to use dashed arrows to associate the transition in an input (in this case $0 \to 1$) that causes the output change.

A *state transition table* is a alternate (but equivalent) description of a sequential logic circuit. Figure 4.2c shows the state transition table for the SR-FF. It has a standardized form that contains all of the possible values of the inputs (along with the current output values) to determine the next output value. In this way, its input structure is similar to that of a truth table in combinational logic circuits, except the input section also includes the current output value.

To describe the SR-FF behavior, let initial inputs $S = 0$ and $R = 0$ and output $Q = 0$, as shown in the first row of the state transition table. Let us examine both the timing diagram and each row of the state transition table:

$(S, R, Q(cur) = 000)$: The next value of the output is $Q(next) = 0$, which is the same as the current state. Its complement is $\overline{Q} = 1$ (as expected). Since the output values remain at these values, this is a *stable state* of the sequential circuit.

$(S, R, Q(cur) = 001)$: $Q(next) = 1$. Since it the same as the current state, this is another *stable state* as long as both $S = 0$ and $R = 0$.

$(S, R, Q(cur) = 010$: The $0 \to 1$ transition on the R input keeps $Q(next) = 0$.

$(S, R, Q(cur) = 011$: The $0 \to 1$ transition on the R input makes $Q(next) = 0$.

$(S, R, Q(cur) = 100$: The $0 \to 1$ transition on the S input makes $Q(next) = 1$.

$(S, R, Q(cur) = 101$: The $0 \to 1$ transition on the S input keeps $Q(next) = 1$.

The transition table does not have the full $8 (= 2^3)$ rows, because it is not allowed to have $S = 1$ and $R = 1$ at the same time.

This SR-FF behavior can be explained by examining the logic circuit. The output Q also connects (or *feeds back*) to the S-OR gate input. Thus, with $Q = 0$ and $S = 0$, the output of the S-OR is $X = 0$. As X passes through the NOT gate, $\overline{Q} = \overline{X} = 1$. The $\overline{Q}$ output also feeds back to an input of R-OR having output Y. With $\overline{Q} = 1$, $Y = 1$, which passes through its NOT gate to produce $Q = \overline{Y} = 0$. This completes the feedback loop that results in a stable set of input and output values. Thus, the SR-FF is in a stable condition when $S = 0$ and $R = 0$.

Lets examine what happens when a transition occurs on the R or S inputs:

$R\ 0 \to 1$: When the R input on the R-OR gate changes from $0 \to 1$, it makes $Y = 1$, which in turn causes $Q = 0$. With $S = 0$ and $Q = 0$, $X = 0$, making $\overline{Q} = 1$, which feeds back to the other input of the R-OR, reinforcing the $R = 1$ value. This is the reset condition that drives $Q \to 0$.

$S\ 0 \to 1$: When the S input on the S-OR gate changes from $0 \to 1$, it makes $X = 1$, which in turn causes $\overline{Q} = 0$. With $R = 0$ and $\overline{Q} = 0$, $Y = 0$, making $Q = 1$, which feeds back to the other input of the S-OR, reinforcing the $S = 1$ value. This is the set condition that drives $Q \to 1$.

In summary, the value of Q *remembers* which input, S or R, was 1 last. This is the *memory feature* of the SR-FF. The following example shows the SR-FF used in a car alarm system, converting a momentary pulse produced by a the car movement sensor into a continuous blaring sound.

EXAMPLE 4.1   Car burglar alarm

When a parked car experiences a sudden bump, a sensitive motion sensor (an accelerometer) produces a short-duration $0 \to 1 \to 0$ electronic pulse. This momentary pulse connects to the S input the SR-FF, as shown in Figure 4.3 and causes its Q output to become and remain at 1. This activates an alarm and keeps it blaring (at least for a while). The alarm is turned off by using a key that provides a $0 \to 1$ transition at the R input that resets $Q = 0$.

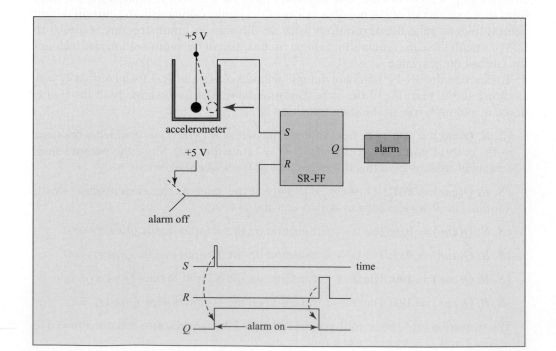

Figure 4.3

SR-FF used in a car alarm.

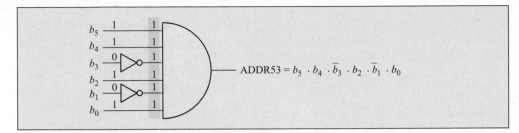

Figure 4.4

Address recognition logic circuit. The 6-input AND gate shown recognizes the 6-bit address 110101 (location 53_{10}).

$$ADDR53 = b_5 \cdot b_4 \cdot \overline{b_3} \cdot b_2 \cdot \overline{b_1} \cdot b_0$$

4.3.1 Addressable Memory

An important component of a computer is its *random access memory* (RAM) in which binary data can be stored and retrieved when needed. We first consider a simple 1-bit memory and then extend it to an 8-bit (byte) memory.

Each location in a digital memory is designated with its own address specified as a binary sequence. To access a particular location in our 1-bit memory, its address is first recognized by using an AND gate. For example, let us consider the memory location 53_{10}, having the 6-bit binary address ADDR53=110101. The six-input AND gate shown in Figure 4.4 recognizes this address when it occurs at its input. That is, ADDR53=1 only when the AND gate input is 110101, otherwise ADDR53=0.

A 1-bit memory location is implemented using a single SR-FF (SR-FF53) having output Q53. To store the binary value INDATA (equal to 0 or 1) into SR-FF53, we first put its address on the address lines. When 53_{10} appears on the address lines, ADDR53=1, and this output connects to two AND gates with the first having output X that connects to the S input of SR-FF53 and other output Y that connects to R, as shown in Figure 4.5. The data INDATA is applied to *all* the SR-FFs in the memory, but only one (SR-FF53) has its address AND gate output equal to one. If INDATA=1, X becomes 1 and SR-FF53 is set, making Q53=1. If INDATA=0, it passes through a NOT gate making $Y = 1$ and SR-FF53 is reset, making Q53=0. When 53_{10} is removed from the address line ADDR53=0, then $X = 0$ and $Y = 0$, thus preventing any changes to occur in SR-FF53. Thus, the INDATA value is stored in SR-FF53.

To retrieve the data value Q53 stored in SR-FF53, we apply ADDR53=110101 to all of the address recognition circuits, making only ADDR53=1. ADDR53 and Q53 are applied to the *data output AND gate* AND53 with output MEM53, as shown in Figure 4.6. If Q53=1 and ADDR53=1, the output MEM53=1. If Q53=0 and ADDR53=1,

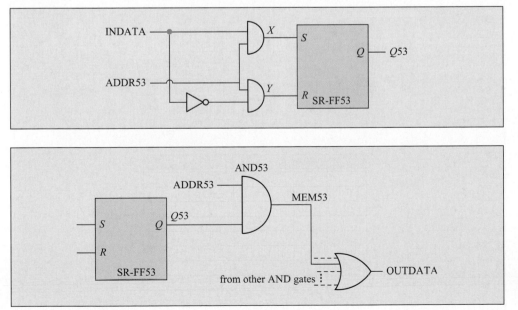

Figure 4.5

Writing the data bit value INDATA into SR-FF53 at memory location ADDR53.

Figure 4.6

Reading the data bit value stored in memory location 53.

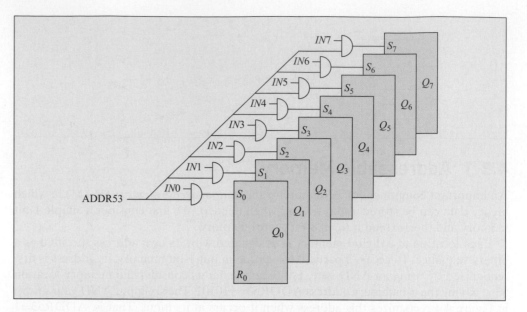

Figure 4.7

A byte in computer memory is organized as a stack of eight SR-FFs.

MEM53=0. All of the other data output AND gates at address $XX_{10} \neq 53$ produce 0 outputs because their ADDRXX=0. The outputs of all data output AND gates are applied to the input of one large OR gate (at least conceptually) to produce the output data OUTDATA. Note that OUTDATA=1 only if Q53=1. That is, the retrieved value OUTDATA has the same binary value as that stored in SR-FF53.

Binary data are typically stored as 8-bit units (bytes). As shown in Figure 4.7, a byte in an 8-bit memory can be considered to be a stack of eight SR-FFs sharing a common address.

4.4 TOGGLE FLIP-FLOP

The toggle flip-flop (T-FF) is the basic counting element of the computer and is shown in Figure 4.8. The T-FF has two inputs, T for toggle and C for clear, and one output Q. The value of Q switches from 0 to 1 or from 1 to 0 (*toggles*) whenever there is a

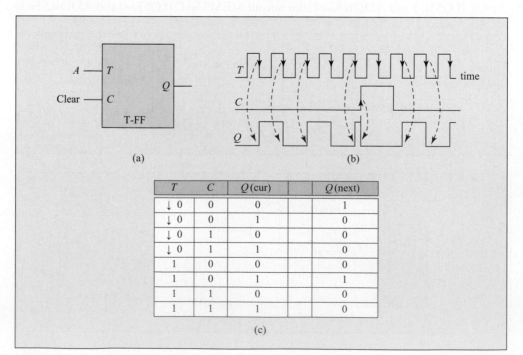

Figure 4.8

Toggle flip-flop (T-FF): (a) logic symbol, (b) timing diagram, and (c) transition table.

T	C	Q(cur)		Q(next)
↓ 0	0	0		1
↓ 0	0	1		0
↓ 0	1	0		0
↓ 0	1	1		0
1	0	0		0
1	0	1		1
1	1	0		0
1	1	1		0

(c)

downward $(1 \rightarrow 0)$ transition at its T input. When a 1 is applied to the C input, $Q = 0$ immediately and maintained at 0 as long as $C = 1$.

The operation of the T-FF is illustrated in the timing diagram and state transition table in Figure 4.8. Figure 4.8a shows the T-FF input (T) connected to a waveform A that has a series of transitions. Figure 4.8b shows the T-FF output changes its value (or, toggles its state) when a $1 \rightarrow 0$ transition occurs at the T input. The in toggling operation of the T-FF occurs in the first two rows of the transition table shown in Figure 4.8b: As long as $C = 0$, a downward transition on T causes Q to toggle (become the complement of the previous value)—hence, the name *toggle flip-flop*.

The timing diagram and transition table show that whenever $C = 1$, $Q = 0$—no matter what occurs on the T input.

The transition in T is important: When T remains at 0, Q does not change. If we attempted to implement a combinational logic circuit from this state transition table, using it as a truth table, then $T = 0$, and $C = 0$ would cause Q to oscillate between 0 and 1 continuously. Clearly, this is not the intended operation, so the digital electronic circuit that implements a T-FF incorporates a transition-sensing device in the T input whose operation is beyond the level of this course.

4.5 COUNTING WITH TOGGLE FLIP-FLOPS

Toggle flip-flops are connected in a chain for counting the downward transitions in pulses. In a chain, the Q output of one T-FF connects to the T input of the next T-FF. A $1 \rightarrow 0$ transition at the T input of any T-FF switches its output Q to the complementary value. The T-FF chain counts the $1 \rightarrow 0$ transitions that occur at the T input of the first T-FF. The T-FF outputs are indexed starting with 0, making Q_0 correspond to the coefficient of 2^0, Q_1 corresponds to the coefficient of 2^1, and so on. The count of downward $(1 \downarrow 0)$ CLOCK transitions is encoded in the T-FFs outputs in the binary form

$$CNT = Q_3 \; Q_2 \; Q_1 \; Q_0$$

For a chain containing M T-FFs, numbered 0 to $M - 1$, the CNT value is determined by

$$CNT = Q_0 + 2Q_1 + 4Q_2 + 8Q_3 + \cdots + 2^{M-1}Q_{M-1} \tag{4.1}$$

Thus, a chain containing M T-FFs produces unique CNT values in the range

$$0 \leq CNT \leq 2^M - 1 \tag{4.2}$$

This is called a *modulo-M counter*. At the $2^{M\text{th}}$ input transition, all Q values reset to 0, making $CNT = 0$.

Modulo-16 binary counter   EXAMPLE   4.2

Figure 4.9 shows four T-FFs connected in a chain. All C inputs are equal to 0. The input to the first T-FF is denoted A, which contains the downward transitions we want to count.

The T-FFs are initialized by resetting all Q values to 0. This is typically done by a short pulse $(0 \rightarrow 1 \rightarrow 0)$ applied simultaneously to all C inputs (not shown in the figure). The initial count is shown as $Q_3 Q_2 Q_1 Q_0 = 0000$ on the lower left.

The $1 \rightarrow 0$ transitions are the ones that cause changes, and these are indicated with downward arrows. The first $1 \rightarrow 0$ transition on A toggles Q_0 from $0 \rightarrow 1$. Q_0 also connects to T on the next T-FF, but its upward transition has no effect because a T-FF changes only on a $1 \rightarrow 0$ transition. After the first $1 \rightarrow 0$ transition on A, $Q_3 Q_2 Q_1 Q_0 = 0001$, corresponding to $CNT = 1$.

The second $1 \rightarrow 0$ transition on A toggles Q_0 from $1 \rightarrow 0$, and this transition toggles Q_1 from $0 \rightarrow 1$. After the second $1 \rightarrow 0$ transition on A, $Q_3 Q_2 Q_1 Q_0 = 0010$, making $CNT = 2$.

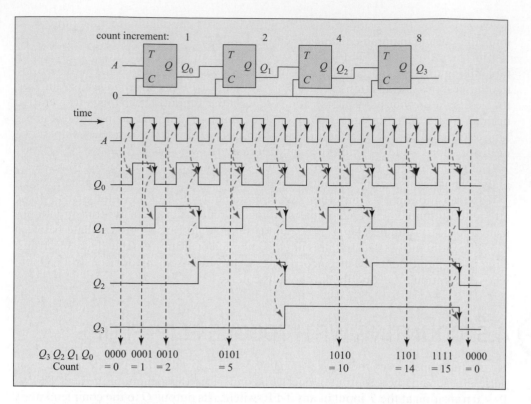

Figure 4.9

Counting with a chain of four T-FFs.

This procedure continues until $Q_3 Q_2 Q_1 Q_0 = 1111$ or $CNT = 15$. The next (16th) $1 \rightarrow 0$ transition on A toggles Q_0 from $1 \rightarrow 0$, and this transition toggles Q_1 $1 \rightarrow 0$, and so on, producing the final stable values $Q_3 Q_2 Q_1 Q_0 = 0000$ for $CNT = 0$. This is the initial state of the T-FF chain, so CNT takes on values from 0 to 15, thus producing a modulo-16 counter.

4.5.1 Modulo-K Counters

This section shows how to implement a modulo-K counter when K is not a power of 2 by including a combinational logic circuit in the counter. The previous section shows that a chain of M T-FFs counts from 0 to $2^M - 1$. If we want a modulo-K with $K < 2^M - 1$, we need to reset the counter when it reaches K.

To form a modulo-K, we need M T-FFs where

$$M \geq \log_2 K = \frac{\log_{10} K}{\log_{10} 2} \tag{4.3}$$

EXAMPLE 4.3   **Number of T-FFs in a modulo-100 counter**

A modulo-100 counter counts from 0 to 99. The number of T-FFs needed in the chain (M) can be determined by either *direct calculation* or using *intuition*.

Direct calculation: To form a modulo-K counter with $K = 100$, we compute

$$M = \text{ceiling}(\log_2 100) = \text{ceiling} \left(\overbrace{\frac{\log_{10} 100}{\log_{10} 2}}^{=2/0.3} \right) = \text{ceiling}(6.7) = 7$$

Thus, we need seven T-FFs in the chain.

Intuition: To count modulo-100, the number of T-FFs need to count at least to that number. Because M T-FFs is a modulo-2^M counter, we need the smallest M such that $2^M \geq 100$. A little experience with powers of 2 provides the answer: because $2^6 = 64$ (<100), $M > 6$. Adding one more T-FF to the chain produces a modulo-2^7 counter. Because $2^7 = 128$ (>100), $M = 7$ is both necessary and sufficient.

A modulo-K counter starts the count at 0 and resets it to 0 at the Kth count. This is accomplished using the following three steps.

STEP 1. Determine the binary pattern that represents K.

STEP 2. Recognize this binary pattern with an AND gate circuit whose inputs connect to $Q_0, Q_1, Q_2, \ldots, Q_{M-1}$. The Q values that equal 1 in the pattern for K connect directly from Q to the AND gate. For the Q values that equal 0, the connection is through a NOT gate to make them 1.

STEP 3. Connect the AND output to the C inputs of all M T-FFs.

When the count equals K, the AND gate output becomes 1, which immediately resets all the T-FFs to zero. The all-zero pattern of the T-FFs is then no longer the pattern that the AND recognizes, making the AND gate output 0. Making $C = 0$ allows the T-FF chain to restart the count at the next downward transition. The pulse at the AND output is very short—so short that the Q values momentarily representing K can be neglected. The next example illustrates this operation.

Modulo-6 counter   **EXAMPLE   4.4**

A modulo-6 counter produces counts 0, 1, 2, 3, 4, 5, 0, 1, 2, . . . To count six downward transitions, we need

$$M = \text{ceiling}(\log_2 6) = \text{ceiling} \left(\overbrace{\frac{\log_{10} 6}{\log_{10} 2}}^{=0.78/0.3} \right) = \text{ceiling}(2.6) = 3$$

or three T-FF's. By intuition, $2^2 = 4$ (too small) and $2^3 = 8$ (necessary and sufficient).

Figure 4.10 shows the three T-FF chain. The modulo-6 counter is implemented using the following three steps.

STEP 1. The binary pattern for 6_{10} is determined as $110_2 (= Q_2 Q_1 Q_0)$.

STEP 2. The outputs of the T-FFs are connected to an AND gate that recognizes this binary pattern. When this pattern occurs, the AND gate output $Y = 1$.

STEP 3. The AND gate output connects to the C inputs of all the T-FFs, resetting them all to zero when $Q_2 Q_1 Q_0 = 110$. This resets the counter to zero and makes $Y = 0$ again, because the binary values become $Q_2 Q_1 Q_0 = 000$.

Figure 4.10 shows that logic diagrams can get very complicated. Straight lines that touch or cross indicate an electrical (and logical) connection is made between them. When a connection is not meant to made, one of the lines is drawn with an arc that passes over the other (or others), as shown in the figure.

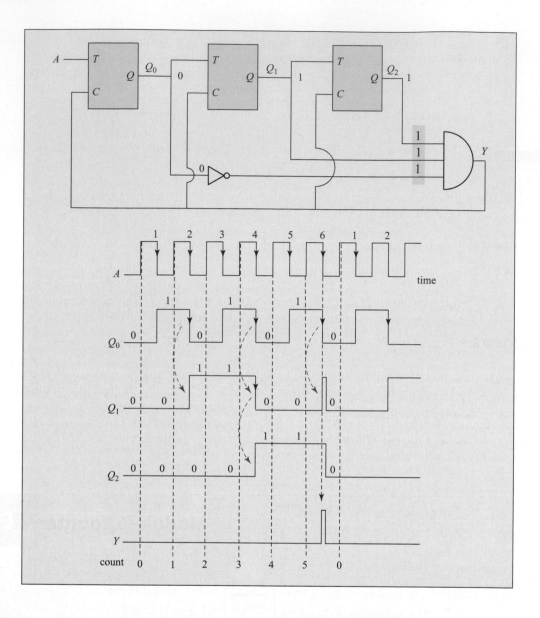

Figure 4.10

A modulo-6 counter circuit.

EXAMPLE 4.5

Modulo-44 counter

To design a modulo-44 counter, we need six T-FFs, as shown in Figure 4.11, because $2^6 = 64$ and $2^5 = 32$ makes five T-FFs insufficient.

The input A contains the $1 \downarrow 0$ transitions that are counted and connects to the first (Q_0) T-FF T input. The Q outputs connect to the T inputs of the next T-FF in the chain. The AND gate recognizes the pattern

$$44_{10} = 32 + 8 + 4 = 101100_2$$

with the logic equation as

$$Y = Q_5 \cdot \overline{Q_4} \cdot Q_3 \cdot Q_2 \cdot \overline{Q_1} \cdot \overline{Q_0}$$

The AND gate output Y connects to the C inputs of all of T-FFs needed to clear them when the count equals 44.

Figure 4.11 shows the T-FF output values at the count of 44. The chain is cleared immediately after this count appears.

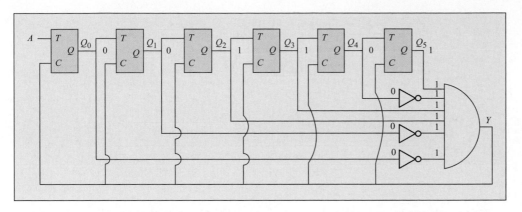

Figure 4.11

Modulo-44 counter in Example 4.5.

The next section describes a method that implements an efficient counter that uses a smaller gate count.

4.5.2 Efficient Modulo-*K* Counter

The previous example of a modulo-6 counter used a 3-input AND gate to recognize the binary pattern for 6 (= 110). While this works, a more efficient counting circuit is implemented by observing the counting sequence and noting that the T-FF Q values that equal 0 do not need to be connected to the AND gate. This is the case because the pattern of 1's in the binary representation of K in a modulo-K counter is sufficient to reset the counting chain at the correct time. The other binary sequences containing the same pattern of 1's occurs at count values greater than K, and hence never occur in the modulo-K counter.

This observation eliminates one or more NOT gates and uses a pattern recognition AND gate with fewer inputs in the efficient counter implementation. The following example illustrates the approach.

Efficient modulo-6 counter implementation **EXAMPLE 4.6**

Consider the values of $Q_2 Q_1 Q_0$ in the counting sequence shown in Figure 4.12. Note that $Q_2 Q_1 = 11$ in the binary pattern for six (=110) and for the pattern for seven (=111).

Because the modulo-6 counter resets at the count of six, **the count never reaches seven.** Thus, the value of Q_0 does not matter. Thus, its NOT gate can be eliminated and an AND gate with fewer inputs is used. Figure 4.13 shows the efficient implementation of the modulo-6 counter.

Count	Q_2	Q_1	Q_0
0	0	0	0
1	0	0	1
2	0	1	0
3	0	1	1
4	1	0	0
5	1	0	1
6	1	1	0
7	1	1	1

Figure 4.12

Values of $Q_2 Q_1 Q_0$ in the counting sequence.

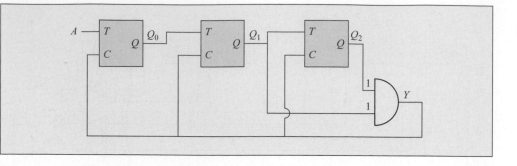

Figure 4.13

Efficient implementation of modulo-6 counter that recognizes only the 1 values in 110 to reset the counting chain.

This result can be generalized to include all T-FF Q values that equal 0 in the pattern for K in a modulo-K counter; Q values that are 0 in the binary value of K do not need to be applied to the pattern-recognition AND gate.

For example, a modulo-85 counter has binary value equal to

$$85_{10} = 1010101$$

If we replace any 0 in this binary value with a 1, the resulting number is greater than 85. For example, replacing the middle 0 produces

$$101\,1\,101 = 85 + 8 = 93_{10}$$

However, this binary pattern is not observed in the modulo-85 counter, because the count resets back to 0 at count 85.

EXAMPLE 4.7

Efficient modulo-44 counter

Figure 4.14 shows an efficient implementation of the modulo-44 counter of Example 4.5. The AND gate recognizes the pattern

$$44_{10} = 32 + 8 + 4 = 101100_2$$

by applying only the bit positions that equal 1 to an AND gate to form the logic equation as

$$Y = Q_5 \cdot Q_3 \cdot Q_2$$

The pattern recognition AND gate used previously in Example 4.5 had six inputs. The efficient implementation uses a 3-input AND gate and eliminates three NOT gates.

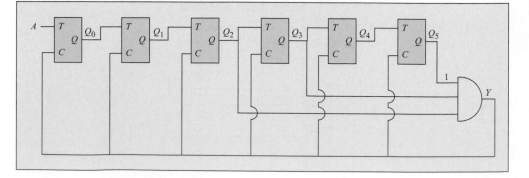

Figure 4.14

Efficient implementation of the modulo-44 counter in Example 4.7.

4.6 Summary

Sequential logic circuits produce an output value that is related not only to the current values of the inputs but also to their past values. The behavior of a sequential circuit is described with a timing diagram or a state transition table. A common sequential circuit, the flip-flop, was presented in its two basic forms: The set-reset flip-flop (SR-FF) is the basic memory unit in the computer, and the toggle flip-flop (T-FF) is the computer's elemental counting element. Combinational and sequential logic were combined to implement efficient counters.

4.7 Problems

4.1 **SR-FF using two NOR gates.** It is not allowed to have $S = 1$ and $R = 1$ at the same time for a SR-FF that is implemented with NOR gates as shown in Figure 4.2. What are the values of Q and $\overline{Q}$ when $S = 1$ and $R = 1$ if by mistake or during power-up?

4.2 **Implementing an SR-FF using two NAND gates.** Implement an SR-FF using two NAND gates and describe its behavior using a timing diagram.

4.3 **Random-access digital memory input.** Design a random-access digital memory that *stores* single bits in two addressable SR-FFs having 3-bit addresses 5 and 7. The memory has the address input lines $A2$, $A1$, $A0$ and data input line DIN and the two SR-FF outputs are Q5 and Q7.

4.4 **Random-access digital memory output.** Design a random-access digital memory that *reads* single bits stored in two addressable SR-FFs having 3-bit addresses 5 and 7. The memory has address input lines $A2$, $A1$, $A0$, and data output line DOUT.

4.5 **Random-access digital memory input/output.** Design a random-access digital memory that can write or read single bits stored in two addressable SR-FFs having 3-bit addresses 5 and 7. The memory has the following inputs and outputs

- Address input lines $A2$, $A1$, $A0$
- Data input line DIN
- Data output line DOUT
- Read/Write input line RW ($RW = 0$ stores DIN into addressed memory. $RW = 1$ puts addressed memory data on DOUT).

4.6 **Counting chain containing two T-FFs.** Sketch the counting chain containing two T-FFs and its timing diagram, starting with $Q_1 Q_0 = 00$. Show the transitions and the count values over a complete count cycle.

4.7 **Modulo-365 counter.** Implement an efficient modulo-365 counter using the following steps.

1. How many T-FFs are needed?
2. What is the binary pattern representing 365?
3. Show that only the 1's in the pattern representing 365 need to be inputs to the recognition AND gate by replacing one of the 0's in the pattern with a 1 and computing the resulting decimal value.
4. Draw the simplified modulo-365 counter circuit.

4.8 Excel Projects

The following Excel projects illustrate the following features:

- Generate a $0 \rightarrow 1 \rightarrow 0$ pulse.
- Simulate Set-Reset and Toggle Flip-Flop operation.
- Recognize a $1 \rightarrow 0$ transition.
- Demonstrate the operation of a counter with an infinite loop.
- Use of Boolean variables to implement a logic equation.

Using a Narrative box (described in Example 13.4), include observations, conclusions, and answers to questions posed in the projects.

4.1 **Generate a two-second pulse.** Modify Example 13.19 to generate a two-second pulse in cell B4 each time a shape labeled *pulse* is clicked.

4.2 **Set Reset Flip-Flop.** Modify Example 13.20 to generate a two-second pulse in cell B4 each time a shape labeled *pulse* is clicked.

4.3 **Three Toggle Flip-Flop chain.** Extend Example 13.21 from two T-FF to three T-FFs. The chain state increments with every click of the *A* input and clears the T-FF chair to zero by clicking on the *Clear* shape. Each click on *A* that maintains the 0 value for one second, which is followed by a 1 value that lasts for one second and then followed by a downward transition back to 0.

4.4 **Modulo-6 counter.** Modify Example 13.22 to implement an efficient modulo-6 counter.

4.5 **Modulo-14 counter.** Modify Example 13.22 to implement a modulo-14 counter by extending the T-FF chain to operate with four T-FFs.

CHAPTER

5

CONVERTING BETWEEN ANALOG AND DIGITAL SIGNALS

LEARNING OBJECTIVES

After completing this chapter, the reader should be able to:

- Understand how to sample analog waveforms without losing information.
- Represent samples using finite-precision numbers.
- Reconstruct analog waveforms from digital samples.

▶ 5.1 INTRODUCTION

Chapter 2 described the role sensors play in producing digital information. This chapter describes a second common source of digital data: sampling analog signals. Such sampling processes occur in smart phones for converting audio waveforms into digital data for transmission and storage in memory and in digital cameras for capturing images.

The main topics in this chapter are covered in the following sequence.

Waveform sampling—Sampling an analog waveform without losing information must consider how quickly the waveform changes in time. Time and image waveforms are characterized by the highest frequency component that is present. All of the information is retained by sampling the waveform at a rate that is higher than twice this highest frequency component. Sampling at a higher rate helps to visualize the waveform from the samples but produces unnecessary data.

Quantization—For digital transmission, storage, or processing, waveform samples must be represented in a binary form used by computers. Quantization is the process of representing the infinite-precision sample values with a finite number of bits. The number of bits employed is a trade-off between the errors introduced and the size of the data that is generated.

Digital-to-analog conversion—Digital data produced by the analog-to-digital conversion process must typically be converted back to analog waveforms. Computer animation generates digital sounds and images using formulas and models, and displaying them requires the same digital-to-analog conversion. Two common techniques are described for producing analog values from digital data.

Practical quantizer designs—This chapter concludes by describing considerations for producing digital data that achieve a desired performance, for example, making a digital system indistinguishable from an analog system.

▶ 5.2 ANALOG-TO-DIGITAL CONVERSION

Consider the digital audio system shown in Figure 5.1 that is a part of your smartphone. Acoustic energy in a microphone produces an analog signal that is a continuous-time waveform. An analog-to-digital converter (ADC) samples the waveforms to produce a sequence of samples that are transformed by a quantizer within the ADC to produce binary data. These data are either stored into a digital memory for later playback (as in recorded music) or transmitted over a digital communication channel (as during a telephone conversation). At the receiving end, a digital-to-analog converter (DAC) reproduces the analog signal that drives a speaker to produce acoustic energy. The goal

Figure 5.1

In a digital audio system, signals are present in both analog and digital forms.

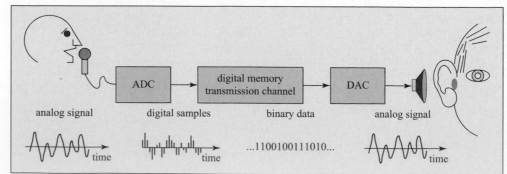

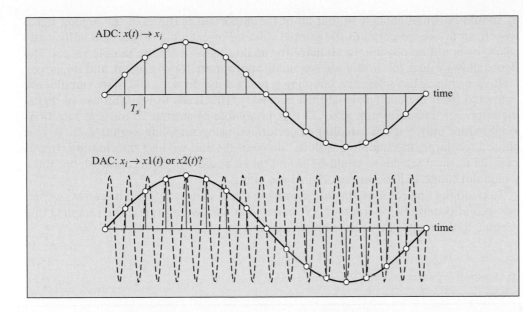

Figure 5.2

Reconstructing an analog waveform from samples presents a problem. Sampling an analog waveform $x(t)$ with sampling period T_s produces a unique set of numbers x_i. Converting this set of numbers back into an analog waveform could produce waveform $x1(t)$ (solid line) or $x2(t)$ (dashed line). Which waveform is correct?

of such a system is to have the reproduction perceptually indistinguishable from the original energy. This chapter describes the various steps that occur in such systems.

5.2.1 Sampling Analog Waveforms

In the sampling process, an analog signal is converted into a sequence of numbers by recording the values of the analog signal at a set of time-sampling points. We find that, although an analog waveform is converted to a unique set of numbers in the sampling process, returning from the samples to an analog waveform is problematic because there are an infinite number of analog signals that could have produced the same set of samples. Figure 5.2 illustrates this problem. So, which analog waveform should we choose? This section considers one-dimensional waveforms that are functions of time, but the same principle holds for sampling images with the function of position.

Figure 5.3 shows three examples of the sampling process with the original analog waveform shown in a solid line and the samples as dots. Typically, the sampling times are regularly spaced, occurring in time every *sampling period* T_s. Intuition tells us that

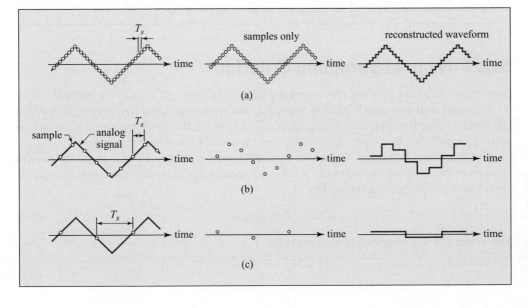

Figure 5.3

Sampling and reconstructing an analog signal. When the sampling points accurately represent the analog signal waveform, the sampling period T_s is appropriate for the signal. (a) T_s is too small, producing more samples than necessary. (b) T_s is sufficiently small. (c) T_s is too large.

T_s should be small enough so that all of the important features of the analog signal waveform are represented in the sample values. For example, when T_s is sufficiently small, it should be possible to *visualize* the analog signal from the sample values. The figure shows values for T_s that are too small, appropriate for the signal, and too large.

If T_s is chosen too small, the sampling is being done densely, and more samples are extracted from the analog signal than necessary. This leads to wasteful use of digital memory. For example, an audio CD that is capable of storing 10 musical selections would store only 5 if the sampling is performed using sampling period $T_s/2$. If T_s is chosen too large, the sampling is done too coarsely, and we lose information. In this case, 20 musical selections could be stored on an audio CD rather than 10, but these would not sound as good as the original recording.

In sampling applications, the term *sampling rate* is denoted f_s and has units *samples per second* (sample/s), or commonly referred to as a *frequency* (Hz). The sampling rate is equal to the reciprocal of the sampling period, as

$$f_s = \frac{1}{T_s} \qquad (5.1)$$

EXAMPLE 5.1 Sampling audio signals

Consider two common cases of sampling audio signals.

Digital telephone. In a digital telephone system, the speech signal is sampled at 8,000 times per second, making $f_s = 8,000$ Hz. The sampling period is

$$T_s = \frac{1 \text{ second}}{8,000 \text{ samples}} = 1.25 \times 10^{-4} \text{ s/sample} = 125 \ \mu\text{s/sample}$$

Audio CD. In an audio CD system, music is sampled at 44,000 times per second, $f_s = 44,000$ Hz. The sampling period is

$$T_s = \frac{1 \text{ second}}{44,000 \text{ samples}} = 0.0000227 \text{ s/sample} = 22.7 \ \mu\text{s/sample}$$

In a stereo system, each channel is sampled at this rate.

> *Factoid:* Our rule of thumb for adequately sampling $x(t)$ to produce x_i is that we should be able to visualize (by eye) the original waveform $x(t)$ from x_i.

5.2.2 Nyquist Sampling Criterion

Since the sampling process can generate huge quantities of data (one sample every T_s seconds) and too much data is wasteful, the question naturally arises: Is there a minimum value for sampling rate f_s that still retains all the information in the original analog signal? We first state the answer and then illustrate it with examples.

The minimum value for f_s is given by the *Nyquist criterion*: If f_{max} is the maximum frequency component present in an analog waveform, the sampling rate must be greater than twice the maximum frequency or

$$f_s > 2 f_{max} \qquad (5.2)$$

Equivalently, the sampling period must be

$$T_s < \frac{1}{2 f_{max}} \qquad (5.3)$$

Telephone-quality speech   EXAMPLE 5.2

Telephone-quality speech is processed by the telephone company to remove the frequencies above the telephone bandwidth. The highest frequency in telephone-quality speech is

$$f_{max} = 3,300 \text{ Hz}$$

The sampling period used in the telephone system is chosen so

$$f_s > 2 f_{max} \quad \text{making} \quad f_s > 6,600 \text{ Hz}$$

The telephone company commonly uses

$$f_s = 8,000 \text{ Hz}$$

for digital telephone applications.

Audio CD system   EXAMPLE 5.3

Let us consider an audio CD system. Most people cannot hear frequencies higher than 20,000 Hz (20 kHz). Thus, storing any higher frequency components is futile, as these components will not be heard or appreciated anyway. Hence, 20 kHz is often taken as an upper bound on high fidelity music, so

$$f_{max} = 20 \text{ kHz}$$

In this case, the Nyquist criterion states that the sampling rate is

$$f_s > 40 \text{ kHz}$$

In practice, audio CD systems employ the sampling rate

$$f_s = 44 \text{ kHz}$$

for producing digitally recorded music. All standard CD players use this rate to play back music.

5.2.3 Aliasing

The Nyquist criterion determines the minimum sampling rate f_s that prevents loss of information. We now describe the problems encountered if the analog signal is not sampled sufficiently often. When $f_s \leq 2 f_{max}$ (that is, if the Nyquist criterion is not obeyed) a new strange frequency will be perceived from the samples called the *alias* frequency f_a.

Let us consider sampling the sinusoidal waveform $\cos(2\pi f_o t)$. When the sample values are viewed, the phenomenon of aliasing causes the eye to see a sinusoid in the sample values having a frequency that is different than that present in the analog signal. Because $\cos(2\pi f_o t)$ contains only one frequency component, the maximum frequency is equal to that frequency or $f_{max} = f_o$. Let this sinusoid be sampled with a sampling rate f_s that is too small (that is, $f_s \leq 2 f_o$). Then the Nyquist criterion is not obeyed, and a sinusoid having an alias frequency f_a is observed, where

$$f_a = |f_s - f_o| \tag{5.4}$$

EXAMPLE 5.4

Demonstration of aliasing

The analog waveform $x(t) = \cos(2\pi f_o t)$ has $f_{max} = f_o$. With sampling rate $f_s > 2f_o$, the sampling period is $T_s = \frac{1}{f_s}$, and the resulting sample values equal

$$x_i = x(t = iT_s) = \cos(2\pi f_o \, i \, T_s)$$

To demonstrate aliasing, increase the analog waveform frequency

$$f_o \rightarrow f_o' = f_o + f_s$$

while keeping the same sampling rate f_s. Clearly, this violates the Nyquist criterion, $f_s < 2f_o'$. The resulting sample values equal

$$x_i' = x'(t = iT_s) = \cos\left(2\pi f_o' \, i \, T_s\right) = \cos(2\pi(f_o + f_s)i \, T_s)$$

Simplifying, we get

$$x_i' = \cos\left(2\pi f_o \, i \, T_s + \overbrace{2\pi f_s i \, T_s}^{=2\pi i}\right)$$

Because adding an integer number of 2π to the argument of a cosine function does not change its value, the last equation simplifies to

$$x_i' = \cos(2\pi f_o \, i \, T_s) = x_i$$

These last values correspond to those that are observed for the original waveform with frequency f_o, even though the analog frequency now equals $f_o + f_s$. Hence, frequency f_o is the alias frequency for $f_o' = f_o + f_s$.

EXAMPLE 5.5

Graphical demonstration of aliasing

Let us consider the sinusoid with frequency $f_o = 2$ Hz,

$$s(t) = \sin(2\pi f_o t) = \sin(4\pi t)$$

This waveform is shown in solid line in Figure 5.4. Because this waveform has only one frequency component ($f_o = 2$ Hz), its maximum frequency is $f_{max} = 2$ Hz. To retain all of the information, the sampling rate must be

$$f_s > 2f_{max} = 4 \text{ Hz} \quad \rightarrow \quad T_s = \frac{1}{f_s} = \frac{1}{4} = 0.25 \text{ s}$$

The sampled time sequence with index i equals the values of the analog waveform at the sampling times:

$$s_i = s(t = iT_s) = \sin(4\pi \, i \, T_s)$$

Consider the value $f_s = 16$ Hz ($T_s = 1/16$ s) shown in Figure 5.4a. These samples form a good perceptual representation of the analog waveform, because the waveform can be visualized from the samples connected by a dashed line. (The human eye performs a linear interpolation between sample points.)

Figure 5.4b shows $f_s = 8$ Hz. No aliasing occurs because $f_s > 2f_o$ ($f_s = 8$ Hz > 4 Hz). The dashed curve *looks* similar to solid curve, but the deviations between the original waveform and linear interpolation are noticeable.

Figure 5.4c shows $f_s = 5$ Hz. No aliasing occurs because $f_s > 2f_o$ ($f_s = 5$ Hz > 4 Hz). The dashed curve differs significantly from solid curve, but techniques exist to recover the solid curve from samples. Although sampling the analog signal at slightly more than twice the highest frequency component in the signal retains the information, it is difficult to visualize the waveform from the samples.

Hence, for a reasonably accurate linear interpolation, the sampling rate must be much higher, at least eight times the highest frequency or four times the Nyquist rate.

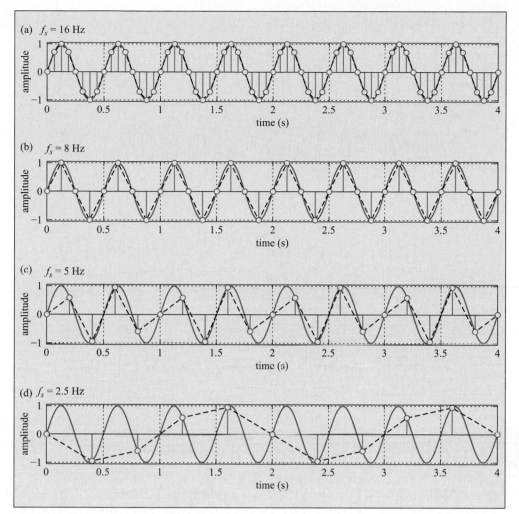

Figure 5.4

Demonstration of aliasing. The analog waveform $s(t) = \sin(2\pi f_o t) = \sin(4\pi t)$, ($f_o = 2\,\text{Hz}$) for $0 \leq t \leq 4\,\text{s}$ (solid curve), is sampled using different sampling rates f_s. The samples shown as circles are connected by dashed lines to form a linear interpolation.

When $f_s < 2 f_{\max}$, aliasing will occur. The alias frequency in the linear interpolation of the sample values is a frequency $f_a < f_s/2$. Consider $f_s = 2.5\,\text{Hz}$ shown in Figure 5.4d. Then, for $f_o = 2\,\text{Hz}$, we have

$$f_a = |f_s - f_o| = |2.5 - 2| = 0.5\,\text{Hz}$$

The waveform having this aliased frequency is shown as a dashed line.

To quantify the alias effect, compare the values at the sample points for both the original and the alias frequencies for $f_s = 2.5\,\text{Hz}$ ($T_s = 2/5\,\text{s}$). Starting at $t = 0$, the sample times occur at the instants

$$t = 0 \quad T_s \quad 2T_s \quad 3T_s \quad 4T_s \quad 5T_s \cdots$$
$$= 0 \quad \tfrac{2}{5}\text{s} \quad \tfrac{4}{5}\text{s} \quad \tfrac{6}{5}\text{s} \quad \tfrac{8}{5}\text{s} \quad 2\text{s} \cdots$$

The values of the original analog cosine waveform at these time instants equal

$$\sin(2\pi f_o \times 0) \quad \sin(2\pi f_o \times T_s) \quad \sin(2\pi f_o \times 2T_s) \quad \sin(2\pi f_o \times 3T_s) \cdots$$
$$\sin(0) \quad \sin(\tfrac{8}{5}\pi) \quad \sin(\tfrac{16}{5}\pi) \quad \sin(\tfrac{24}{5}\pi) \cdots$$
$$0 \quad -0.951 \quad -0.588 \quad 0.588 \cdots$$

The values of the alias signal at these time instants equal

$$\sin(2\pi f_a \times 0) \quad \sin(2\pi f_a \times T_s) \quad \sin(2\pi f_a \times 2T_s) \quad \sin(2\pi f_a \times 3T_s) \cdots$$
$$\sin(0) \quad \sin(\tfrac{2}{5}\pi) \quad \sin(\tfrac{4}{5}\pi) \quad \sin(\tfrac{6}{5}\pi) \cdots$$
$$0 \quad -0.951 \quad -0.588 \quad 0.588 \cdots$$

Note that the sample values of both the original and the alias signal are equal.

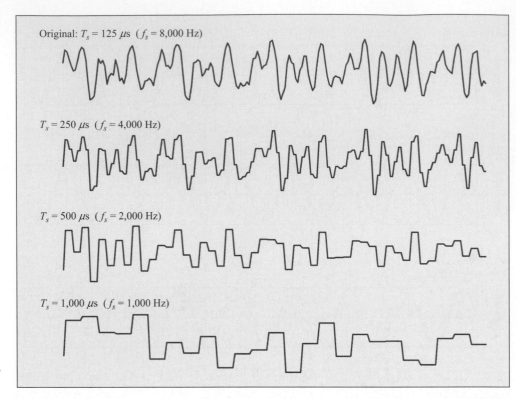

Figure 5.5

Demonstration of aliasing in microphone speech.

EXAMPLE   **5.6**   ## Aliased microphone speech waveform

Figure 5.5 shows a speech waveform that was acquired using the microphone on a laptop. The signal is sampled at $f_s = 8,000$ Hz. The sampling period T_s was doubled in each successive waveform to demonstrate aliasing. As T_s increases, more details in the speech waveform are lost. When $T_s = 1,000\,\mu s$, the speech is unintelligible.

Anti-Aliasing Filters

To prevent unintended aliasing, a waveform is low-pass filtered to reduce frequency components that are greater than half of the sampling frequency f_s using an *anti-aliasing filter*. Ideally, this filter should pass all frequency components $f < f_s/2$ and eliminate all components $f \geq f_s/2$. Such higher frequency components are typically present in the noise contained in the waveform.

5.3 SPATIAL FREQUENCIES

Images contain spatial frequencies. Consider a Cartesian (x, y) coordinate system that contains a gray-scale image with x index $0 \leq i \leq n_x - 1$ and y index $0 \leq j \leq n_y - 1$. Figure 5.6 shows images with $n_x = 256$ and $n_y = 256$. The pixel in the upper left-hand corner has an index of $(0,0)$. A gray-scale image contains pixel values $v_{i,j}$ in the range $0 \leq v_{i,j} \leq A_{\max}$, where 0 corresponds to black and $A_{\max}$ to white.

Horizontal spatial frequencies vary in the x direction, and vertical frequencies vary in the y direction. The sinusoidal period is related to the size of the image in that direction. For example, horizontal frequencies have a fundamental period containing n_x points.

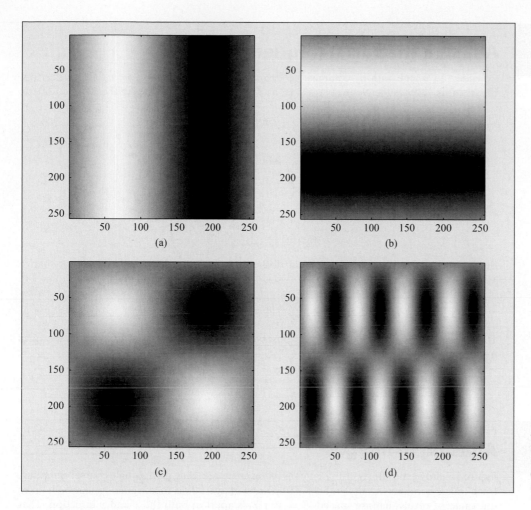

Figure 5.6

Spatial frequencies: (a) $f_x = 1$ cycles/row, $f_y = 0$ cycles/column; (b) $f_x = 0$, $f_y = 1$; (c) $f_x = 1$, $f_y = 1$; and (d) $f_x = 4$, $f_y = 1$.

A horizontal frequency $0 \leq f_x \leq f_{x,\max}$ is given by the x vector as

$$v_i = \frac{A_{\max}}{2} \left[1 + \sin\left(\frac{2\pi f_x i}{n_x} \right) \right] \quad \text{for } 0 \leq i \leq n_x - 1 \tag{5.5}$$

adding 1 to the sine values makes them vary between 0 and 1, making $0 \leq v_i \leq A_{\max}$, necessary to display the gray-scale image.

The image is sampled at every pixel. Hence, following the Nyquist criterion, the maximum spacial frequency produces at least two samples per period. Therefore, with n_x pixels, the maximum frequency equals

$$f_{x,\max} < \frac{n_x}{2} \text{ cycles/row} \tag{5.6}$$

Similarly, a vertical frequency $0 \leq f_y \leq f_{y,\max}$ is given by the y vector as

$$v_j = \frac{A_{\max}}{2} \left[1 + \sin\left(\frac{2\pi f_y j}{n_y} \right) \right] \quad \text{for } 0 \leq j \leq n_y - 1 \tag{5.7}$$

The maximum vertical spacial frequency is

$$f_{y,\max} < \frac{n_y}{2} \text{ cycles/column} \tag{5.8}$$

EXAMPLE 5.7

Aliasing in spacial frequencies

Consider an image with n_x columns. To demonstrate aliasing, let $f_x = n_x + 1$ cycles/row. Clearly, this violates the Nyquist criterion, as $f_{x,\max} < n_x/2$. The values in the x direction are

$$v_i = \frac{A_{\max}}{2}\left[1 + \sin\left(\frac{2\pi f_x i}{n_x}\right)\right]$$

$$= \frac{A_{\max}}{2}\left[1 + \sin\left(\frac{2\pi(n_x+1)i}{n_x}\right)\right]$$

$$= \frac{A_{\max}}{2}\left[1 + \sin\left(\overbrace{\frac{2\pi n_x i}{n_x}}^{=2\pi i} + \frac{2\pi i}{n_x}\right)\right]$$

Because adding an integer number of 2π to the argument of a sine function does not change its value, the last equation simplifies to

$$v_i = \frac{A_{\max}}{2}\left[1 + \sin\left(\frac{2\pi i}{n_x}\right)\right]$$

These last values correspond to those observed for $f_x = 1$ cycles/row, which is the alias frequency.

EXAMPLE 5.8

Aliasing in images

Figure 5.7 shows the effects of spatially sampling an image using increasingly coarser sampling. Each image contains $(n_x, n_y) = (500, 500)$ points. Image 5.7a is the original. Image 5.7b takes the value of pixels that are spaced $T_s = 10$ pixels apart in both the x and y direction. That pixel value sets the values of a $T_s \times T_s$ square in the image in order to maintain the size of the original image.

As T_s increases, the detail of finer (*higher spatial frequency*) objects, such as the whiskers, begin to degrade. When the viewer *squints*, it forms a spatial low-pass filter that removes the higher frequencies in all of the images, and they begin to look the same, although without fine details.

▶ 5.4 QUANTIZATION

Analog-to-digital conversion in practice involves two steps:

Sample-and-hold: Maintaining the sample value at a constant level for a sufficient time interval to allow the conversion process to be completed.

Quantization: Representing the sampled analog value with a finite number of bits.

5.4.1 Sample-and-Hold Operation

Because the quantization process takes time to accomplish, the analog sample value must be *held* constant during the conversion process. Figure 5.8 shows that the sample value is held at voltage value that occurs at the sampling time, as

$$V_a = V_{\text{IN}}(t_{\text{SaH}}), \quad \text{for } t \geq t_{\text{SaH}} \tag{5.9}$$

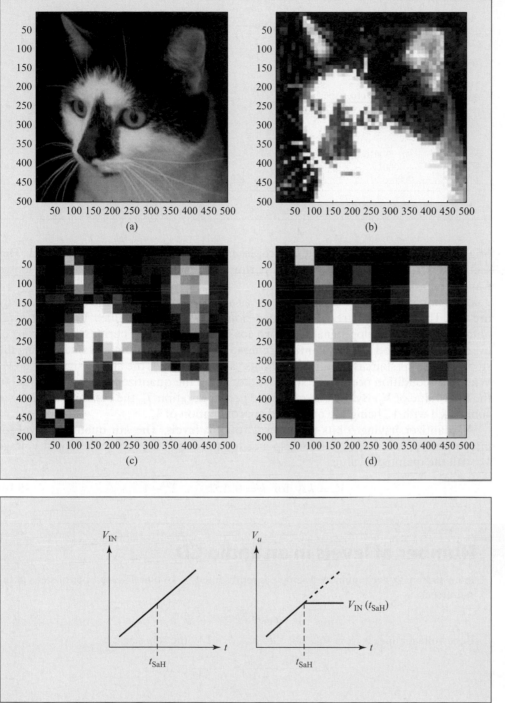

Figure 5.7

Sampling an image: (a) original 500 × 500 pixels; (b) $T_s = 10$ pixels; (c) $T_s = 25$ pixels; and (d) $T_s = 50$ pixels.

Figure 5.8

Sampling and holding $V_{IN}(t)$ to produce the infinite-precision value V_a to be quantized, with $V_a = V_{IN}(t_{SaH})$.

Having a constant, infinite-precision, voltage value V_a, we can proceed to convert it into a finite-precision approximation V_q, which is described next.

5.4.2 Staircase Method for Quantization

Quantization is the process of transforming a sample value having infinite precision, such as $s_i = 3.1415926\ldots$, into a finite-precision value, such as $sq_i = 3.14$, that is expressed with a finite number of bits, such as the 7-bit binary code word 0110010. The number of bits must be sufficiently large to provide a reasonable approximation to the

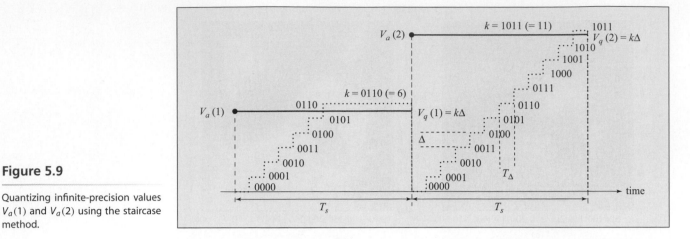

Figure 5.9

Quantizing infinite-precision values $V_a(1)$ and $V_a(2)$ using the staircase method.

original sample value with the error representing an added noise to the signal. This section describes how quantization is performed in your smartphone and for sensor waveform processing in robotics.

One simple method of explaining how a sample value is quantized is shown in Figure 5.9. The figure illustrates the conversion of two sample values $V_a(1)$ and $V_a(2)$ obtained from an analog signal from two consecutive sampling periods. At the start of every sample period T_s, the quantizer generates a staircase pattern of values having the step-size Δ. This staircase continues until its value V_q exceeds the analog signal value V_a. When this condition occurs, the staircase stops, and the quantized value is set equal to the final value of V_q. By the end of sample period duration T_s, the conversion has been completed with V_q being the quantized approximation of V_a.

A quantizer having b bits produces 2^b output levels. The kth quantizer level for integer $k = 0, 1, 2, \ldots, 2^b - 1$ corresponds to approximating the analog input voltage V_a with the quantized value

$$V_q = k\Delta \quad \text{for} \quad k = 0, 1, 2, \ldots, 2^b - 1 \tag{5.10}$$

EXAMPLE 5.9 Number of levels in an audio CD

In an audio CD, each quantized sample is represented by 16 bits. The number of steps in the staircase is

$$n_{\text{steps}} = 2^{16} = 65{,}536$$

If the voltage range is from $V_{\text{min}} = 0$ V to $V_{\text{max}} = +2$ V, the step size is

$$\Delta = \frac{V_{\text{max}} - V_{\text{min}}}{n_{\text{steps}} - 1} = \frac{2\text{ V}}{65{,}535} \approx 10\ \mu\text{V}$$

The quantizer produces a binary code that is the base-2 representation of integer k—not a voltage value. To reconstruct the voltage value at playback from a binary stream, the number of bits in the codeword n_b, sample time T_s, and step size Δ must be known.

To generate a staircase with 2^b requires a fast clock. If T_s is the sample period, a new step must be produced every T_Δ seconds, where

$$T_\Delta = \frac{T_s}{2^b} \tag{5.11}$$

EXAMPLE 5.10

Staircase quantizer

Microcontrollers (MCUs) typically include 10-bit ADCs as part of the chip design. Each ADC produces $2^{10} = 1,024$ levels. An MCU typically uses a voltage supply $V_s = 5$ V, which determines the voltage range of the ADC. The corresponding step size is

$$\Delta = \frac{V_s}{2^b - 1} = \frac{5\,\text{V}}{1023} = 0.0049\,\text{V}$$

An easier approximation to remember is to use $2^{10} \approx 1,000$ to produce

$$\Delta \approx 5\,\text{V}/1,000 = 5\,\text{mV}$$

The ADC produces sample value k with $0 \le k \le 1023$. If $k = 500$, the corresponding quantized value of V_a equals

$$V_q = k\Delta = 500 \times 0.0049\,\text{V} = 2.45\,\text{V}$$

The approximation would produce the value $V_q = 500 \times 5\,\text{mV} = 2.5\,\text{V}$.

If a staircase conversion must occur within $T_s = 1$ ms, the time to generate each step in a 10-bit quantizer is

$$T_\Delta = \frac{T_s}{2^b} = \frac{10^{-3}\,\text{s}}{1024} = 9.77 \times 10^{-7}\,\text{s} \approx 1\,\mu\text{s}$$

5.4.3 Setting Number of Quantizer Bits

A discrete voltage value produced by a quantizer, x_q, is encoded as the binary count of the number of steps used in the conversion, as shown in Figure 5.9. If a small number of bits is used, a staircase pattern having a small number of large-sized steps is needed to cover the desired voltage range. The large step size can produce large errors between the analog value x_a and its approximation x_q. Using the staircase quantizer described previously, the quantized value x_q is always larger than the corresponding analog value x_a. Then the quantizer error is determined by the step size Δ by

$$x_q - x_a < \Delta \tag{5.12}$$

Quantizing a sinusoid using 3 bits is shown in Figure 5.10a. This figure shows a graph (x_a, x_q) that indicates the transformation produced by the staircase quantizer in converting x_a into x_q. The resulting quantization is very coarse, and the resulting errors in the quantized signal introduce significant deviations from the pure sinusoid, resulting in a harsh sound.

If b bits are used to generate the staircase pattern, the number of steps equals $n_{\text{steps}} = 2^b$. For example, a 10-bit quantizer produces $n_{\text{steps}} = 2^{10} = 1,024$.

The staircase pattern starts at 0 V and the size of the step between levels equals Δ V. The range of the quantizer extends from the minimum voltage $V_{\text{min}} = 0$ to the maximum voltage $V_{\text{max}} = (n_{\text{steps}} - 1)\Delta$.

EXAMPLE 5.11

Quantized audio waveforms

Figure 5.11 shows the waveforms when using a different number of bits for encoding a speech waveform produced by a laptop microphone. The quantized waveforms were played back on the speakers. Although it is noticeably distorted, even 1-bit encoding produces intelligible speech.

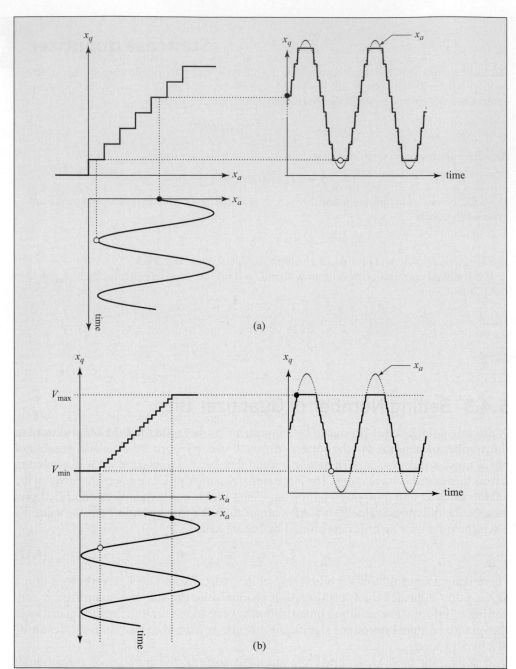

Figure 5.10

Quantizing a sinusoidal waveform. An analog value x_a is quantized to value x_q (a) using 3 bits (8 levels), and (b) using 4 bits (16 levels). Note that the insufficient quantizer range causes severe clipping.

EXAMPLE 5.12 **Quantized digital images**

The effects of using different number of bits for encoding an image are shown in Figure 5.12. The lower limit sets the darkest shade of gray in the image and the upper limit sets the lightest shade. As the number of bits is reduced, there are fewer gray levels in the image.

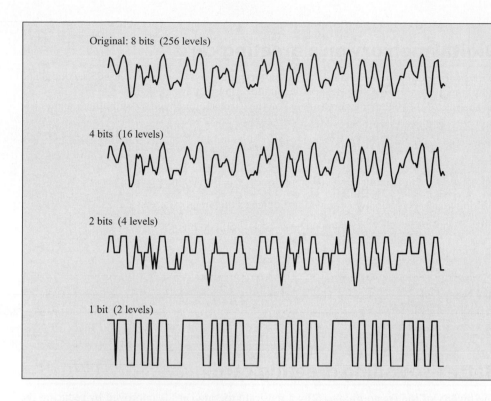

Figure 5.11

Varying the number of quantization bits in an audio waveform.

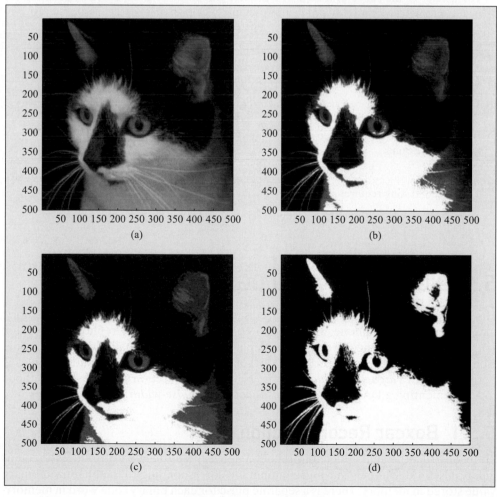

Figure 5.12

Varying the number of bits in displaying an image: (a) image gray scale quantized to 8 bits (256 levels), (b) 3 bits (8 levels), (c) 2 bits (4 levels) and (d) 1 bit (2 levels).

EXAMPLE 5.13

Digital memory on a greeting card

A musical greeting card contains a 1.5 V battery and plays a song that lasts 5 seconds. The sampling rate was $f_s = 5,000$ Hz, and the quantizer has 64 levels.

The size of the digital memory (in bits) on the card is found by first computing the number of bits per sample as

$$\log_2 64 = 6 \text{ bits/sample}$$

This gives a bit rate equal to

$$6 \text{ bits/sample} \times 5,000 \text{ samples/second} = 30,000 \text{ bits/second (bps)}$$

Multiplying this bit rate by the song duration gives the memory size as

$$30,000 \text{ bps} \times 5 \text{ s} = 150,000 \text{ bits}$$

The quantizer step size is computed from the battery voltage as

$$\Delta = \frac{1.5 \text{ V}}{63} = 0.024 \text{ V} = 24 \text{ mV}$$

EXAMPLE 5.14

Better-sounding greeting card

The quality of the sound produced by a musical greeting card is improved by reducing the quantizer Δ. If the card in the previous example is improved by reducing the step size to $\Delta = 10$ mV, the number of quantizer steps increases to

$$n_{\text{steps}} = \frac{1.5 \text{ V}}{0.01 \text{ V}} = 150$$

The number of bits b to provide 150 levels is found from

$$\log_2 150 = \frac{\log_{10} 150}{\log_{10} 2} = 7.22$$

Because 7 bits are insufficient, $b = 8$ is required for this quantizer. The resulting bit rate using the same f_s equals

$$8 \text{ bits/sample} \times 5,000 \text{ samples/second} = 40,000 \text{ bps}$$

The 5-second song requires a memory size equal to

$$40,000 \text{ bps} \times 5 \text{ s} = 200,000 \text{ bits}$$

▶ 5.5 DIGITAL-TO-ANALOG CONVERSION

Having stored an audio signal in quantized samples as a collection of bits in a memory, like that in your smartphone, we would now like to hear an analog replica of the audio. If the system is designed properly, the replica will be perceptually indistinguishable from the original. This reconstruction of the analog signal from the digital samples is performed with a *digital-to-analog converter* (DAC). This section describes two methods for implementing a DAC: *boxcar reconstruction* and *pulse-width modulation*.

5.5.1 Boxcar Reconstruction

To reconstruct the analog signal from the samples, the *boxcar DAC* produces a rectangular pulse of duration T_s having an amplitude corresponding to the quantized binary value stored in memory. There is a separate pulse for each binary code word in memory,

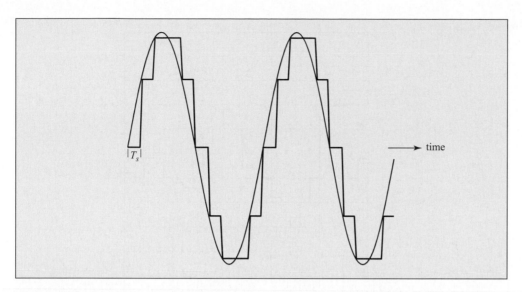

Figure 5.13

A boxcar reconstruction of a sinusoidal waveform. Original waveform is shown in blue; the boxcar DAC reconstruction is in red.

and each pulse lasts for the duration of the original sampling period T_s. When plotted over time, these pulses produce a step-like constant or *boxcar* approximation to the original analog waveform. Figure 5.13 shows a reconstruction of a sinusoidal waveform from the digital values stored in memory, along with knowledge of sampling time T_s and step size Δ. Note that these values could have been obtained by sampling a sinusoidal waveform or *synthesized* by computing the sinusoidal values from the trigonometric function.

ADC—DAC · EXAMPLE 5.15

Consider the analog waveform reconstructed from digital samples shown in Figure 5.14. The same sampling rate and quantizer were used in both sampling and reconstruction.

The sampling rate f_s is determined from the duration of the constant-valued segments that are observed to last two small divisions on the oscilloscope trace. With each large division corresponding to 0.01 ms, the sample period is

$$T_s = \frac{2}{5} \times 0.01 \text{ ms} = 0.04 \text{ ms}$$

The sample rate is the reciprocal of the sample period, so

$$f_s = \frac{1}{T_s} = \frac{1}{4 \times 10^{-5}} = 2.5 \times 10^4 \text{ Hz}$$

The quantizer step size Δ is determined from the smallest vertical increment that is displayed, which is observed to be one small division. With each large division being 0.5 V and containing five small divisions, we have

$$\Delta = \frac{1}{5} \times 0.5 \text{ V} = 0.1 \text{ V}$$

If the boxcar DAC uses the same Δ and a 6-bit quantizer, consider the reconstruction of the boxcar approximation from the binary values

$$001001 \ 111001$$

To do the decoding, the sequence is first grouped into 6-bit codes to yield

$$\underbrace{001001}_{=9} \ \underbrace{111001}_{=57}$$

Then these values are scaled by the step size to produce

$$9\Delta = 0.9 \text{ V} \quad \text{and} \quad 57\Delta = 5.7 \text{ V}$$

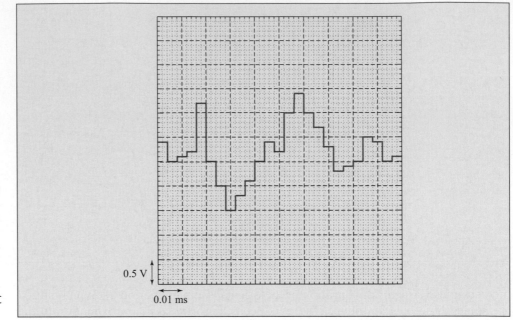

Figure 5.14

Sample waveform produced by DAC as displayed by an oscilloscope.

0.5 V

0.01 ms

5.5.2 Pulse-Width Modulation

Pulse-width modulation (PWM) is an alternate, less expensive technique that varies the ON/OFF time of a binary (digital) waveform to produce a desired average (analog) value. This technique is commonly used to dim LED displays and turn motors at variable speeds.

Figure 5.15 shows pulse trains with variable duty cycles that produce variable average voltages. The period of the PWM waveform T_{PWM} is constant and is usually much shorter than sampling time T_s to allow an average voltage to be determined. For example, if $T_s = 1$ ms, a reasonable value would be $T_{PWM} = 0.1$ ms.

A PWM system that operates with b bits divides the T_{PWM} period into sub-intervals with each having duration δT, so

$$\delta T = \frac{T_{PWM}}{2^b - 1} \tag{5.13}$$

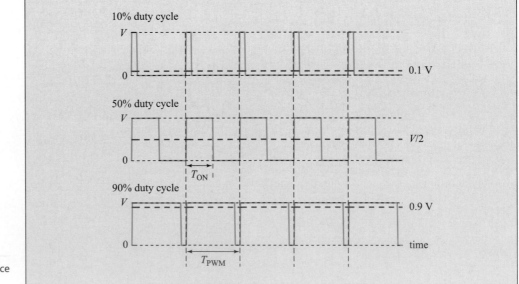

Figure 5.15

PWM waveform duty cycles produce a variety of average voltages.

For example, for $n_b = 4$, $\delta T = T_{PWM}/15$. For $T_{PWM} = 0.25$ ms, $\delta T = 0.017$ ms (17 μs).

A desired average voltage is obtained by maintaining the ON state for duration T_{ON} by specifying the number of intervals k, as

$$T_{ON} = k\delta T \tag{5.14}$$

where $0 \leq k \leq 2^b - 1$. The average voltage produced by the PWM waveform for a specified k value equals

$$V_q(k) = \frac{T_{ON}}{T_{PWM}} V_{max} = \frac{k\delta T}{T_{PWM}} V_{max} \tag{5.15}$$

The effective PWM step size equals the smallest increment in V_q, as

$$\Delta_{PWM} = \frac{\delta T}{T_{PWM}} V_{max} = \frac{V_{max}}{2^b - 1} \tag{5.16}$$

The average voltage for a specified k value is equal to

$$V_q(k) = k\Delta_{PWM} \tag{5.17}$$

which varies from 0 V ($k = 0$) to V_{max} ($k = 2^b - 1$).

The *duty cycle* of a binary waveform is the percentage of the period that the waveform spends in the ON state. The duty cycle of the PWM waveform is given by

$$\text{Duty cycle} = \frac{T_{ON}}{T_{PWM}} \times 100\% \tag{5.18}$$

PWM voltage $\quad$ EXAMPLE 5.16

A 5-bit PWM system operates with $T_{PWM} = 2$ ms and $V_{max} = 5$ V. The sub-interval duration equals

$$\delta T = \frac{T_{PWM}}{2^5 - 1} = \frac{2\,\text{ms}}{31} = 0.065\,\text{ms}\ (65\,\mu s)$$

The effective step size equals

$$\Delta_{PWM} = \frac{\delta T}{T_{PWM}} V_{max} = \frac{0.065\,\text{ms}}{2\,\text{ms}}\,5\,\text{V} = 0.16\,\text{V}$$

The value of k can be specified between 0 and 31 to yield an average voltage, as

$$V_q(k) = k\Delta_{PWM} = 0.16\,k\text{V}$$

When $k = 10$, $V_q = 1.6$ V.

To produce $V_q = 2.4$ V the value of k equals

$$k = \frac{V_q}{\Delta_{PWM}} = \frac{2.4\,\text{V}}{0.16\,\text{V}} = 15$$

The duty cycle value is

$$\text{Duty cycle} = \frac{k\delta_T}{T_{PWM}} \times 100\% = \frac{15 \times 0.065\,\text{ms}}{2\,\text{ms}} \times 100\% = 49\%$$

By controlling the duty cycle and having a sufficiently high frequency of repetition, the PWM waveform can produce V_q values equivalent to the boxcar DAC. Figure 5.16 shows the averages over several cycles produce samples of a sinusoidal waveform.

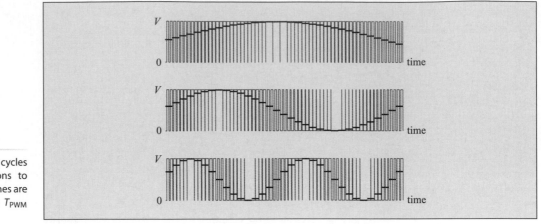

Figure 5.16

Pulse trains with variable duty cycles produce good approximations to sinusoidal waveforms. Solid lines are averages computer over two T_{PWM} periods.

5.6 Summary

This chapter described and analyzed the procedure that transforms an analog signal, such as speech or music, into a sequence of numbers using an analog-to-digital converter (ADC). Analog-to-digital conversion was found to occur in two steps: First the analog waveform is sampled, and then the samples are quantized by being represented by a finite number of bits. The Nyquist criterion states that the sampling rate must be greater than twice the highest frequency present in the signal in order to prevent information loss. A sampling rate lower that the Nyquist rate produces alias signals.

Quantization converts the infinite-precision analog values into a finite-precision digital values. The quanti-

zation error increases with step size Δ. A staircase model illustrated the quantization process. Successive approximation performs quantization in b steps rather than 2^b steps in the staircase method.

Two methods of digital-to-analog conversion were described. The boxcar digital-to-analog converter (DAC) reconstructs the analog waveform using piece-wise constant values. Pulse-width modulation (PWM) is an efficient DAC that uses a digital waveform with a variable duty cycle to from an analog voltage.

5.7 Problems

5.1 Sampling audio waveforms. A talking book generates a digital audio waveform that was generated by sampling the original audio waveform $2,000$ times a second. What is the maximum allowed frequency for proper operation?

5.2 Cell phone audio waveforms. The audio signals on your cell phone occupy the frequency range from 300 Hz to 3,400 Hz. What sampling frequency would you specify for digitizing the voice waveform?

5.3 Cell phone HD Voice waveforms. The HD Voice systems for cell phone systems occupy the frequency range from 50 Hz to 7,000 Hz and permit high-quality speech to be transmitted. What sampling frequency would you specify for digitizing the voice waveform?

5.4 Aliasing. Let the sampling frequency $f_s = 10,000$ Hz with the result that the alias frequency $f_a = 1,000$ Hz is observed. What two values of the original frequency f_o could have produced this alias frequency?

5.5 Maximum spacial frequencies. Your old cellphone display has a 180 rows and 320 columns. What are the maximum horizontal and vertical spacial frequencies present in an icon image that would produce an un-aliased image on that display?

5.6 2014_{10} in binary. What is the binary representation of 2014_{10}?

5.7 01011010_2 in decimal. What is the decimal representation of 010101010_2?

5.8 Magic cards. *Magic cards* allow you to guess a number between 0 and 15 that a person was thinking of by answering whether the number appeared on a set of four cards. These cards appear in Figure 5.17 with one number replaced by dashes in each card. What are the missing numbers?

5.9 Video game audio. Your computer game uses digitized audio stored as 10 bits per sample. How many voltage levels does this represent? If the audio is reproduced

Figure 5.17

Magic cards used in Problem 5.8.

over a range of 0 V to 5 V, what is step size Δ? If the audio samples were generated 44,000 times per second in each of two stereo channels, how many bits per second does the system produce to give your game sound?

5.10 **Binary addresses on the Internet.** Current binary addresses of Websites use 4 bytes. What is the number of possible unique addresses?

5.11 **CD audio duration.** An audio CD stores 650 megabytes (650 MB) of data, where 1 MB = 2^{20} bytes = 1,048,576 bytes, and 1 byte is an 8-bit data unit. The sampling rate $f_s = 44$ kHz is used with 16-bit quantization. What duration of stereo music (two separate waveforms) can be stored on a CD? Give answer in minutes.

5.12 **ADC with staircase quantizer.** An 8-bit ADC performs a conversion every sampling period $T_s = 0.1$ ms. What is the period of the staircase T_Δ?

5.13 **Boxcar DAC output waveform.** A digital memory produces the following bit sequence that goes to a 8-bit boxcar DAC that ranges between 0 V and 5 V, and produces a reconstruction every $T_s = 0.1$ ms. Sketch the reconstructed analog waveform produced by the following bits, providing amplitude and time values in the sketch.

$$1010\ 0000\ 0110\ 0111\ 0000\ 0111$$

5.14 **Musical greeting card with boxcar DAC.** A musical greeting card plays a 10-second segment of a song when it is opened. If the original audio was processed to remove all frequencies at and above 3 kHz and 5 bits quantize the samples, what is the minimum number of bits stored on the card digital memory?

5.15 **PWM duty cycle.** A PWM waveform extends from 0 V to 5 V and has a period $T_{PWM} = 10$ ms. What is the duty cycle of the PWM waveform that produces a 1 V average?

5.16 **PWM DAC on-time.** Let $V_{max} = 5$ V and $T_{PWM} = 2$ ms. What value of T_{ON} produces $V_{ave} = 3.4$ V?

5.17 **PWM DAC value.** Let $V_{max} = 5$ V and $T_{PWM} = 2$ ms in a PWM DAC that uses 8-bits. What V_q is closest to $V_{ave} = 3.4$ V?

5.8 Excel Projects

The following Excel projects illustrate the following features:

- Computing and plotting sinusoidal sequences.
- Demonstration of aliasing.
- Demonstration of quantization.
- DAC using boxcar approximations.
- PWM operation with an animation.

Using a Narrative box (described in Example 13.4), include observations, conclusions, and answers to questions posed in the projects.

5.1 **Sampling 2 periods of a sinusoid.** Using Example 13.23 as a guide, compose a worksheet to plot two periods of a sinusoidal waveform that are sampled 16 times per period.

5.2 **Sampling a sinusoid at 32 points.** Using Example 13.23 as a guide, compose a worksheet to plot one period of a sinusoidal waveform that is sampled 32 times per period.

5.3 **Sampling a sinusoid at 4 points.** Using Example 13.23 as a guide, compose a worksheet to plot four periods of a sinusoidal waveform that is sampled four times per period.

5.4 **Demonstration of aliasing.** Using Example 13.24 as a guide, demonstrate the aliasing that occurs when we make $f_o = 0.9$ when $f_s = 1$. Explain the difference in the observed aliased waveforms between $f_o = 0.9$ and $f_o = 1.1$.

5.5 **Quantizing waveforms.** Using Example 13.25 as a guide, compose a worksheet to quantize the waveform

$$\sin(2\pi i/24) + \sin(2\pi i/32)\,\text{V} \text{ for } 0 \le i \le 64$$

using code words with $n_b = 4$ bits.

5.6 **Boxcar DAC.** Using the results of Project 5.5 and Example 13.26 as a guide, reconstruct the quantized waveform sq_i from the codewords having $n_b = 4$ bits.

5.7 **PWM DAC.** Using Example 13.27 as a guide, compose a worksheet to form a PWM reconstruction using the code words formed in Project 5.5.

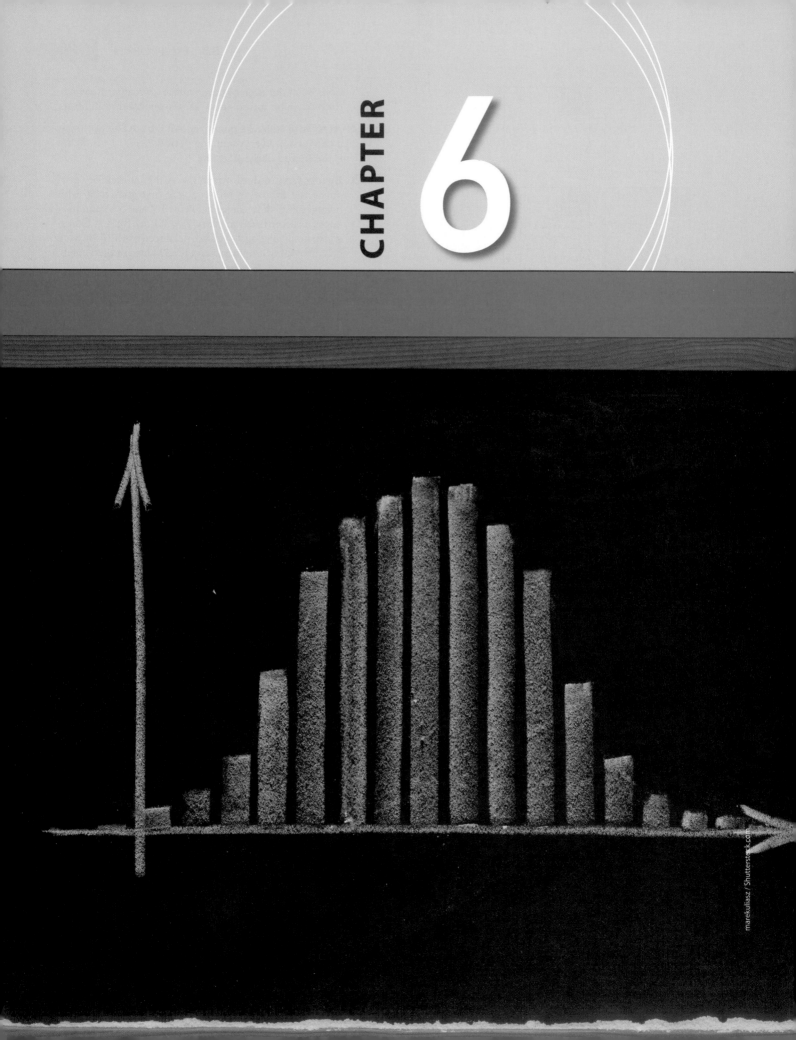

MODELING RANDOM DATA AND NOISE

LEARNING OBJECTIVES

After completing this chapter, the reader should be able to:

- Understand how probability models data sources and treats noise.
- Understand the consequences of signal-to-noise ratio on data transmission.
- Use pseudo-random number generators to generate random data and simulate noise.

▶ 6.1 INTRODUCTION

To describe random noise that occurs in signals, we introduce some topics from probability. Probability is the mathematical tool that helps us describe, interpret, and predict the statistical behavior of random events, specifically those that contribute noise in transmitting data signals. Probability theory is a vast subject that covers many types of randomness. We necessarily limit the topics to those that help us understand how to cope with noise.

Most data analysis software packages contain a *pseudo-random number generator* (PRNGs) that produces pseudo-random numbers. The two most common are random numbers that are uniformly distributed over [0,1) and Gaussian random numbers that are distributed according to the *bell-shaped curve*. This chapter describes each PRNG and their application. The *signal-to-noise ratio (SNR)* is an important parameter in transmitting data in the presence of noise and is equal to the signal energy over the noise energy. The main result of this chapter is that deterministic signals add in amplitude and noise signals add in variance. This is the secret to processing signals in the presence of noise for reliable data transmission.

▶ 6.2 USING PROBABILITY TO MODEL UNCERTAINTY

Probability is the branch of mathematics that treats randomness. The first practitioners of probability were *probably* gamblers. In the mid 1600's, a problem involving gambling led two famous mathematicians, Pascal and Fermat, to develop a mathematical theory of probability to describe games of chance. Their method, known as the *relative frequency approach*, is intuitive, still applicable, and is used in this text.

Typical data exhibit a known trend that also contains random deviations, collectively called *noise*. Depending on the measurement procedure, these deviations can have one or more causes:

- *Electronic noise*, such as hiss on weak radio stations.

- *Transmission errors* due to computer glitches.

- *Sensor inaccuracies* caused by aging components or weak batteries.

- *Fluctuations caused by dynamic environments*, as when trying to predict hurricane paths.

- *Human mistakes*—typos do happen.

When we call such deviations *random*, we assume that they are not only *unpredictable* but they also follow laws of probability. Even though the exact values of the deviations cannot be determined, probability provides some information about their *statistics*, such as their average value and average variation.

To apply probability to our data communication problem, the following five elements are helpful.

1. The **experiment** is an activity that produces unpredictable results.

2. The **outcome** indicates the possible results that the experiment can produce.

3. The **random variable** *(RV)* is a rule that assigns a numerical value to each outcome (because it is easier to deal with the statistics of numbers than more general outcomes). A RV is denoted as an upper-case letter, such as X, which assigns a numerical value to the experimental outcome. In contrast, lower-case letters will refer to non-random or *deterministic* values.

We consider the following two RVs that will be used to generate all other RVs:

Y describes the RV that is uniformly distributed over the range [0,1).

G describes the RV that has a normalized Gaussian distribution with zero mean and unit variance.

4. An **event** is a numerical condition of the RV indicated with square brackets. Events include

- $[X = a]$: the RV X produces realizations (random numbers) equal to the non-random value a.

- $[X > 0]$: the RV X produces realizations that have positive values.

- $[a \leq X < b]$: the RV X produces realizations that lie in the interval $[a, b)$.

5. The **probability measure** of an *event*. The probability can take values between 0 and 1 for an experimental outcome.

- $Prob[.] = 0$ indicates that the event will likely not occur, such as hitting the exact center of a bullseye.

- $Prob[.] = 1$ indicates that the event will likely always occur.

- $Prob[.] = 0.5$ indicates that the event is likely to occur half the time an experiment is performed.

6.3 HISTOGRAMS

A *histogram* displays the frequency of occurrence of a set of random numbers that are realized when performing an experiment that produces RV X. Statistical analysis deals with the *average behavior* of RVs, and histograms show the possible values that have occurred.

A histogram is formed by defining a set of bins and then counting the number of times a realization (random number) falls within each bin. Each bin is defined by a numerical interval, such as $[a, b)$. The square bracket denotes a *closed limit* (that is, if $X = a$, that bin is incremented). The parenthesis indicates an *open limit* (that is, if $X = b$, that bin is not incremented, rather the next bin defined by $[b, c)$ is incremented).

Because the values are random, the count in each bin is also random. Our intuition indicates that larger numbers of data points lead to more accurate estimates. Hence, when there are many random numbers, the count within each bin with large counts tend to be accurate and repeatable. The next example illustrates the reliability of bin counts as the quantity of data points increases.

Histogram of uniformly distributed random numbers EXAMPLE 6.1

The random numbers Y_i generated by the uniform PRNG fall within the interval [0,1). We form bins that cover this interval using equally sized, non-overlapping segments producing 50 bins:

$$\text{bin } 1 \rightarrow [0, 0.02)$$

$$\text{bin } 2 \rightarrow [0.02, 0.04)$$

$$\text{bin } 3 \rightarrow [0.04, 0.06)$$

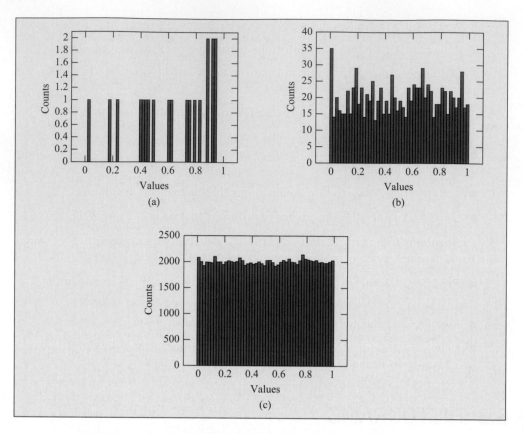

Figure 6.1

Histogram with 50 bins displaying the results of uniformly distributed pseudo-random numbers, Y_1 to Y_n: (a) $n = 20$, (b) $n = 1,000$, and (c) $n = 100,000$.

and so on until

$$\text{bin } 50 \rightarrow [0.98, 1.0)$$

Note that the half-open intervals $[.,.)$ ensure that each Y value falls into only one bin and all Y values in $[0,1)$ produce counts by falling into some bin.

Figure 6.1 shows histograms for Y_i for $i = 1, 2, \ldots, n$ when $n = 20$, $n = 1,000$, and $n = 100,000$. When random numbers are said to be uniformly distributed over $[0,1)$, we expect the bins to yield counts that are approximately equal for the bins that cover $[0,1)$. Yet, the figure shows that this will happen—and then only approximately—when n is very large. Another way to look at it is that a histogram will be accurate only when the count in a narrow bin is large, say greater than 100.

> *Factoid:* To obtain a representative and repeatable histogram a large quantity of numbers is needed.

6.4 PROBABILITY DENSITY FUNCTIONS

When n becomes very large (approaches infinity) and the bin size is made small, the histogram approximates a function called the *probability density function* (PDF). In the detection problem, the most appropriate event describes the RV producing a value that lies within an interval, such as $[a \leq X < b]$, with the probability of the event being denoted $P[a \leq X < b]$. Indeed, such an interval can be considered to be just one of the bins in a histogram.

The probability density function (PDF), denoted mathematically as $p_X(x)$, is the function describing RV X at point x (a *dummy* variable) that is produced in the limit as both the number of trials n approaches infinity and the bin interval δ centered on x approaches 0. If x is the center of a bin interval and n_x is the number of random numbers that fall into that bin, then the PDF is defined by

$$p_X(x) = \lim_{n \to \infty} \lim_{\delta \to 0} \frac{n_x}{n\delta} \tag{6.1}$$

When $p_X(x)$ is a function that does not contain discontinuities and the limits are approached correctly (as mathematicians caution), the limit exists and produces the desired PDF.

The PDF has the following three properties.

1. The PDF is greater than or equal to zero for all x. Thus, $p_X(x) \geq 0$ for all x.

2. The area under the $p_X(x)$ curve equals one.

3. The probability that X produces a realization that falls within interval $a \leq x < b$ equals the area under the $p_X(x)$ curve between points $x = a$ and $x = b$.

6.4.1 Moments

In addition to the PDF, a RV is also described by simpler values called *moments*. The two most important moments are the *mean* and *variance*, which are described in this section.

Mean

For the RV X described by PDF $p_X(x)$, the theoretical value of the mean μ_X is the x point ($x = \mu_X$) at which the curve representing the distribution of mass would *balance*.

The μ_X value is commonly approximated by computing the arithmetic average of the random number sequence called the *sample average Ave(X)*. If we observe n random numbers $X_1, X_2, \ldots, X_n$, the sample average is computed as

$$\text{Ave}(X) = \frac{1}{n} \sum_{i=1}^{n} X_i \quad (\approx \mu_X) \tag{6.2}$$

"Ave(X)" starts with an upper-case A because, being a function of RVs X_1 to X_n, it too is a RV whose value depends on the data values actually observed when performing the experiment. The sample average Ave(X) approaches the theoretical non-random value μ_X when n becomes very large (that is, when we have a lot of data). The larger the n, the closer the value of Ave(X) will likely be to μ_X.

Variance

The other important moment is the variance σ_X^2. Its square root σ_X is the *standard deviation* (SD) that measures the spread of the RVs about the mean value μ_X.

The sample variance is computed from the square of the individual RVs minus the square of the mean

$$\text{Var}(X) = \text{Ave}(X^2) - (\text{Ave}(X))^2 \tag{6.3}$$

If the RV X has zero mean, $\mu_x = 0$, which is a common occurrence in this book, its sample variance equals

$$\text{Var}(X) = \text{Ave}(X^2) = \frac{1}{n} \sum_{i=1}^{n} X_i^2 \tag{6.4}$$

As with the sample average and mean, Var(X) $\to \sigma_X^2$ as n increases.

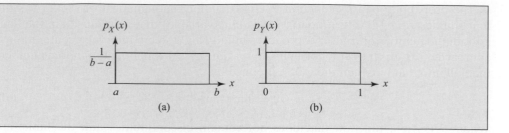

Figure 6.2

Uniform PDF $p_X(x)$. (a) General form for RV X lying in the range $[a, b)$. (b) Specific form for the common RV Y in the range $[0,1)$.

The sample standard deviation (SD) is defined by several different formulas. For our purposes, we use the simplest form, which is the square root of the sample variance. Thus,

$$SD(X) = \sqrt{Var(X)} \tag{6.5}$$

> **Factoid:** In computing estimates, larger values for n (that is, more data) are likely to produce more accurate results. Alternatively, accurate estimates should not be expected when only small values of n are available.

6.4.2 Uniform PDF

The uniformly distributed RV X that has the PDF:

$$p_X(x) = \begin{cases} \frac{1}{b-a} & \text{for } a \leq x < b \\ 0 & \text{otherwise} \end{cases}$$

Figure 6.2 shows the uniform PDF $p_X(x)$ that governs the RV X that produces random numbers that are uniformly distributed over $[a, b)$. The most commonly used form of the uniform RV is Y, which lies in the range $[0,1)$

The uniform PDF is usually assumed when we know from basic principles that the RV is uniformly distributed between two limits, such as rounding errors.

EXAMPLE 6.2 Uniform PDF

The uniform PDF is used to model several important applications in this book:

Uniform pseudo-random number generator: Data analysis software packages always include a pseudo-random number generator (PRNG) that produces random numbers that are uniformly distributed over the interval $[0,1)$. In this case, $a = 0$ and $b = 1$. We reserve the RV Y to denote the random value produced by this PRNG, and its PDF is denoted $p_Y(x)$.

Rounding error: The residual when rounding a real number to an integer is often modeled as a RV R that is uniformly distributed over $-0.5 \leq x < 0.5$. In this case, $a = -0.5$ and $b = 0.5$. For example, rounding $3.14 \rightarrow 3$ has residual $R = 3 - 3.14 = -0.14$.

Quantization error: When an infinite-precision analog value V_a is quantized to finite-precision V_q having step-size Δ, then the error

$$\epsilon_q = V_q - V_a$$

is modeled as a RV ϵ_q that is uniformly distributed over $0 \leq x < \Delta$. In this case, $a = 0$ and $b = \Delta$. The height is then $1/\Delta$.

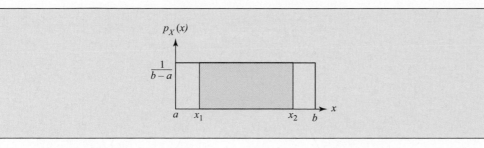

Figure 6.3

Shaded area is the probability that $x_1 \leq X < x_2$, denoted $P[x_1 \leq X < x_2]$.

Computing the area under the uniform PDF over $-\infty \leq x < \infty$, we need to consider only the interval $[a, b)$ over which $p_X(x)$ is greater than zero. The area is then the nonzero height multiplied by the corresponding interval, as

$$\text{Area}(a, b) = \overbrace{\frac{1}{b-a}}^{\text{height}} \times \overbrace{b - a}^{\text{width}} = 1 \qquad (6.6)$$

Figure 6.3 shows the probability of the event $[a \leq x_1 \leq X < x_2 < b]$ is computed as

$$\text{Area}(x_1, x_2) = \overbrace{\frac{1}{b-a}}^{\text{height}} \times \overbrace{x_2 - x_1}^{\text{width}} = \frac{x_2 - x_1}{b-a} \qquad (6.7)$$

Computing μ_Y and σ_Y^2

EXAMPLE 6.3

From the definition of μ_Y, the PDF $p_Y(x)$ would balance about the point $x = 0.5$. Hence, $\mu_Y = 0.5$.

Let the Y produce the $10 \,(= n)$ realization of $Y_1, Y_2, \ldots, Y_{10}$ as

$$0.774, 0.787, 0.699, 0.350, 0.301, 0.562, 0.824, 0.070, 0.918, 0.372$$

Computing the average by summing and dividing by 10, we find the sample average equals

$$\text{Ave}(Y) = \frac{1}{10} \sum_{i=1}^{10} Y_i = 0.566$$

This result is *approximately equal to* 0.5. As n is increased, $\text{Ave}(Y)$ would tend to be closer to μ_Y.

For RV Y uniformly distributed over [0,1), the variance $\sigma_Y^2 = 1/12 = 0.0833$. The SD σ_Y then equals

$$\sigma_Y = \sqrt{\sigma_Y^2} = \sqrt{\frac{1}{12}} = 0.3$$

The random numbers Y_i fall uniformly over the interval [0,1). Using the previous realizations, the sample variance equals

$$\text{Var}(Y) = 0.0696$$

This result is approximately equal to the expected 0.0833. As n is increased, $\text{Var}(Y)$ would tend to be closer to σ_Y^2.

6.4.3 Gaussian PDF

The *Gaussian* probability density function, also known as the familiar bell-shaped curve. The Gaussian PDF is important because it occurs in physical processes, such as thermal

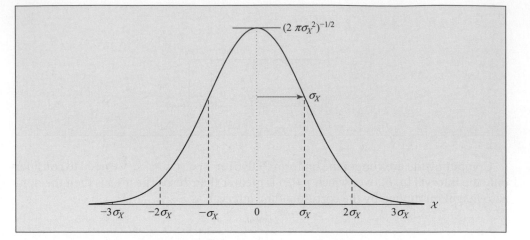

Figure 6.4

Gaussian probability density function $p_X(x)$, which is also known as the bell-shaped curve. The common zero-mean case is shown.

noise. We model thermal noise N_i as a Gaussian random number, for the following reasons.

- Noise is *random* because we cannot predict the values that will occur and we will never see the exactly same noise sequence again.

- Noise is *Gaussian* because each noise sample can be considered to be the resulting sum of many independent random contributing factors.

The bell-shaped curve describing RV X is specified by the mean μ_X and SD σ_X. Figure 6.4 shows the general form of a Gaussian PDF that produces RV X given by

$$p_X(x) = \frac{1}{\sqrt{2\pi}\,\sigma_x}\, e^{-\frac{(x-\mu_X)^2}{2\sigma_X^2}} \tag{6.8}$$

One important result for the Gaussian RV X is that 95% of the realizations fall in the range

$$[\mu_x - 1.96\sigma_X, \mu_x + 1.96\sigma_X]$$

This result will be used for determining the ability of detectors to differentiate signal from noise.

The most common version of the Gaussian RV is the *standard* Gaussian RV, which is denoted in this text by G, having zero mean ($\mu_G = 0$) and unit variance ($\sigma_G^2 = 1$, $\sigma_G = 1$). Thus,

$$p_G(x) = \frac{1}{\sqrt{2\pi}}\, e^{-\frac{x^2}{2}} \tag{6.9}$$

Our main goal for the Gaussian PDF is to describe thermal noise, denoted by the RV N, which has zero mean $\mu_N = 0$ and SD σ_N, with PDF

$$p_N(x) = \frac{1}{\sqrt{2\pi}\,\sigma_N}\, e^{-\frac{x^2}{2\sigma_N^2}} \tag{6.10}$$

▶ 6.5 PROBABILITY MASS FUNCTION

The previous PDF describes a RV that produces a realization that can have any value within an interval, like [0,1) or $(-\infty, \infty)$. When the RV X produces values that are members of a set of discrete values which is denoted as $\{x_i\}$, we need an alternate description. This case occurs in communication problems when a set of specific values

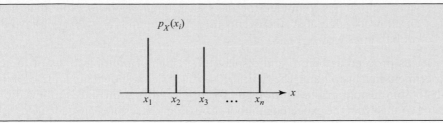

Figure 6.5

Probability mass function $p_X(x_i)$.

is transmitted. Then X can take on a particular value x_i with probability greater than zero. The assignment of these probability values is described by the *probability mass function* (PMF), which is denoted $p_X(x_i)$ and defined by

$$p_X(x_i) = \text{Prob}[X = x_i] \tag{6.11}$$

The PMF is a set of values defined at a set of discrete values defined by x_i, for $i = 1, 2, 3, \ldots, n$, as shown in Figure 6.5.

The coin-flip experiment   EXAMPLE 6.4

A coin is flipped and produces either a head or tail outcome. Consider assigning the RV X the value $+1$ when the flip outcome is a head and -1 when it is a tail. Intuition tells us that a fair coin will produce each of the two possible outcomes (-1 and $+1$) with equal probability. Hence, the PMF is given by

$$p_X(1) = 0.5$$

$$p_X(-1) = 0.5$$

$$p_X(x) = 0, \text{ for } x \neq -1, 1$$

Figure 6.6 shows this PMF.

The PMF has the following three properties:

1. $p_X(x_i) \geq 0$ for all x_i.

2. Since one of the allowable values must occur at each trial, it follows that

$$\sum_{x_i} p_X(x_i) = 1 \tag{6.12}$$

3. Given two non-random constants a and b where $a \leq b$, the probability of the event that $a \leq X < b$ equals

$$\text{Prob}[a \leq X < b] = \sum_{a \leq x_i < b} p_X(x_i) \tag{6.13}$$

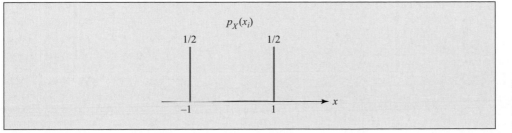

Figure 6.6

Probability mass function $p_X(x_i)$ for the coin-flip experiment.

where the sum is computed for all x_i such that $a \leq x_i < b$. Note that, if $x_i = b$ is an allowed value of X, the events $[a \leq X < b]$ and $[a \leq X \leq b]$ have different probabilities.

Let us proceed as before and display the results of multiple trials of the coin-flip experiment in a histogram. The typical 100-trial series is shown in Figure 6.7a produces the 100-bin histogram shown in Figure 6.7b. The histogram shows a deviation in the actual counts from the expected number of 50 occurrences of each, which is a "statistical fluctuation" due to the small number of trials. The graph of observed outcomes indicates that runs of six or more of the same outcome are not that unusual.

Increasing the number of coin-flip trials to a million produces the 100-bin histogram shown in Figure 6.8a. It shows that each possible value occurs approximately 1/2 of the time. As before, we reduce the bin size by a factor of ten to produce the 1000-bin histogram shown in Figure 6.8b. The narrower bin produces a histogram with values concentrated about the two possible integer values of −1 and +1.

Continuing to increase the number of trials (n_t) to infinity and to reduce the bin size to zero, the histogram form converges to the PMF defined at each possible discrete value of X that can occur. For the coin-flip experiment, the PMF has non-zero values only for $x = -1$ and $x = +1$. If the number of times $X = -1$ is denoted by n_{-1} and the number of times $X = +1$ by n_{+1} in n_t trials, the PMF is computed at these two points as

$$\text{Prob}[X = -1] = \lim_{n_t \to \infty} \frac{n_{-1}}{n_t} \quad \left(= \frac{1}{2} \text{ for a fair coin} \right) \tag{6.14}$$

$$\text{Prob}[X = +1] = \lim_{n_t \to \infty} \frac{n_{+1}}{n_t} \quad \left(= \frac{1}{2} \text{ for a fair coin} \right)$$

The PMF values at all other values of x equal zero.

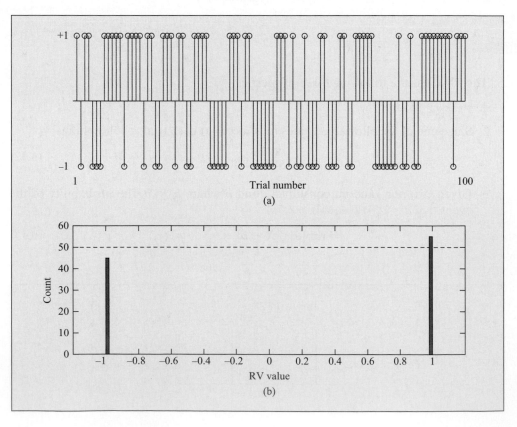

Figure 6.7

Coin flip experiment. (a) RV values observed from 100 trials. (b) 100-bin histogram where the dashed line indicates a count of 50.

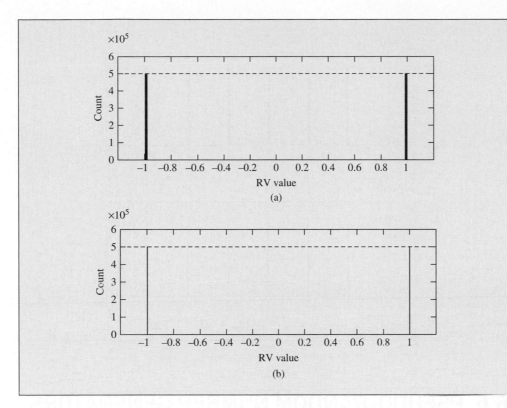

Figure 6.8

Histograms from one million trials of coin-flip experiment: (a) 100 bins and (b) 1000 bins.

PMF describing a die-toss experiment

EXAMPLE 6.5

In a die-toss experiment, a die is tossed and the outcome is the top face. The RV T is the count of dots on the top face. Then T can only exhibit the integer values from 1 to 6. Multiple trials of the die-toss experiment produces the data displayed in a histogram. The 100-bin histogram for a typical 120-trial series is shown in Figure 6.9. Intuition tells us that each of the possible outcomes is equally likely to occur. The histogram shows a variation in the actual counts from the expected number of 20 ($= 120/6$) occurrences of each due to the small number of trials.

Increasing the number of trials to a 1.2 million produces the 100-bin histogram shown in Figure 6.10. It shows that each possible value occurs approximately 1/6 of the time. As before, we reduce the bin size by a factor of 10 to produce the histogram shown in the figure. This narrower bin produces a histogram with values concentrated about the possible integer values. Continuing to increase the number of trials to infinity and to reduce the bin size to zero, the histogram converges to a set Dirac delta functions describing the PDF.

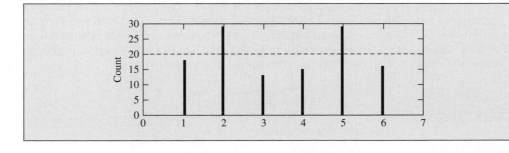

Figure 6.9

100-bin histogram of RV values from 120 trials of die-toss experiment. The dashed line indicates count equal to 20.

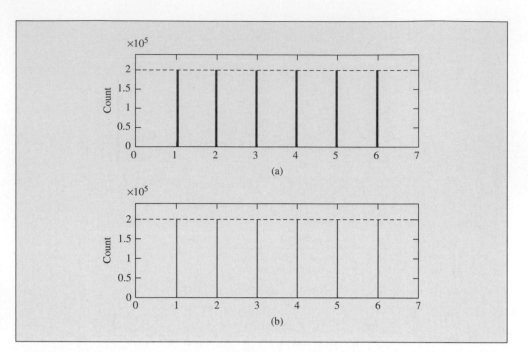

Figure 6.10

Histograms from 1.2 million trials of die-toss experiment where the dashed line indicates count equal to 200,000 for (a) 100-bin histogram and (b) 1000-bin histogram.

▶ 6.6 PSEUDO-RANDOM NUMBER GENERATORS

Most data analysis software packages contain a *pseudo-random number generator* (PRNGs) that generates pseudo-random numbers. This section describes the two most common PRNGs.

A PRNG replaces the actual experiment of flipping a coin or tossing a die with a computer program called a *simulation*. Because a computer program follows a fixed set of instructions known as the *algorithm*, the results cannot be truly random. The random numbers a PRNG produces, however, do exhibit reasonable *unpredictability*. This section describes the two most common PRNGs.

6.6.1 Uniform PRNG

The RV Y produces realizations that are uniformly distributed over the range $[0,1)$. The Uniform PRNG generates random numbers that uniformly distributed over the same range.

EXAMPLE 6.6 Simulating a fair coin flip

Let us examine how a computer can simulate the flip of a fair coin. We can simulate the behavior of a coin flip by using the PRNG Y_i values lying in the $[0,1)$ interval by dividing the interval into two equal sub-intervals and assigning a head outcome to the first sub-interval and a tail to the second. Although any partitioning that produces two equal lengths, Figure 6.11 shows the simplest partition of $[0,1)$ into two equal sub-intervals to be

$$[0, 1) \to \overbrace{[0, 0.5)}^{\text{head}} \overbrace{[0.5, 1)}^{\text{tail}}$$

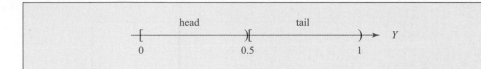

Figure 6.11

Simulating a fair coin flip with a PRNG.

This partition into two half-open intervals has the following desirable properties.

- It covers the entire [0,1) interval. Hence, any random number will fall into *some* interval. This prevents the case when a possible Y value does not fall into any interval.
- It covers the entire [0,1) interval with no overlaps. Hence, every possible value falls into *only one* of the sub-intervals. This prevents any ambiguity when $Y = 0.5000\ldots$.

Uniform PRNG   EXAMPLE 6.7

Let us illustrate the four elements to model a problem using probability by observing the values produced by the Excel uniform PRNG RAND().

1. The experiment is making a call to the uniform PRNG that produces a random number.
2. The outcome is random value in the range [0,1).
3. The RV Y corresponds to a numerical value that is produced.
4. To determine the PDF $p_Y(x)$, we perform the experiment n times, producing RVs

$$Y_1, Y_2, Y_3, Y_4, Y_5, \ldots, Y_n$$

Figure 6.12 shows typical values for $n = 100$. Note all of the values fall within the range [0,1), as expected.

To generate histograms, we set the bin size very small ($\delta = 0.01$) and perform many experiments by making n large. Figure 6.13 shows histograms that are observed for $n = 100$ and $n = 10^6$.

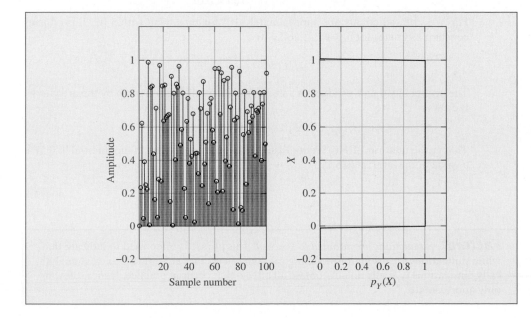

Figure 6.12

One hundred values of the uniformly distributed random numbers.

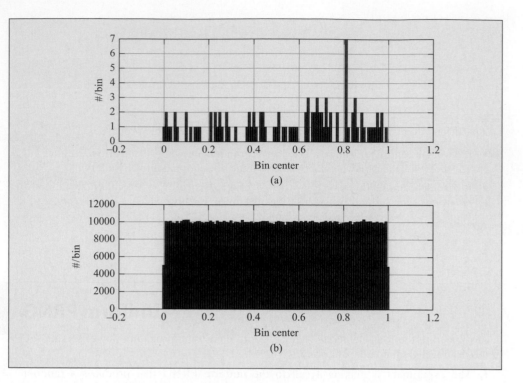

Figure 6.13

Histograms for the uniform PRNG with $\delta = 0.01$: (a) $n = 100$ and (b) $n = 10^6$.

Note that the more accurate histogram generated with $n = 10^6$ random numbers exhibits an approximately uniform level over $[0,1)$. A finite number of randomly generated numbers will never produce equal bin counts—except by rare accident. Because $\delta > 0$ and n is finite, the value of $p_Y(x)$ is approximated as

$$p_Y(x) \approx \frac{n_{a_i}}{n\delta}$$

For $n = 100$, the counts vary from 0 to 5 for bins lying within $[0,1)$, producing unreliable estimates for $0 \le x < 1$ ranging from $p_Y(x = 0.02) = 0$ to

$$p_Y(x = 3.8) \approx \frac{5}{100 \times 0.01} = 5$$

For $n = 10^6$, all counts are approximately 10^4 for bins lying within $[0,1)$, producing constant estimates for $0 \le x < 1$ equal to

$$p_Y(x) \approx \frac{10^4}{10^6 \times 0.01} = 1$$

The probability that Y produces a value in the interval $[0,0.5)$ is given by the area under the curve

$$P[0 \le Y < 0.5] = 0.5$$

That is, the uniform PRNG is equally likely to produce a value in the interval $[0,0.5)$ as a value in the interval $[0.5,1)$.

> **Factoid:** *Simulations are doomed to succeed.* This phrase is often used to indicate that, while simulations are useful for modeling random effects that are difficult to treat analytically, simulations typically do not contain all effects that occur in practice. Hence, they are only approximate.

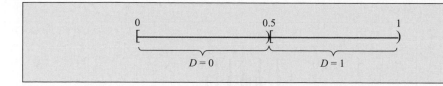

Figure 6.14

Using the PRNG to simulate binary data.

Generating random binary data   EXAMPLE 6.8

As a model of generating random binary data, consider flipping a coin and observing if it comes up heads or tails. To simulate the coin flip, we use a PRNG that generates a pseudo-random numbers uniformly distributed in the range [0,1). This interval is first divided into two equal non-overlapping parts with one part corresponding to a $D = 0$ and the other a $D = 1$. The arbitrary (but convenient) division into $[0, 0.5)$ and $[0.5, 1)$ is shown in Figure 6.14. The random number produced by the uniform PRNG being less than 0.5 makes $D = 0$, otherwise $D = 1$.

6.6.2 Gaussian PRNG

Most signal processing software contains a PRNG that produces the *standard* Gaussian RV G having with zero mean ($\mu_G = 0$) and unit variance ($\sigma_G^2 = 1$). The corresponding Gaussian PDF is

$$p_G(x) = \frac{1}{\sqrt{2\pi}} \, e^{-\frac{x^2}{2}} \tag{6.15}$$

Figure 6.15 shows typical values produced by a Gaussian PRNG for $n_x = 100$. The samples do not exhibit discrete values but vary continuously. Hence, a Gaussian PDF is appropriate. We note that almost all of the values fall within $[-3, 3]$. This verifies the observation that 99.7% of the Gaussian values fall within $[-3\sigma, 3\sigma]$ and 95.4% of the Gaussian values fall within $[-2\sigma, 2\sigma]$.

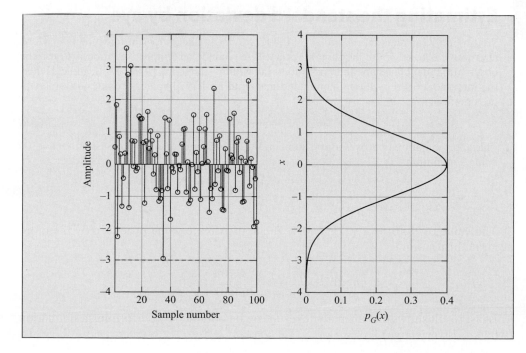

Figure 6.15

One hundred values of the standard RV G_i, where $\mu_G = 0$ and $\sigma_G = 1$, produced by a PRNG and the corresponding Gaussian curve. Note that 99.7% of the random numbers fall between $\pm 3\sigma_G$.

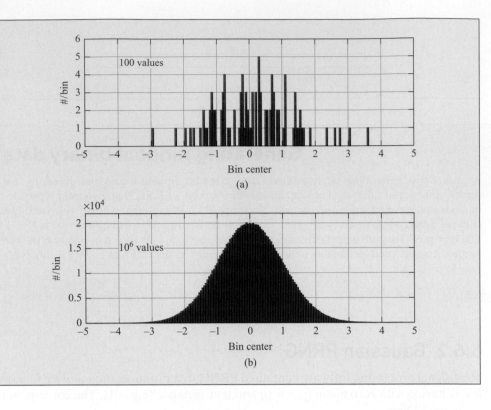

Figure 6.16

Histograms for the Gaussian PRNG with $\delta = 0.05$: (a) $n = 100$ and (b) $n = 10^6$.

To generate a histogram, we set the bin size very small ($\delta = 0.05$) and perform many experiments. Figure 6.16 shows histograms for $n = 100$ and $n = 10^6$.

Note that the more accurate histogram generated with $n = 10^6$ random numbers exhibits the familiar bell-shaped curve. Note that Gaussian random numbers almost all fall within $[-3\sigma_G, 3\sigma_G] = [-3, 3]$. This feature is useful for estimating the SD from a set of Gaussian random numbers.

EXAMPLE 6.9 Estimating the standard deviation by eye

We observe Gaussian data N_i for $i = 0, 1, 2, \ldots, n_x - 1$ and we want to estimate σ_N^2. One alternative is to compute the sample variance. A second easier approximation is to use the observation that Gaussian random numbers fall within $[-3\sigma_N, 3\sigma_N]$ 99.7% of the time. Hence, the interval defined by the minimum and maximum observed values provides a simple approximation, as

$$\max(N_i) - \min(N_i) \approx 6\sigma_N \tag{6.16}$$

This approximation improves as the number of random numbers n_x increases, with $n_x = 100$ being a minimum.

The value of $p_G(0) = 1/\sqrt{2\pi} = 0.4$. The approximate value produced by the more accurate histogram is

$$p_G(0) \approx \frac{n_0}{n_x \, \delta} = \frac{2 \times 10^4}{(0.05)(10^6)} = 0.4$$

This last result shows that an analytic result can be obtained by performing a simulation (when n_x is very large).

6.7 RANDOM NUMBER ARITHMETIC

The following chapters describe how a detected data signal is processed in the presence of random noise. A processor manipulates the detected data

$$X_i = s_i + N_i \text{ for } i = 0, 1, 2, \ldots, n_x - 1$$

comprised of a known (non-random) signal sequence s_i plus the random noise sequence N_i. The processor computes output value V with

$$V = \sum_{i=0}^{n_x-1} s_i X_i = \overbrace{\sum_{i=0}^{n_x-1} s_i^2}^{\text{non-random}} + \overbrace{\sum_{i=0}^{n_x-1} s_i N_i}^{\text{random}} \tag{6.17}$$

The value V is a RV that contains both a non-random component and a random component that is modeled as a Gaussian random variable. The goal here is to determine the mean of V (μ_V), its variance σ_V^2, and its PDF $p_V(x)$.

This processor operation involves three arithmetic manipulations that we consider separately.

1. Adding a non-random constant value to a RV.

2. Multiplying a RV by a non-random constant.

3. Adding independent RVs.

6.7.1 Adding a Constant to a RV

Consider adding a non-random constant to a random number, as shown in Figure 6.17. Let G_i be a Gaussian RV and add a non-random constant a to each realization to produce RV $Y1_i$. Then

$$Y1_i = a + G_i \tag{6.18}$$

The mean of $Y1_i$ has been shifted from 0 to a. Hence,

$$\mu_{Y1} = a + \overbrace{\mu_G}^{=0} = a \tag{6.19}$$

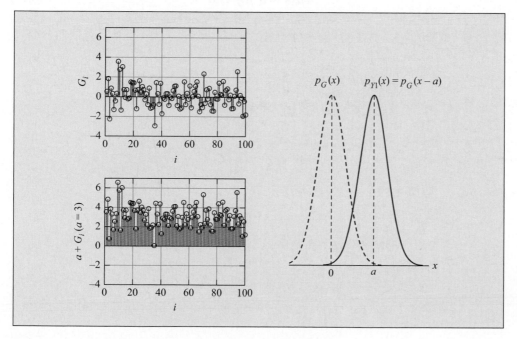

Figure 6.17

Probability density functions $p_G(x)$ and $p_{Y1}(x)$ when $Y1_i = a + G_i$. The results for $a = 3$ are shown.

The form of $p_{Y1}(x)$ is the same as that of $p_G(x)$—except for a shift. Hence, the variance, which is a measure of the deviation from the mean, is unchanged:

$$\sigma_{Y1}^2 = \sigma_G^2 \tag{6.20}$$

The form of the PDF of $Y1_i$ is then

$$p_{Y1}(x) = p_G(x - a) = \frac{1}{\sqrt{2\pi}\,\sigma_G} e^{-\frac{(x-a)^2}{2\sigma_G^2}} \tag{6.21}$$

EXAMPLE 6.10 — Non-random signal plus Gaussian random noise

The signal s is a non-random constant. Gaussian noise N having mean $\mu_N = 0$ and variance σ_N^2 is added to s to produce the detected signal X, as

$$X = s + N$$

The detected signal X is random (because of N) and has mean, variance, and PDF computed as

$$\mu_X = s + \mu_N = s$$

$$\sigma_X^2 = \sigma_N^2$$

The PDF of X is Gaussian (because N is Gaussian) with the form specified by its mean and variance as

$$p_X(x) = \frac{1}{\sqrt{2\pi}\,\sigma_N} e^{-\frac{(x-s)^2}{2\sigma_N^2}}$$

6.7.2 Multiplying a RV by a Constant

Consider multiplying a random number by a constant a, as shown in Figure 6.18. Let G_i be a Gaussian RV that is multiplied by the constant a to produce RV $Y2_i$. Then

$$Y2_i = aG_i \tag{6.22}$$

The mean of $Y1_i$ has been not been shifted. Hence,

$$\mu_{Y2} = a \overbrace{\mu_G}^{=0} = 0 \tag{6.23}$$

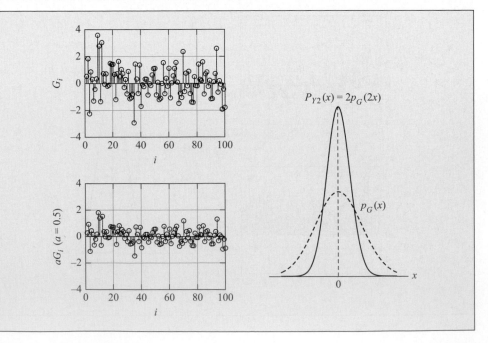

Figure 6.18

Probability density function $p_G(x)$ and $p_{Y2}(x)$ when $Y2_i = 0.5G_i$.

The form of $p_{Y2}(x)$ is a stretched (when $a > 1$) or contracted (when $a < 1$) version of $p_G(x)$. To maintain the area under the curve to equal 1, the amplitude of $p_{Y2}(x)$ is scaled by $1/a$, which is a reduced (for $a > 1$) or increased (for $a < 1$) version of $p_G(x)$. The variance, which is a measure of the deviation from the mean, is computed as

$$\sigma_{Y2}^2 = a^2 \sigma_G^2 \qquad (6.24)$$

The PDF of $Y2_i$ has a Gaussian form that is specified by its mean and variance as

$$p_{Y2}(x) = \frac{1}{\sqrt{2\pi}\, a\sigma_G} e^{-\frac{x^2}{2(a\sigma_G)^2}} \qquad (6.25)$$

Multiplying a detected sample value by a non-random constant

EXAMPLE 6.11

The detected signal is

$$X = s + N$$

The noise N is a Gaussian RV with mean $\mu_G = 0$ and variance σ_N^2. If we multiply X by 2, we get

$$2X = 2s + 2N$$

The mean, variance, and PDF of $2X$ equal

$$\mu_{2X} = 2s + 2\mu_N = 2s$$

$$\sigma_{2X}^2 = 4\sigma_X^2 = 4\sigma_N^2$$

$$p_{2X}(x) = \frac{1}{\sqrt{2\pi}\,(2\sigma_N)} e^{-\frac{(x-2s)^2}{2(4\sigma_N^2)}} = \frac{1}{\sqrt{8\pi}\,\sigma_N} e^{-\frac{(x-2s)^2}{8\sigma_N^2}}$$

6.7.3 Adding Independent RVs

The RVs X_1 and X_2 are said to be *independent* if knowing the value produced by X_1 tells us nothing about the value produced by X_2. Each time a PRNG produces two random numbers (say X_1 and X_2), the values are assumed to be independent because neither value can easily be predicted from the other. Independence simplifies the calculations involving sums of RVs.

Central limit theorem

EXAMPLE 6.12

The *central limit theorem* (CLT) states that, under conditions that typically hold in practice, the sum of n independent RVs approaches a Gaussian distribution as n becomes large, *no matter what the distributions of the RVs themselves*. We apply the CLT to argue that thermal noise has a Gaussian distribution.

Consider a resistor R at room temperature that contains a huge number of thermally excited electrons moving inside it. If there are n electrons and I_k is the random current due to the motion of the kth electron, the total instantaneous (random) current I is the sum of the individual currents. The random instantaneous voltage V measured across the terminals of resistor R is then given by

$$V = RI = R\sum_{k=1}^{n} I_k$$

The current I is the sum of many electrons that move randomly and independently. Hence, the central limit theorem indicates that the distribution of V approaches the Gaussian distribution with the variance σ_N^2 being the noise power. *No matter what the distribution of the original RVs (the individual currents I_k), the random voltage V can be accurately described by the Gaussian distribution by the central limit theorem.*

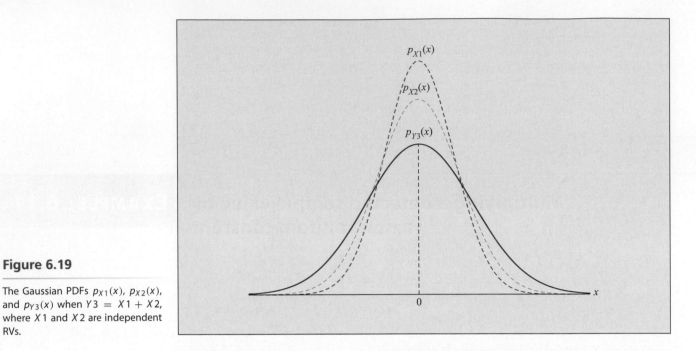

Figure 6.19

The Gaussian PDFs $p_{X1}(x)$, $p_{X2}(x)$, and $p_{Y3}(x)$ when $Y3 = X1 + X2$, where $X1$ and $X2$ are independent RVs.

> *Factoid:* Your SAT score resulted from the averaging effect of many different and mostly independent influences, such as your study habits, memory, stress level, and so on. The CLT is the reason that the histograms of SAT scores typically exhibit a bell-shaped curve.

Let $X1$ and $X2$ be independent Gaussian RVs having zero means ($\mu_{X1} = \mu_{X2} = 0$), variances σ_{X1}^2 and σ_{X2}^2, and $Y3$ equal their sum. Thus,

$$Y3 = X1 + X2 \tag{6.26}$$

The mean of $Y3$ is the sum of the means of $X1$ and $X2$, as

$$\mu_{Y3} = \mu_{X1} + \mu_{X2} = 0 \tag{6.27}$$

Because $X1$ and $X2$ are independent RVs, the variance of $Y3$ equals the sum of their variances, as

$$\sigma_{Y3}^2 = \sigma_{X1}^2 + \sigma_{X2}^2 \tag{6.28}$$

Because $X1$ and $X2$ are Gaussian, the PDF of $Y3$ is also Gaussian and specified by its mean and variance values, as

$$p_{Y3}(x) = \frac{1}{\sqrt{2\pi\sigma_{Y3}^2}} e^{-\frac{x^2}{2\sigma_{Y3}^2}} = \frac{1}{\sqrt{2\pi\left(\sigma_{X1}^2 + \sigma_{X2}^2\right)}} e^{-\frac{x^2}{2(\sigma_{X1}^2 + \sigma_{X2}^2)}} \tag{6.29}$$

Note that the σ_{Y3} was brought into the square root and became σ_{Y3}^2. The PDFs of $X1$, $X2$, and $Y3$ are shown in Figure 6.19.

6.8 Summary

This chapter introduced elements of probability theory that are useful for understanding noise and data detection. The main motivation for studying probability is that both the desired information encoded in the signal and the noise are most intuitively described in probabilistic terms (that is, in terms of a random outcome of some experiment). The probabilistic description of an estimation problem was developed in terms of a suitable experiment. A RV is a rule that assigns a numerical value to the desired information, and a probability measure describes the likelihood of observing a range of values.

6.9 Problems

6.1 Sample mean, variance, and SD. Let the RV X produce the $n_x = 10$ realizations:

$$1.7, -0.8, -0.6, 0.5, 0.3, -0.5, 0.8, -0.1, 0.9, -0.7$$

Compute the sample mean, variance, and SD.

6.2 Probability that Y lies in [0.25,0.75). What is the probability that the RV Y that is uniformly distributed over $[0,1)$ will lie in the range $[0.25, 0.75)$?

6.3 Probability that Y lies in [0.25,2). What is the probability that the RV Y that is uniformly distributed over $[0,1)$ will lie in the range $[0.25,2)$?

6.4 Simulating random die toss. Starting with realizations of Y, how would you form T, which is the RV that simulates the result of a fair die toss that equals the number of dots showing on the top face? Sketch the PMF of T.

6.5 Adding 1 to Y. Starting with RV Y that is uniformly distributed over $[0,1)$, we add 1 to each value we observe to form

$$X = 1 + Y$$

Compute μ_X, σ_X^2, and sketch $p_X(x)$.

6.6 Adding 1 to G. Starting with standardized Gaussian RV G, add 1 to each value to form

$$X = 1 + G$$

Compute μ_X, σ_X^2, and sketch $p_X(x)$.

6.7 Multiplying Y by 2. Starting with RV Y being uniformly distributed over $[0,1)$, multiply each value to form

$$Z = 2Y$$

Compute μ_Z, σ_Z^2, and sketch $p_Z(x)$.

6.8 Multiplying G by 2. Starting with standardized Gaussian RV G, multiply each value by 2 to form

$$Z = 2G$$

Compute μ_Z, σ_Z^2, and sketch $p_Z(x)$.

6.9 Summing two independent standardized Gaussian RVs. Starting with independent standardized Gaussian RVs G_1 and G_2, form

$$V = 2(G_1 + 1) - (G_2 - 1)$$

Compute μ_V, σ_V^2, and $p_V(x)$.

6.10 Excel Projects

The following Excel projects illustrate the following features:

- Generating random numbers using Excel functions RAND() and NORM.S.INV(RAND()).

- Simulating experimental outcomes.

- Simulating Gaussian random noise.

- Adding noise to data.

Using a Narrative box (described in Example 13.4), include observations, conclusions, and answers to questions posed in the projects.

6.1 Random numbers using RAND(). Extend Example 13.28 to generate $n_Y = 25$ random numbers Y_i for $0 \le i \le 24$, and compute the sample averages. Observe how the sample averages change as a new set of random numbers is generated.

6.2 Two Random Dice. Extend Example 13.30 to simulate the toss of two dice, $D1_i$ and $D2_i$ for $0 \le i \le 24$, and add the observed values together. Observe how the sums change as another set of random numbers is generated.

6.3 Histogram of uniformly distributed random numbers. Using Example 13.32 as a guide, generate a histogram of ten thousand random numbers produced by RAND().

6.4 Random Gaussian noise. Extend Example 13.33 to generate $n_N = 25$ random Gaussian numbers N_i for $0 \le i \le 24$ and compute the sample averages. Combine columns B and C to form N_i with a single formula using the PRNG. Observe how the sample averages change as another set of random numbers is generated.

6.5 Noisy digital die-toss signals. Using Example 13.34 as a guide, compose a worksheet that generates 25 samples of random data corresponding to a die toss that is corrupted by additive random Gaussian noise with $\sigma_N = 0.2$

6.6 Thresholding noisy binary signals. Extend Example 13.35 to form 20 samples of random binary data that is corrupted by additive random Gaussian noise with specified σ_N. Attempt to determine the data from S_i by applying a threshold $\tau = 0.5$ to estimate the data value. Repeatedly generate new data. Determine the σ_N value that typically produces 1 error in the 20 values of S_i.

6.7 Histogram of noisy binary signals. Using Example 13.32 as a guide, generate a histogram of ten thousand noisy binary signals having a specified variance.

CHAPTER

7

DETECTING DATA SIGNALS IN NOISE

LEARNING OBJECTIVES

After completing this chapter, the reader should be able to:

- Apply data processing techniques to extract data from signals.
- Understand the purpose of complementary signals for robust data transmission.
- Use a pseudo-random number generator to generate binary data and to simulate noise.
- Appreciate why the most important parameter in data transmission is the signal-to-noise ratio equal to the signal energy divided by the noise variance.
- Investigate the performance of data transmission signals in noise through simulations.

⊳ 7.1 INTRODUCTION

This chapter describes how data are transmitted reliably in the presence of noise—even from great distances, such as from Mars. Data signal transmission is a special case in which the signal values are known at both the transmitter and receiver (that is, a specified set of values are employed for transmitting a 0 and another set for transmitting a 1). This chapter describes how a processor matched to this set of values determines the bit value that was transmitted. Two signals having complementary values produce a maximal difference in the matched processor output. Reliable transmission is typically achieved at the cost of longer transmission times. To reduce the probability of error for reliable data transmission, the signal energy is increased by appropriate signal design.

The main topics in this chapter are covered in the following sequence.

Data transmission model—A source produces binary data that is encoded using a signal that is transmitted over a channel. During transmission, the channel introduces random noise, and a processor in the receiver tries to extract the signal that was transmitted. Random number generators are employed to simulate random binary data produced by the source and the noise introduced by the channel.

Signal processing—The conventional processor that operates on received signals forms the sum of weighted signal values. Knowing the signal values that were transmitted leads to a processor whose weights are matched to these signal values.

Signal design—When a matched processor operates on received signals, we design the two signals that encode data values 0 and 1 to form a complementary pair (that is, the signal values of one are the negative values of the other). This choice maximizes the matched-processor output differentiation of these two signals for signals having a particular energy.

Processing signals in noise—When a matched processor operates on complementary signals that are corrupted with noise, the reliability of detecting the correct (transmitted) signal value is determined solely by the ratio of signal energy to noise variance, which is known as the *signal-to-noise ratio (SNR)*.

Probability of error—Because noise is random, data transmission system performance is measured in terms of the *probability of error*. No data transmission system operates perfectly, causing this probability to be greater than zero. A simulation procedure is described to estimate the probability of error by performing multiple transmissions in the presence of noise and counting the number of errors.

⊳ 7.2 DATA TRANSMISSION MODEL

Figure 7.1 shows the block diagram description of binary data transmission that includes the following elements.

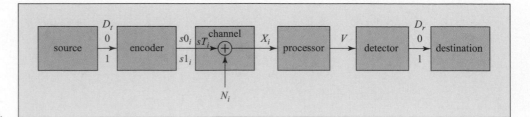

Figure 7.1

Model of a data transmission system.

Source produces a binary data value D_t that is equal to either a 0 or 1 and is to be transmitted to a destination. This chapter considers the case of transmitting a single bit that is randomly generated. The upper-case letter indicates D_t is a random quantity (that is, its value is not predictable) that obeys the laws of probability.

Encoder encodes the data bit values using transmission signal waveform $s(t)$. A *signal sequence* is an indexed set of sample values denoted

$$s_i \quad \text{for} \quad i = 0, 1, 2, \ldots, n_x - 1 \tag{7.1}$$

where i is the index and n_x is the signal duration. This sequence is obtained by converting the signal waveform $s(t)$ by sampling with period T_s, or

$$s_i = s(t)|_{t=iT_s} \quad \text{for} \quad i = 0, 1, 2, \ldots, n_x - 1 \tag{7.2}$$

where s_0 corresponds to $s(t)$ at time $t = 0$ and $s(t)$ extends over $0 \le t < n_x T_s$.

The encoder assigns signals according to the D_t bit value with

$$D_t = 1 \rightarrow s1_i \quad \text{for} \quad i = 0, 1, 2, \ldots, n_x - 1 \tag{7.3}$$

$$D_t = 0 \rightarrow s0_i \quad \text{for} \quad i = 0, 1, 2, \ldots, n_x - 1$$

Data signals are special in that both the encoder at the transmitter end of the channel and processor at the receiving end know their values. One constraint on these signals is that both $s0_i$ and $s1_i$ must have the same energy $\mathcal{E}_s$, which is computed as the sum of their squared values:

$$\mathcal{E}_s = \sum_{i=0}^{n_x-1} s0_i^2 = \sum_{i=0}^{n_x-1} s1_i^2 \tag{7.4}$$

Channel transports the transmitted signal sT_i for $i = 0, 1, 2, \ldots, n_x - 1$ from the transmitter to receiver, where

$$sT_i = s1_i \quad \text{if} \quad D_t = 1 \tag{7.5}$$

$$= s0_i \quad \text{if} \quad D_t = 0$$

A channel can be a length of wire or a medium that supports wireless transmission. In the transmission process, the channel adds random noise N_i to sT_i to form the signal at the channel output X_i for $i = 0, 1, 2, \ldots, n_x - 1$, where

$$X_i = s1_i + N_i \quad \text{if} \quad D_t = 1 \tag{7.6}$$

$$X_i = s0_i + N_i \quad \text{if} \quad D_t = 0$$

X_i is random and upper case, because N_i being random makes the sum random as well.

Processor uses the observed sequence X_i to compute an output value V. Because noise makes X_i random, V is also random.

Detector applies V to a threshold detector to determine the received bit value, producing the receiver value $D_r = 0$ or 1.

Destination accepts data value D_r for accomplishing the desired task.

Data Processing Task

The task at hand is to process X_i to determine the data value D_t that it encodes. Because X_i contains random noise, there is a possibility of making a mistake. There are two types of errors that can occur:

$$\text{If} \quad D_t = 1 \quad \text{and} \quad D_r = 0 \quad \rightarrow \quad \text{false negative} \tag{7.7}$$

$$\text{If} \quad D_t = 0 \quad \text{and} \quad D_r = 1 \quad \rightarrow \quad \text{false positive}$$

Figure 7.2

Error model when binary data are transmitted over a noisy channel.

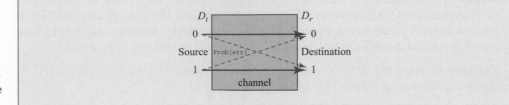

With random noise, we need to compute the probability of making an error. Figure 7.2 shows a simple model that describes data communication over a noisy channel. Prob[error] indicates the probability that an error occurs. In a practical system, probability of error should very small, such as Prob[error] $< 10^{-3}$ or less than one error in 1,000 transmissions.

This chapter describes why it is not theoretically possible to design a system that is perfect (that is, one that has Prob[error] $= 0$). The engineering problem is to design a system that has an acceptably small Prob[error] by designing the transmission signals and processing them appropriately.

7.3 PROCESSING DATA SIGNALS

Our description of signal processing procedures progresses in the following steps.

STEP 1. We first describe the basic processing operation employed by most digital receivers. The general approach is termed a *linear processor* that stores samples of the received signal X_i for $i = 0, 1, 2, \ldots, n_x - 1$ and scales them by a set of coefficients to produce a single output value V.

STEP 2. The linear processor that achieves the best performance is called the *matched processor*, where the coefficients are tuned to the data signals values. This processing is also known as *template matching*, *correlation detection*, and *matched filtering*. This is the most reliable method of extracting known data values from a detected signal corrupted with additive Gaussian noise.

STEP 3. Motivated by the matched processor, we design a pair of data signal sequences, $s1_i$ and $s0_i$, that produce maximally different output values V to differentiate a signal that encodes a 1 from that encoding a 0.

This section considers the ideal noiseless case to illustrate the processing that occurs. The next section extends the processing to signals that contain noise.

7.3.1 Linear Processor

Detected signal X_i for $i = 0, 1, 2, \ldots, n_x - 1$ that contains noise must be processed to enhance the information-carrying components and reduce the noise. The method used in practice is called a *linear processor*, which includes a sample memory that stores n_x values X_i for $i = 0, 1, 2, \ldots, n_x - 1$ (that is, all of the data samples that are in $s1_i$ or $s0_i$). These stored sample values are multiplied by delay-specific coefficients c_i for $i = 0, 1, 2, \ldots, n_x - 1$, and the products are summed to form an output value V according to

$$V = \sum_{i=0}^{n_x-1} c_i \, X_i \tag{7.8}$$

Figure 7.3 shows the structure of a general linear processor that operates on the detected input sequence X_i to produce output V. The blocks indicated with δ represent memory units that store an input value for one sample period. For example, at time i, the input

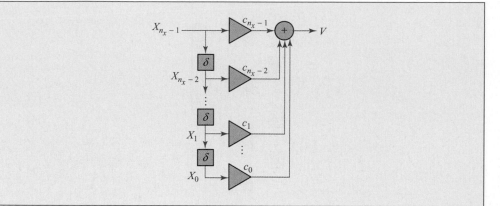

Figure 7.3

Block diagram of the general linear processor at the time that it observes all transmitted signal values X_i for $i = 0, 1, 2, \ldots, n_x - 1$.

to a memory unit is X_i, while the memory unit output equals the past value X_{i-1}. The value of X_i is multiplied by coefficient c_i with the multipliers indicated by triangles. The products at the multiplier outputs are added together at a summing junction to produce V.

This processor structure is a natural choice for data signal processing:

- The transmitted signal has a finite duration (n_x).

- Some signal values are more important than others, and larger magnitude coefficients are assigned to the more important signal sample values.

- The processor uses all n_x values to produce a single output value V that the detector uses to decide the received data value.

Linear processor EXAMPLE 7.1

The observed signal sequence X_i for $i = 0, 1, 2, \ldots, n_x - 1$ ($n_x = 3$) equals

$$X_0 = 2, \; X_1 = -1, \; \text{and} \; X_2 = 1$$

A linear processor operates on $n_x = 3$ samples with coefficients

$$c_0 = 1, c_1 = 2, \; \text{and} \; c_2 = 3$$

The linear processor produces output value V, as

$$V = \sum_{i=0}^{2} c_i \, X_i = c_0 \, X_0 + c_1 \, X_1 + c_2 \, X_2 = 1(2) + 2(-1) + 3(1) = 3$$

The coefficient magnitude $|c_i|$ measures the *importance* of the contribution of signal value X_i to the value of V. The largest magnitude coefficient c_2 indicates that X_2 has more weight in determining the V value than the other X_i values.

7.3.2 Matched Processor

The general linear processor applies weights to the observed input values through the coefficients. *In the presence of a constant random noise level, it is intuitive that X_i values that have larger magnitudes are detected more reliably than those with smaller magnitudes.* Hence, larger magnitude signals should be given more weight by assigning larger coefficients to them. This is the motivation for the *matched processor*.

To implement a matched processor, the general processor undergoes the following modification: The filter coefficients c_i are *tuned* to the values of the signal that it is

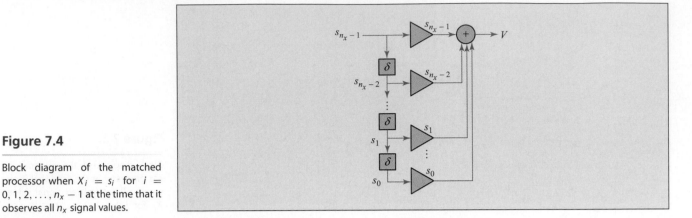

Figure 7.4

Block diagram of the matched processor when $X_i = s_i$ for $i = 0, 1, 2, \ldots, n_x - 1$ at the time that it observes all n_x signal values.

designed to detect. We start by designing a signal sequence s_i for $i = 0, 1, 2, \ldots, n_x - 1$ having known values to transmit data. We can design the processor matched to s_i by specifying $c_i = s_i$. The processor output V then given as

$$V = \sum_{i=0}^{n_x-1} s_i \, X_i \tag{7.9}$$

Figure 7.4 shows the structure of a matched processor when $X_i = s_i$ for $i = 0, 1, 2, \ldots,$ $n_x - 1$ at the time that it observes all n_x signal values to produce output V.

When the signal to which the matched processor is *tuned* occurs (that is, when $X_i = s_i$ for $i = 0, 1, 2, \ldots, n_x - 1$),

$$V = \sum_{i=0}^{n_x-1} s_i \, \overbrace{X_i}^{=s_i} = \sum_{i=0}^{n_x-1} s_i^2 = \mathcal{E}_s \tag{7.10}$$

So, when the processor matched to a signal actually detects this signal, the output is large—equal to the sum of the squares of the signal values. This large positive value equals the signal energy $\mathcal{E}_s$. It can be shown that the matched processor is the best we can do to detect signals in the presence of noise.

EXAMPLE 7.2   Matched processor in the ideal noiseless case

Let the signal sequence with $n_x = 3$ be

$$s_0 = 2, s_1 = -1, \text{ and } s_2 = 1$$

The three coefficients matched to this signal are

$$c_0 = 2, c_1 = -1, \text{ and } c_2 = 1$$

When the observed signal is $X_i = s_i$ for $i = 0, 1, 2, \ldots, n_x - 1$, the matched processor produces output value V, as

$$V = \sum_{i=0}^{2} c_i \, X_i = \sum_{i=0}^{2} s_i \, X_i = s_0 \, s_0 + s_1 \, s_1 + s_2 \, s_2 = (2)^2 + (-1)^2 + (1)^2 = 6$$

The energy in the signal s_i is $\mathcal{E}_s = 6$.

Let us apply a different signal having the same energy to this matched processor. Consider the time-reversed signal

$$s_0' = 1, s_1' = -1, \quad \text{and} \quad s_2' = 2$$

Because s_i' has the same three values as s_i (but in a different order) their energies are equal, as

$$\mathcal{E}_s = \mathcal{E}_{s'} = 6$$

When $X_i = s_i'$ is processed by the matched processor tuned to s_i, the output V' is

$$V' = \sum_{i=0}^{2} s_i \overbrace{X_i}^{=s_i'} = s_0 s_0' + s_1 s_1' + s_2 s_2' = (2)(1) + (-1)(-1) + (1)(1) = 4$$

Note that $V' < \mathcal{E}_s$. This demonstrates that a matched processor tuned to a signal generates the largest possible value when that signal is observed, for signals having the same energy.

7.3.3 Complementary Signals

The previous section shows that a matched processor tuned to a signal generates the largest possible signal when that signal is observed. We want to apply this observation to signals that encode the transmitted binary value D_t with $s1_i$ encoding a transmitted 1 ($D_t = 1$) and $s0_i$ encoding a 0 ($D_t = 0$). We would like to specify these signals so that the matched processor output produces very different values for a detector to easily differentiate the occurrence of these two signals. It turns out that the largest difference in matched processor output occurs when $s1_i$ and $s0_i$ are *complementary*, that is

$$s0_i = -s1_i \ \text{ for } \ i = 0, 1, 2, \ldots, n_x - 1 \tag{7.11}$$

We form a complementary pair from specified signal s_i for $i = 0, 1, 2, \ldots, n_x - 1$ by defining

$$s1_i = s_i \ \text{ for } \ i = 0, 1, 2, \ldots, n_x - 1 \quad \text{when } D_t = 1 \tag{7.12}$$

$$s0_i = -s_i \ \text{ for } \ i = 0, 1, 2, \ldots, n_x - 1 \quad \text{when } D_t = 0$$

Doing this provides two signals having the same energy, as $\mathcal{E}_{s1} = \mathcal{E}_{s0} = \mathcal{E}_s$.

Signal energy of sinusoidal complementary signals   EXAMPLE 7.3

Figure 7.5 shows one period of a sinusoidal sequence

$$s_i = A \sin(2\pi i / 32) \ \text{ for } \ i = 0, 1, 2, \ldots, n_x - 1 (n_x = 32)$$

that forms the pair of complementary signals

$$s1_i = s_i$$

$$s0_i = -s_i$$

The energy in a sinusoidal sequence having amplitude A and an integral number of periods equals

$$\mathcal{E}_s = \sum_{i=0}^{n_x-1} s_i^2 = \sum_{i=0}^{31} A^2 \sin^2(2\pi i / 32) = A^2 \sum_{i=0}^{31} \frac{1 + \cos(4\pi i / 32)}{2} = 16A^2 \quad (= n_x A^2 / 2)$$

This is true because the sum of sinusoidal values over an integral number of periods equals zero. In our case,

$$\sum_{i=0}^{31} \cos(4\pi i / 32) = 0$$

This leaves n_x values of $A^2 / 2$ to produce the sum $\mathcal{E}_s = n_x A^2 / 2$.

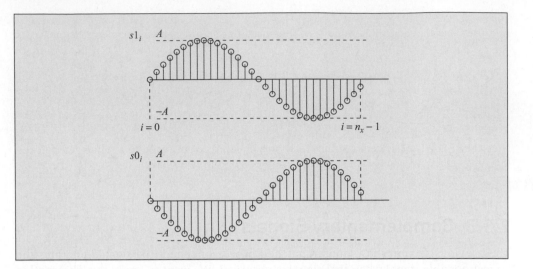

Figure 7.5

Complementary data signals for $s_i = A\sin(2\pi i/32)$ for $i = 0, 1, 2, \ldots, n_x - 1$ with $n_x = 32$.

To show the utility of the matched processor, consider the ideal (noiseless) cases when $X_i = s1_i$ and $X_i = s0_i$.

$X_i = s1_i$: The output V of matched processor when $X_i = s1_i = s_i$ is denoted by $V_{|X=s1}$ (in words, V given that $X = s1$) and is computed as

$$V_{|X=s1} = \sum_{i=0}^{n_x-1} s1_i \overbrace{X_i}^{=s1_i} = \sum_{i=0}^{n_x-1} s1_i^2 = \mathcal{E}_s \qquad (7.13)$$

$X_i = s0_i$: The output V when $X_i = s0_i = -s_i$ is denoted by $V_{|X=s0}$ and equals

$$V_{|X=s0} = \sum_{i=0}^{n_x-1} s0_i \overbrace{X_i}^{s0_i=-s_i} = -\sum_{i=0}^{n_x-1} s_i^2 = -\mathcal{E}_s \qquad (7.14)$$

which is a negative value with large magnitude.

Note that when $\mathcal{E}_s$ is large, the matched processor output $V_{|X=s1} = \mathcal{E}_s$ can be easily differentiated from $V_{|X=s0} = -\mathcal{E}_s$.

EXAMPLE 7.4   Processing complementary signals

Let $s_i = 1$ for $0 \leq i \leq 3$. We form complementary signals

$$s1_i = s_i = 1 \text{ for } 0 \leq i \leq 3$$

$$s0_i = -s_i = -1 \text{ for } 0 \leq i \leq 3$$

The energy of s_i equals

$$\mathcal{E}_s = \sum_{i=0}^{3} s_i^2 = \sum_{i=0}^{3} (1)^2 = 4$$

The matched processor produces the outputs:

$$V_{|X=s1} = \sum_{i=0}^{n_x-1} s_i \overbrace{X_i}^{=s1_i} = \sum_{i=0}^{n_x-1} \overbrace{s_i}^{=1} \overbrace{s1_i}^{=1} = \sum_{i=0}^{3} 1 = 4$$

and

$$V_{|X=s0} = \sum_{i=0}^{n_x-1} s_i \overbrace{X_i}^{=s0_i} = \sum_{i=0}^{n_x-1} \overbrace{s_i}^{=1} \overbrace{s0_i}^{=-1} = \sum_{i=0}^{3} (-1) = -4$$

> *Factoid:*   A data system does not need to use complementary signals, but one that does will work better (that is, produce fewer errors) when noise is present.

7.4 PROCESSING SIGNALS IN NOISE

Noise is typically considered to be a *Gaussian random process* (that is, it obeys the famous *bell-shaped curve*, as shown in Figure 7.6). We model the noise sample N_i as a Gaussian random number.

- N_i is *random* because we cannot predict its value.

- N_i is *Gaussian* because each noise sample can be considered to be the sum of many independent random factors.

The bell-shaped curve for noise has mean value $\mu_N = 0$ and standard deviation (SD) σ_N with the *variance* equal to σ_N^2. One important observation for Gaussian N_i is that observed noise values tend to cluster around it mean value (in this case, zero) with 95% of the values falling within the range

$$[-1.96\sigma_N, 1.96\sigma_N] \tag{7.15}$$

> *Factoid:*   Note: The SD σ_N has the same units as the signal values s_i, while the variance σ_N^2 has the same units as signal energy $\mathcal{E}_s$.

7.4.1 Signal-to-Noise Ratio

A parameter that describes both the quality of the received signal and the probability of detecting the correct data value is given by the *signal-to-noise ratio (SNR)*, which is equal to the ratio of the signal energy $\mathcal{E}_s$ to the noise variance:

$$\text{SNR} = \frac{\mathcal{E}_s}{\sigma_N^2} \tag{7.16}$$

Electrical engineers express this value in decibel (dB) units, as

$$\text{SNR}_{\text{dB}} = 10 \log_{10} \left(\frac{\mathcal{E}_s}{\sigma_N^2} \right) \tag{7.17}$$

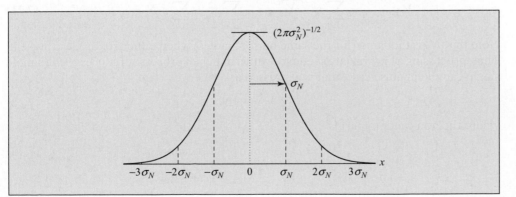

Figure 7.6

Gaussian distribution describing noise sample values.

EXAMPLE 7.5

Signal-to-noise ratio

Let the specified signal s_i for $i = 0, 1, 2, \ldots, n_x - 1$ have signal energy

$$\mathcal{E}_s = \sum_{i=0}^{n_x-1} s_i^2 = 100$$

If Gaussian noise with $\sigma_N = 2$ is added to the signal, 95% of the N_i values fall within $[-3.92, 3.92]$. The signal-to-noise power is

$$\text{SNR} = \frac{\mathcal{E}_s}{\sigma_N^2} = \frac{100}{4} = 25$$

and the value in decibels is

$$\text{SNR}_{\text{dB}} = 10 \log_{10} \left(\frac{\mathcal{E}_s}{\sigma_N^2} \right) = 10 \log_{10} 25 = 14 \, \text{dB}$$

7.4.2 Matched Processor Output

We can now examine the output of the matched processor when noise is present in the two signals that can occur.

$X_i = s1_i + N_i$: In this case,

$$V_{|X=s1+N} = \sum_{i=0}^{n_x-1} s_i X_i = \sum_{i=0}^{n_x-1} s_i (s1_i + N_i) \qquad (7.18)$$

Setting $s1_i = s_i$, we have

$$V_{|X=s1+N} = \sum_{i=0}^{n_x-1} s_i \overbrace{X_i}^{=s_i+N_i} = \overbrace{\sum_{i=0}^{n_x-1} s_i^2}^{=\mathcal{E}_s} + \overbrace{\sum_{i=0}^{n_x-1} s_i N_i}^{=N^*} = \mathcal{E}_s + N^* \qquad (7.19)$$

where N^* is a random number that represents the random Gaussian noise component in the matched processor output.

$X_i = s0_i + N_i$: In this case,

$$V_{|X=s0+N} = \sum_{i=0}^{n_x-1} s_i X_i = \sum_{i=0}^{n_x-1} s_i (s0_i + N_i) \qquad (7.20)$$

Setting $s0_i = -s_i$, we have

$$V_{|X=s0+N} = \sum_{i=0}^{n_x-1} s_i \overbrace{X_i}^{=-s_i+N_i} = -\sum_{i=0}^{n_x-1} s_i^2 + \sum_{i=0}^{n_x-1} s_i N_i = -\mathcal{E}_s + N^* \qquad (7.21)$$

Note that N^* is the same (for the same noise samples) when either $s0_i$ or $s1_i$ is present. Thus, processing a known data sequence with a matched processor reduces to the simple single-sample in noise case, that is expressed as

$$V = s^* + N^* \qquad (7.22)$$

where the non-random part is

$$s^* = \mathcal{E}_s \quad \text{when} \quad s1_i \text{ occurs} \qquad (7.23)$$

$$s^* = -\mathcal{E}_s \quad \text{when} \quad s0_i \text{ occurs}$$

and the random part is the single value derived from processing the set of noise samples

$$N^* = \sum_{i=0}^{n_x-1} s_i N_i \tag{7.24}$$

When complementary signals are used, V is applied to a threshold detector to determine the received data value D_r:

$$\text{If } V \geq 0 \rightarrow D_r = 1 \tag{7.25}$$

$$\text{If } V < 0 \rightarrow D_r = 0$$

An analysis using probability theory shows three important features that describe the matched processor output V.

1. V is a random number that has a Gaussian distribution.

2. The mean of V depends on which signal was transmitted as

$$\mu_V = \mathcal{E}_s \quad \text{when} \quad X_i = s1_i + N_i \tag{7.26}$$

$$\mu_V = -\mathcal{E}_s \quad \text{when} \quad X_i = s0_i + N_i$$

3. The variance of V equals

$$\sigma_V^2 = \sigma_N^2 \sum_{i=0}^{n_x-1} s_i^2 = \sigma_N^2 \mathcal{E}_s \tag{7.27}$$

and is the same value independent of whether $s1_i$ or $s0_i$ was transmitted.

Figure 7.7 shows the Gaussian distribution of V when $\mathcal{E}_s \gg \sigma_V$ and when $\mathcal{E}_s \approx \sigma_V$. An error occurs when the V value falls on the wrong side of threshold $\tau = 0$. This happens when a signal that produces $V = +\mathcal{E}_s$ under ideal noiseless conditions is corrupted with a specific set of noise values that causes the matched processor to produce $V < 0$.

When $\mathcal{E}_s \gg \sigma_V$, we should expect almost error-free performance with V always falling on the correct side of the threshold τ. However, this is not always possible, because the tails of a Gaussian distribution theoretically extend in the negative direction to $x = -\infty$ and in the positive direction to $x = +\infty$. Even when $\mathcal{E}_s \gg \sigma_V$, it is theoretically possible for rare V values to fall on the wrong side of threshold τ, no matter how large we make $\mathcal{E}_s$.

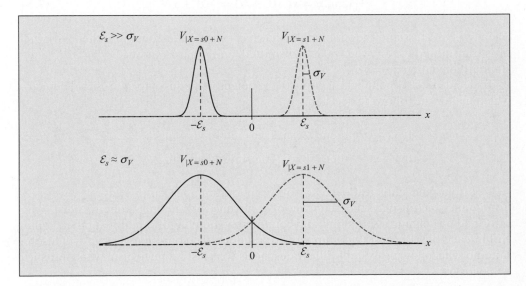

Figure 7.7

The Gaussian distribution of V when $\mathcal{E}_s \gg \sigma_V$ and when $\mathcal{E}_s \approx \sigma_V$.

7.4.3 Detecting Signals in Noise

We start with the simplest signal with $s0_i$ and $s1_i$ consisting of only one sample having magnitude s. Thus,

$$s1 = s \text{ if } D_t = 1 \quad \text{and} \quad s0 = -s \text{ if } D_t = 0 \tag{7.28}$$

The signal energy is

$$\mathcal{E}_s = s^2 \tag{7.29}$$

The received signal is corrupted with the single noise sample N:

$$X = s + N \quad \text{if } s1 \text{ is transmitted} \tag{7.30}$$

$$X = -s + N \quad \text{if } s0 \text{ is transmitted}$$

where N is a zero-mean Gaussian random number having variance σ_N^2.

The signal-to-noise ratio is computed as a ratio of signal energy to noise variance, as

$$\text{SNR} = \frac{\mathcal{E}_s}{\sigma_N^2} = \frac{s^2}{\sigma_N^2} \tag{7.31}$$

The matched processor computes the output

$$V = \sum_{i=0}^{n_x-1} s_i X_i = s X \tag{7.32}$$

In this simple case, the result depends on what signal was transmitted, as

$$V_{|X=s1+N} = (\overbrace{s1}^{=s} + N) = s^2 + sN \tag{7.33}$$

or

$$V_{|X=s0+N} = s(\overbrace{s0}^{=-s} + N) = -s^2 + sN \tag{7.34}$$

Then V is a Gaussian random number that has mean value equal to either $+s^2$ when $s1$ is transmitted or $-s^2$ when $s0$ is transmitted, and it has a variance $\sigma_V^2 = s^2\sigma_N^2$ and a standard deviation $\sigma_V = s\sigma_N$.

To decide if the X contains an $s1$ or $s0$, the detector applies V to a threshold detector, which makes the following decision.

$$\text{If } V \geq 0 \quad \text{then } s1 \text{ was transmitted, making } D_r = 1 \tag{7.35}$$

$$\text{If } V < 0 \quad \text{then } s0 \text{ was transmitted, making } D_r = 0$$

To display the statistical behavior of a random number (in this case V), we use a *histogram*, which displays the frequency of occurrence of random numbers. A histogram is formed by defining a set of bins and counting the number of random numbers falling within each bin. Because the numbers are random, the count in each bin will also be random. The following example illustrates the use of a histogram to display the statistical behavior of V.

EXAMPLE 7.6

Simple signal with noise

To illustrate the problem, consider a source that transmits two hundred data signals: one hundred 0's (each using $s0 = -10$) followed by one hundred 1's (each using $s1 = 10$). Consider first the high SNR case, followed by a low SNR case.

Figure 7.8a shows the two hundred signals corrupted by Gaussian noise with $\sigma_N = 1$ (the high SNR case). The signal energy is

$$\mathcal{E}_s = s^2 = 100$$

and the signal-to-noise ratio

$$\text{SNR} = \frac{\mathcal{E}_s}{\sigma_N^2} = \frac{100}{1} = 100$$

and in decibel units, it is

$$\text{SNR}_{\text{dB}} = 10 \log_{10} 100 = 20 \text{ dB}$$

These two hundred signals are each applied to the matched processor that computes V as

$$V = sX = 10X$$

When $X = s0 + N = -10 + N$, the signal $s0$ is constant, and the random noise value N is different for each observed X value. In this case, the matched-processor output for each observed X value equals

$$V_{|X=s0+N} = -100 + 10N = -\mathcal{E}_s + 10N$$

Because N has a Gaussian distribution, $V_{|X=s0+N}$ also has a Gaussian distribution with mean $-\mathcal{E}_s$ and SD $10\sigma_N$. These one hundred $V_{|X=s0+N}$ values generate the histogram component that is centered about $x = -\mathcal{E}_s$ in Figure 7.8a.

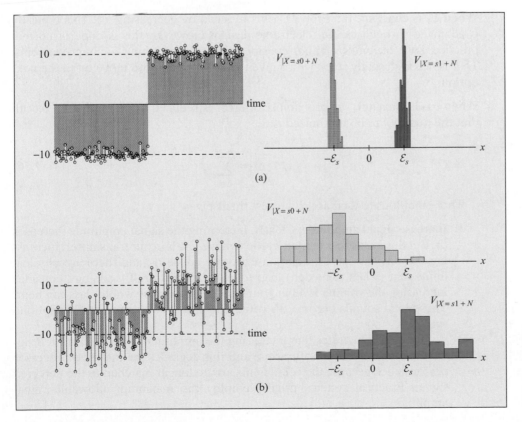

Figure 7.8

Two hundred data signals with noise and histogram of matched-processor output V: (a) High SNR (20 dB) and (b) Low SNR (0 dB).

When $X = s1 + N = 10 + N$, the matched-processor output for each observed X value equals

$$V_{|X=s1+N} = 100 + 10N = \mathcal{E}_s + 10N$$

These one hundred $V_{|X=s1+N}$ values generate the histogram component that is centered about $x = \mathcal{E}_s$.

The histogram in Figure 7.8a indicates that a threshold detector using threshold value $\tau = 0$ would correctly separate the $s0$ and $s1$ signals almost all the time and, thus, have a Prob[error] close to zero.

Figure 7.8b shows the same data values and signal energy ($\mathcal{E}_s = 100$) as in Figure 7.8a but with more noise added ($\sigma_N = 10$). The signal-to-noise ratio in this higher-noise case equals

$$\text{SNR} = \frac{\mathcal{E}_s}{\sigma_N^2} = \frac{100}{100} = 1$$

and in decibel units, it is

$$\text{SNR}_{\text{dB}} = 10 \log_{10} 1 = 0 \text{ dB}$$

The histograms of $V_{|X=s0+N}$ and $V_{|X=s1+N}$ in Figure 7.8b show values that overlap the zero value. A threshold detector using $\tau = 0$ would produce many errors in this case. Because the noise N is random, the errors also occur randomly, and the system performance is measured by its Prob[error]. The system operating in the high noise case would have an unacceptably large Prob[error] (≈ 0.16).

Hence, the SNR is an important factor that determines the reliability of data transmission systems in terms of its Prob[error].

The previous example illustrates that we can reduce the number of errors by increasing the signal-to-noise ratio. The SNR can be improved either by increasing $\mathcal{E}_s$ or by reducing σ_N^2 using the following methods.

When $\mathcal{E}_s$ is constant, the Prob[error] is reduced by decreasing σ_N. This typically is achieved through careful electronic design. However, this is done in normal practice, so it becomes a minor consideration when looking for improvements. That is, we are usually stuck with a given value for σ_N, so we move on to the next option.

When σ_N is constant, the probability of error is reduced by increasing $\mathcal{E}_s$. Recall that the quantity to be maximized is

$$\mathcal{E}_s = \sum_{i=0}^{n_x-1} s_i^2 \tag{7.36}$$

Three methods to increase $\mathcal{E}_s$ suggest themselves.

1. **Increase signal amplitude:** Clearly, increasing the signal amplitude increases $\mathcal{E}_s$. But this increase is limited by the voltage levels in our transmitter. Increasing these levels causes radiation exposure concerns, and there are physical limits that (when exceeded) lead to signal distortion. By analogy, consider increasing the volume level in your headphones so the whole class can hear. These load sounds become distorted and can even destroy the headphone speakers.

   Figure 7.9 illustrates that increasing the amplitude of the transmitted signal eventually leads to saturation, and this degrades the matched-processor performance because the coefficients are no longer matched to the detected signal. Practical systems usually employ the maximum allowable signal amplitude.

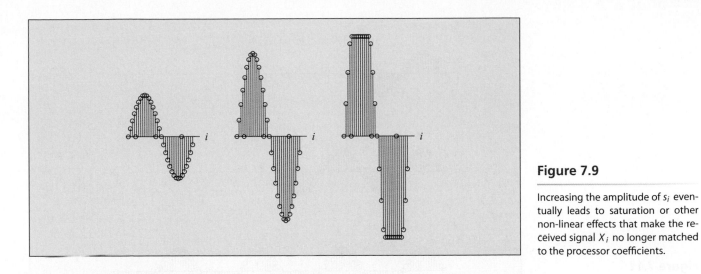

Figure 7.9

Increasing the amplitude of s_i eventually leads to saturation or other non-linear effects that make the received signal X_i no longer matched to the processor coefficients.

2. **Form binary signals:** Examining Eq. (7.36) and applying the constraint

$$|s_i| \leq V_{max} \tag{7.37}$$

engineers developed *binary signals* in which s_i takes on only one of two values, either $+V_{max}$ or $-V_{max}$, as

$$s_i = \pm V_{max} \text{ for } i = 0, 1, 2, \ldots, n_x - 1 \tag{7.38}$$

Conventional signals can be converted to binary signals by extending positive values to $+V_{max}$ and negative values to $-V_{max}$, as shown in Figure 7.10.

The energy of a binary signal is

$$\mathcal{E}_s = \sum_{i=0}^{n_x-1} s_i^2 = \sum_{i=0}^{n_x-1} V_{max}^2 = n_x V_{max}^2 \tag{7.39}$$

Since the signal is always at a maximum magnitude, the binary signal has the maximum possible energy that can be placed on a channel. Recall, a sinusoidal sequence having amplitude V_{max} has energy $\mathcal{E}_s = n_x V_{max}^2/2$, which is half that achievable with a binary signal.

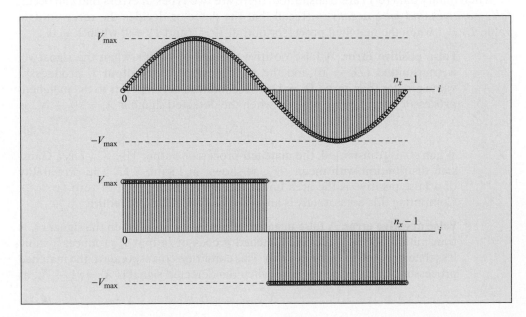

Figure 7.10

Increasing s_i to the amplitude limits imposed by channel maximizes $\mathcal{E}_s$. A sinusoidal s_i (top) has $\mathcal{E}_s = n_x V_{max}^2/2$. A binary signal is formed by maximizing amplitudes to $\pm V_{max}$ (bottom) and has $\mathcal{E}_s = n_x V_{max}^2$.

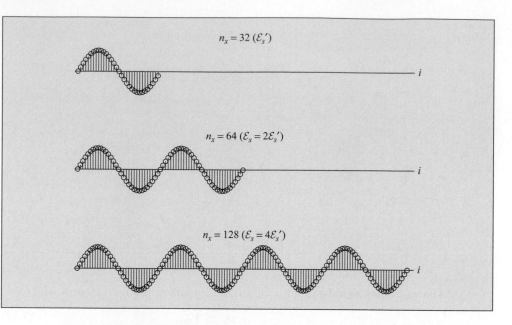

Figure 7.11

Increasing signal duration n_x lowers data transmission rate.

3. **Increase signal duration:** Finally, if a binary signal already uses the maximum allowed magnitudes, we can still increase $\mathcal{E}_s$ by making the signal longer by increasing n_x, as indicated by Eq. (7.39). However, this means that it takes longer to transmit the signal s_i for $i = 0, 1, 2, \ldots, n_x - 1$, which encodes each binary data value. Figure 7.11 illustrates that increasing the duration of the transmitted signal leads to lower data transmission rates.

Practical systems are often forced to use this longer transmission option. This is the reason that faint data signals (such as digital image data from Mars) take a long time to transmit.

▶ 7.5 PROBABILITY OF ERROR

The probability of error is not simple to compute analytically for Gaussian random numbers, so this section provides a qualitative explanation. The following section describes a simulation method to provide a quantitative estimate of the Prob[error].

When binary data (0/1) are transmitted, there are two types of errors that can occur: *false positives and false negatives.* Recall that the detector decides the received data value $D_r = 1$ when the matched processor output $V > 0$ and $D_r = 0$ when $V \leq 0$.

False positive error. A false positive (FP) error occurs when the signal $s0_i$ is transmitted ($D_t = 0$) and the matched-processor output V produces a value greater than zero ($D_r = 1$). This condition corresponds to the matched-processor output being positive when the detected signal is $X_i = s0_i + N_i$, as

$$V_{|X=s0+N} > 0 \qquad (7.40)$$

When $s0_i$ is transmitted, the matched-processor output $V_{|X=s0+N}$ has a Gaussian distribution with mean $-\mathcal{E}_s$, as shown in Figure 7.12. The probability of a false positive is the area under this curve to the right of zero ($x > 0$). Computing this area exactly is an analytically difficult procedure.

False negative error. A false negative (FN) error occurs when the signal $s1_i$ is transmitted ($D_t = 1$) and the matched processor output V produces a value less than or equal to zero ($D_r = 0$). This condition corresponds to the matched processor output being negative when the detected signal is $X_i = s1_i + N_i$, as

$$V_{|X=s1+N} \leq 0 \qquad (7.41)$$

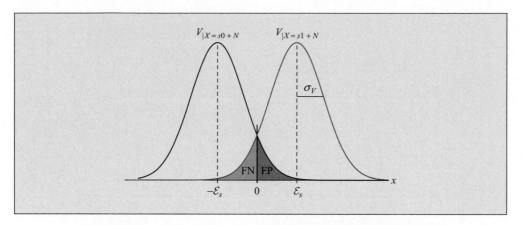

Figure 7.12

Probability of error equals the shaded areas FP (false positive) and FN (false negative) under the Gaussian curves.

When $s1_i$ is transmitted, the matched processor output $V_{|X=s1+N}$ has a Gaussian distribution with mean $\mathcal{E}_s$, as shown in Figure 7.12. The probability of a false negative is the area under this curve to the left of zero ($x \leq 0$).

Qualitatively, the Prob[error] equals the sum of the FP and FN areas and reducing these areas leads to a smaller Prob[error]. The careful reader would have noticed the FN and FP areas are equal when the threshold value $\tau = 0$. Shifting the τ value in the positive direction reduces FP while increasing FN.

The probability of error can be reduced either by reducing σ_V (making the bell-shaped curves narrower) or by increasing $\mathcal{E}_s$ (spreading the curves farther apart). The following example illustrates how increasing the signal duration n_x reduces the Prob[error].

Increasing n_x to reduce the probability of error   EXAMPLE 7.7

This example examines the Prob[error] qualitatively by determining the change in the Gaussian distributions as the signal duration n_x increases. The signal amplitude is set to be the largest possible value ($a = V_{\max}$) to maximize $\mathcal{E}_s$. The noise level σ_N is held constant with σ_N also equal to a. The results are shown in Figure 7.13.

For $n_x = 1$: Let the signal encode the data with a single sample using signal $s = a$. The detected data X is the sum of the signal and noise sample N, as

$$X = a + N \quad \text{if} \quad D_t = 1$$
$$X = -a + N \quad \text{if} \quad D_t = 0$$

The signal energy is $\mathcal{E}_s = a^2$, and the SNR equals

$$\text{SNR} = \frac{\mathcal{E}_s}{\sigma_N^2} = \frac{a^2}{\sigma_N^2}$$

For $\sigma_N = a$, SNR $= 1$.

The matched-processor output V has a Gaussian distribution described by

$$\mu_V = \mathcal{E}_s = a^2 \quad \text{if} \quad D_t = 1$$
$$= -\mathcal{E}_s = -a^2 \quad \text{if} \quad D_t = 0$$

The SDs of both Gaussian distributions are the same and equal to

$$\sigma_V = \sigma_N = a$$

For $n_x = 1$, the two Gaussian curves overlap significantly indicating a large Prob[error].

For $n_x = 2$: Let the signal encode the data with two samples using $s_0 = s_1 = a$. The detected sequence X_i is the sum of s_i and noise N_i, as

$$X_i = a + N_i \quad \text{for} \quad i = 0, 1 \quad \text{if} \quad D_t = 1$$
$$X_i = -a + N_i \quad \text{for} \quad i = 0, 1 \quad \text{if} \quad D_t = 0$$

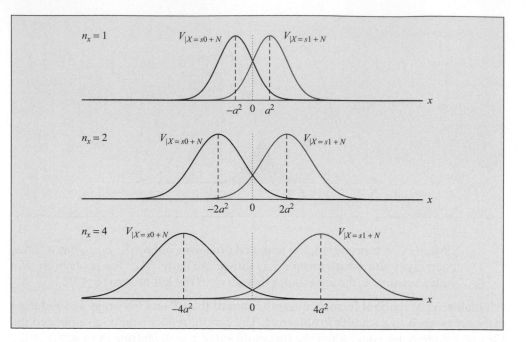

Figure 7.13

Probability of error reduction by increasing n_x.

The signal energy is $\mathcal{E}_s = 2a^2$, and the SNR equals

$$\text{SNR} = \frac{\mathcal{E}_s}{\sigma_N^2} = \frac{2a^2}{\sigma_N^2} = 2$$

The matched-processor output V has a Gaussian distribution described by

$$\mu_V = \mathcal{E}_s = 2a^2 \quad \text{if} \quad D_t = 1$$

$$\mu_V = -\mathcal{E}_s = -2a^2 \quad \text{if} \quad D_t = 0$$

Using Eq. (7.27), $\sigma_V^2 = \mathcal{E}_s \sigma_N^2$, gives the SD of both distributions of V as

$$\sigma_V = \sqrt{\mathcal{E}_s}\,\sigma_N = \sqrt{2}\,a\,\sigma_N = \sqrt{2}a^2$$

The two Gaussian curves overlap less for $n_x = 2$ than for $n_x = 1$, indicating a smaller Prob[error].

For $n_x = 4$: Let the signal encode the data with four samples $s_0 = s_i = s_2 = s_3 = a$. The detected sequence X_i is the sum of s_i and noise N_i, as

$$X_i = a + N_i \quad \text{for} \quad 0 \le i \le 3 \quad \text{if} \quad D_t = 1$$

$$X_i = -a + N_i \quad \text{for} \quad 0 \le i \le 3 \quad \text{if} \quad D_t = 0$$

The signal energy is $\mathcal{E}_s = 4a^2$, and the SNR equals

$$\text{SNR} = \frac{\mathcal{E}_s}{\sigma_N^2} = \frac{4a^2}{\sigma_N^2} = 4$$

The matched-processor output V has a Gaussian distribution described by

$$\mu_V = \mathcal{E}_s = -4a^2 \quad \text{if} \quad D_t = 0$$

$$\mu_V = \mathcal{E}_s = 4a^2 \quad \text{if} \quad D_t = 1$$

Using Eq. (7.27), $\sigma_V^2 = \mathcal{E}_s \sigma_N^2$ gives the SD of V as

$$\sigma_V = \sqrt{\mathcal{E}_s}\,\sigma_N = 2\,a\,\sigma_N = 2\,a^2$$

The two Gaussian curves overlap even less for $n_x = 4$, indicating a further reduction in the Prob[error].

Figure 7.13 shows that increasing n_x increases $\mathcal{E}_s$ (separating the two Gaussian curves) but also increases σ_V (broadening each Gaussian curve). Generalizing this result, as n_x increases, the Gaussian curves separate linearly with n_x and broaden with $\sqrt{n_x}$. Thus, increasing n_x provides a net reduction in the Prob[error]. Hence, any desired Prob[error] (> 0) can be achieved by specifying a sufficiently large n_x. This explains the ability to transmit data from Mars—which requires very large n_x. The trade-off is that increasing n_x lowers the data rate—which explains why it takes so long to acquire a digital image from Mars.

The next section describes a simulation procedure that estimates the Prob[error] by varying n_x, the amplitude of s_i, or the noise variance σ_N^2.

7.6 ESTIMATING PROBABILITY OF ERROR WITH SIMULATIONS

Because noise is random, error performance is measured in terms of the *probability of error*, denoted Prob[error]. Computing Prob[error] analytically is difficult, because Gaussian functions produce area functions (integrals) that are difficult to evaluate. This section describes a simulation procedure to estimate the Prob[error] by performing multiple transmissions in the presence of noise and counting the number of errors.

While SNR is an important measure, what we really want to know is *"How many transmission errors can I expect to occur?"* Although we want no errors, such a system in practice would be too slow, because to accomplish this, the signal duration would need to be too large. An optimal system is one in which errors are rare, and these rare errors can be corrected using error-correction techniques.

Because errors are caused by random disturbances, we need to observe the system performance *statistically* by counting errors when we transmit a large number of signals. To illustrate this approach, let n_t denote the total number of transmissions and N_e (a random number) denote the number of errors that are observed. The *error probability* is estimated as

$$\text{Prob[error]} \approx \frac{N_e}{n_t} \tag{7.42}$$

We explore this problem by describing the following components.

- A source that generates random binary data.

- The Gaussian random noise model.

- Signal design to minimize the Prob[error].

- System simulation to verify performance.

7.6.1 Modeling Random Data

A source that produces random binary data is typically modeled by using a pseudo-random number generator (PRNG) that produces random numbers that lie uniformly within the interval [0, 1) (the *uniform PRNG*). The closed interval, which is denoted with a square bracket on the left, indicates that 0 is a value in the set that can occur, while the open interval with the parenthesis on the right indicates that 1 is not in the set, although any non-negative number < 1 is possible.

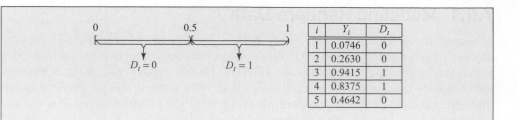

Figure 7.14

Y_i for $i = 0, 1, 2, \ldots, n_Y - 1$ ($n_Y = 200$) and histogram with 50 bins formed by 10^5 random numbers produced by the uniform PRNG. Each bin contains approximately 2,000 counts.

EXAMPLE 7.8   Histogram of uniform PRNG

Let Y_i denote the random numbers generated by the uniform PRNG. We form bins that cover this interval using equally sized non-overlapping segments producing 50 bins:

$$\text{bin } 1 \to [0, 0.02)$$

$$\text{bin } 2 \to [0.02, 0.04)$$

$$\text{bin } 3 \to [0.04, 0.06)$$

and so on until

$$\text{bin } 50 \to [0.98, 1.0)$$

Note that the half-open intervals [.,.) ensure that each Y_i value falls into only one bin and all Y_i values in [0,1) produce counts.

Figure 7.14 shows 200 Y_i values and the histogram formed by 10^5 random numbers. When random numbers are said to be uniformly distributed over [0,1), we expect the bins to yield counts that are approximately equal for the bins that cover [0,1), and the figure verifies this. With 50 bins and 10^5 random numbers, each bin contains approximately 2,000 counts.

EXAMPLE 7.9   Simulating random data bits

To simulate random bits, we use the uniform PRNG that generates a pseudo-random number that is uniformly distributed over [0, 1). We divide this interval into two equal non-overlapping parts. The convenient division into [0, 0.5) and [0.5, 1) is shown in Figure 7.15. The importance of dividing an interval using half-open intervals is two-fold:

- The entire interval is covered (that is, all numbers are accounted for).
- Any random number falls into only one interval.

If $0 \leq Y_i < 0.5 \to D_t = 0$ and if $0.5 \leq Y_i < 1 \to D_t = 1$.

Figure 7.15

Using the uniform PRNG to simulate binary data. Each random number Y_i yields a random data bit.

i	Y_i	D_t
1	0.0746	0
2	0.2630	0
3	0.9415	1
4	0.8375	1
5	0.4642	0

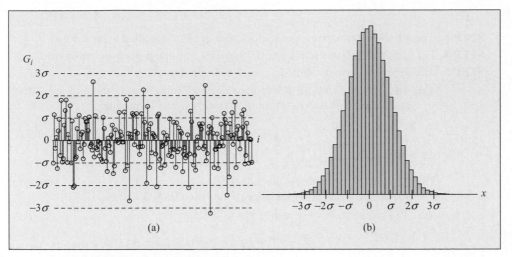

Figure 7.16

Gaussian PRNG. (a) G_i for $i = 0, 1, 2, \ldots, n_G - 1$ ($n_G = 200$). (b) Histogram with 51 bins formed by one million random numbers.

7.6.2 Modeling Random Noise

Noise is typically considered to be a *Gaussian random process* (that is, it obeys the famous *bell-shaped curve* with values that tend to be concentrated about its mean value). Noise samples are simulated with a Gaussian PRNG that produces random numbers having a Gaussian distribution. The most common is the *standardized Gaussian random number* denoted G_i, which has a zero mean ($\mu_G = 0$) and unit variance ($\sigma_G^2 = 1$).

These G_i random numbers simulate zero-mean Gaussian random noise N_i having variance σ_N^2 simply by multiplying G_i by the non-random value σ_N, as

$$N_i = \sigma_N G_i \tag{7.43}$$

Hence, N_i produces a realization (a random number) that has a zero mean and variance σ_N^2.

Histogram of Gaussian random numbers   EXAMPLE 7.10

The standardized Gaussian random numbers G_i have zero mean and unit variance. Figure 7.16 shows 200 G_i values and the histogram formed by 10^5 G_i values. The bell-shaped form of the histogram distribution is evident only when the number of random numbers is large (10^6 in this example). Note that Gaussian random numbers tend to fall close to the mean value ($=0$) and almost all fall within $[-3\sigma_G, 3\sigma_G]$ ($= [-3, 3]$).

Simulating data transmission   EXAMPLE 7.11

A simulation was performed using the following six steps.

STEP 1. The basic signal sequence s_i for $i = 0, 1, 2, \ldots, n_x - 1$ was specified and its energy $\mathcal{E}_s$ computed. The signals used in the simulation include

$$s = 1 \quad (n_x = 1, \ \mathcal{E}_s = 1)$$

$$s_i = \sin(2\pi i/16) \quad \text{for } i = 0, 1, 2, \ldots, n_x - 1 \quad (n_x = 16, \ \mathcal{E}_s = 8)$$

$$s_i = 1 \quad \text{for } i = 0, 1, 2, \ldots, n_x - 1 \quad (n_x = 16, \ \mathcal{E}_s = 16)$$

STEP 2. Complementary signals $s1_i$ ($= s_i$) and $s0_i$ ($= -s_i$) for $i = 0, 1, 2, \ldots, n_x - 1$ were formed.

STEP 3. The number of trials was specified. Each trial had a specified value for σ_N.

STEP 4. For each trial, a large number n_t (10^6) of data transmissions were performed.

STEP 5. The following were computed for each transmission.

(a) The transmitted bit D_t (0/1) was randomly chosen using the uniform PRNG, and the corresponding transmitted signal sT_i was formed as

$$sT_i = s1_i = s_i \quad \text{for } D_t = 1$$
$$= s0_i = -s_i \quad \text{for } D_t = 0$$

(b) The Gaussian random noise sequence N_i for $i = 0, 1, 2, \ldots, n_x - 1$ having the specified σ_N was generated using a Gaussian PRNG and Eq. (7.43).

(c) The received signal sequence was formed as

$$X_i = sT_i + N_i \quad \text{for } i = 0, 1, 2, \ldots, n_x - 1$$

(d) The X_i was applied to the matched processor tuned to s_i to produce output V.

(e) The V value was applied to the threshold detector with $\tau = 0$ to determine D_r as

$$\text{If } V \leq 0 \rightarrow D_r = 0, \text{ otherwise } D_r = 1$$

(f) If $D_r \neq D_t$, the error count N_e was incremented.

STEP 6. At the end of each trial, the probability of error was estimated as

$$\text{Prob[error]} = \frac{N_e}{n_t}$$

This simulation produced the following results.

Figure 7.17 shows the Prob[error] decreases as $\mathcal{E}_s/\sigma_N^2$ increases. Note the following values from the graph.

$$\text{Prob[error]} = 0.37 \quad \text{for} \quad \mathcal{E}_s/\sigma_N^2 = 0.1 \quad \text{(the first value in the curve)}$$

and

$$\text{Prob[error]} = 0.16 \quad \text{for} \quad \mathcal{E}_s/\sigma_N^2 = 1$$

It does not matter whether $\mathcal{E}_s$ is increased by increasing amplitude or by increasing n_x. The factor that determines the Prob[error] is solely the SNR $= \mathcal{E}_s/\sigma_N^2$.

Figure 7.18 compares the Prob[error] versus σ_N for three signal types having the same amplitude.

1. Single sample $s = 1$ having $\mathcal{E}_s = s^2 = 1$.
2. Sinusoidal sequence $s_i = \sin(2\pi i/16)$ for $i = 0, 1, 2, \ldots, n_x - 1$ with $n_x = 16$, having $\mathcal{E}_s = n_x/2 = 8$.
3. Binary sequence $s_i = 1$ for $0 \leq i \leq 15$, having $\mathcal{E}_s = n_x = 16$

Note that at the same σ_N value, the Prob[error] decreases as $\mathcal{E}_s$ increases, as expected. As σ_N increases and begins to swamp the signal, the Prob[error] $\rightarrow 0.5$. This is the probability of error that is expected if we simply guess what data value was transmitted (without even detecting X_i), because we would be correct half the time.

In practice, the requirements for the error probability depend on the application. One successful system may have Prob[error] $< 10^{-3}$ or less than 1 error in 1,000 transmissions on average, while critical systems typically require Prob[error] $< 10^{-6}$. Practical data transmission systems tolerate the achievable Prob[error] by employing error-correcting codes, as discussed in Chapter 10, to correct rarely occurring errors.

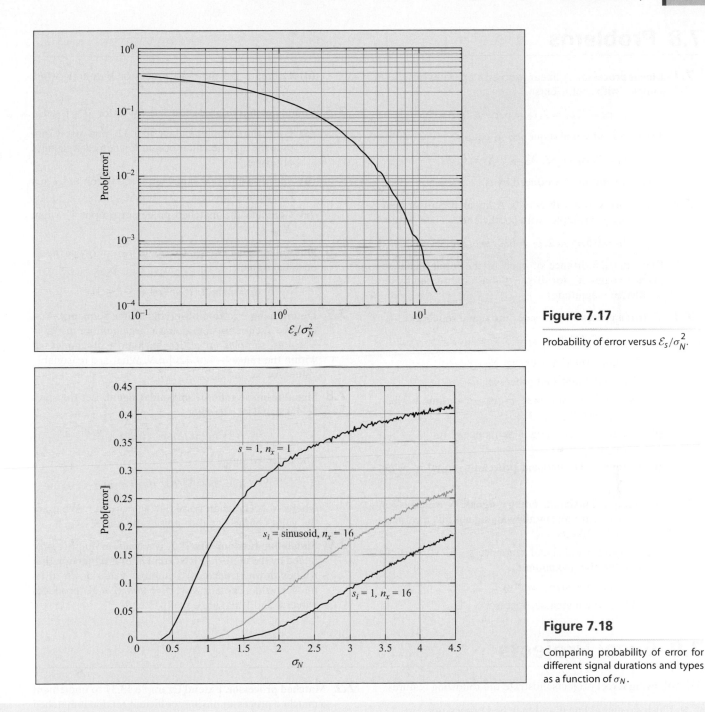

Figure 7.17

Probability of error versus $\mathcal{E}_s/\sigma_N^2$.

Figure 7.18

Comparing probability of error for different signal durations and types as a function of σ_N.

7.7 Summary

This chapter described processing data signals in the presence of noise. In data transmission applications, the signal values are known, having a specified set of values for transmitting a 0 and another set for transmitting a 1. Matched processors exploit this knowledge by tuning its coefficients to these known values. Two signals having complementary values encode binary data values to produce a maximal difference in the matched-processor output. Simulations of data transmissions over a noise channel indicate that the factor determining the Prob[error] is the signal-to-noise ratio $\mathcal{E}_s/\sigma_N^2$. Binary signals that have amplitudes at the voltage limits ($\pm V_{\max}$) maximize the signal energy $\mathcal{E}_s$ for a fixed value of signal duration n_x. For a channel having a constant noise variance σ_N^2, the Prob[error] can be reduced to any desired level greater than zero by increasing the signal duration n_x—but at the expense of transmission time.

7.8 Problems

7.1 Linear processor. A linear processor operates on $n_x = 5$ samples with coefficients

$$c_0 = 1, c_1 = 2, c_2 = 3, c_3 = 2, c_4 = 1$$

The observed signal sequence X_i equals

$$X_0 = 2, X_1 = -3, X_2 = 1, X_3 = 0, X_4 = -1$$

Compute output V produced by X_i.

7.2 Linear processor with $c_i = 0$. A linear processor operates on $n_x = 5$ samples with coefficients

$$c_0 = 0.5, c_1 = 2, c_2 = 0, c_3 = -3, c_4 = -1$$

Rank the importance of each of the input signal sequence values X_i for $0 \le i \le 4$ as determined by the coefficient magnitudes.

7.3 Matched processor. Consider the signal sequence

$$s_0 = 1, s_1 = 2, s_2 = 3, s_3 = 4, s_4 = 5$$

(a) Compute the signal energy $\mathcal{E}_s$.

(b) Design the matched processor.

(c) Compute the matched processor output V when $X_i = s_i$ for $0 \le i \le 4$.

(d) Specify a different signal sequence s_i' for $0 \le i \le 4$ having the same $\mathcal{E}_s$.

(e) Compute the matched processor output V' when $X_i = s_i'$.

7.4 Designing a maximum energy signal. A transmission channel is constrained by allowing signals that have magnitudes $|s_i| \le 2$ Volts.

(a) Design a valid signal sequence s_i for $0 \le i \le 4$ that has the maximum $\mathcal{E}_s$.

(b) Compute the signal energy $\mathcal{E}_s$.

(c) Design the matched processor.

(d) Compute the matched processor output V when $X_i = s_i$.

7.5 Complementary signals. Let $s_i = i + 1$ for $0 \le i \le 4$.

(a) Form the complementary pair $s1_i$ and $s0_i$. Compute the matched filter output V when each signal is observed.

(b) Specify a different signal sequence s_i' for $0 \le i \le 4$ having the same $\mathcal{E}_s$.

(c) Compute the matched processor output V' when $X_i = s_i'$.

7.6 Sinusoidal signal energy. Compute the energy of the sinusoidal signal

$$s_i = 2\sin(2\pi i/8) \text{ for } 0 \le i \le 31$$

7.7 Determining σ_N^2 from observing a time sequence. You observe a sequence containing a large number of Gaussian noise samples and notice that almost all samples fall within the range $[-120, 120]$ mV. What are reasonable values for σ_N and σ_N^2?

7.8 Signal-to-noise ratio of sinusoidal signal. Let the sinusoidal signal be given by

$$s_i = 2\sin(2\pi i/8) \text{ for } 0 \le i \le 31$$

The detected signal is

$$X_i = s_i + N_i \text{ for } 0 \le i \le 31$$

where N_i is Gaussian noise having $\sigma_N = 0.1$. What are the values of the signal-to-noise ratio and SNR_{dB}?

7.9 Simulating random bits. If Y_i is a random number produced by the uniform PRNG and *floor* is a function that rounds down (truncates) a number value down to its integer value, explain why $D = floor(2 * Y_i)$ produces values 0 and 1 that are equally likely.

7.9 Excel Projects

The following Excel projects illustrate the following features.

- Illustrate digital linear and matched processors.

- Compose a VBA Macro to estimate the probability of error by performing a large number of trials.

Using a Narrative box (described in Example 13.4), include observations, conclusions, and answers to questions posed in the projects.

7.1 Linear processor. Extend Example 13.36 to implement a linear processor having coefficient set

$$c_i = 2i + 1 \text{ for } i = 0, 1, 2, \ldots, n_x - 1 \ (n_x = 10)$$

Enter a Gaussian random input sequence X_i for $0 \le i \le n_x - 1$, and compute output V.

7.2 Matched processor. Extend Example 13.37 to implement a matched processor having coefficient set that is matched to input signal

$$s_i = \sin(2\pi i/16) \text{ for } i = 0, 1, 2, \ldots, n_x - 1 \ (n_x = 16)$$

and compute output V when $X_i = s_i$ for $i = 0, 1, 2, \ldots, n_x - 1$.

7.3 Complementary signals. Extend Example 13.38 to implement a matched processor having coefficients matched to input signal

$$s_i = \sin(2\pi i/16) \text{ for } i = 0, 1, 2, \ldots, n_x - 1 \ (n_x = 16)$$

Generate the pair of complementary signals $s1_i$ and $s0_i$, and compute outputs $V|_{X=s1}$ and $V|_{X=s0}$.

7.4 **Matched processor output when observing signals in noise.** Modify Example 13.39 to implement a matched processor having coefficients matched to input signal

$$s_i = \sin(2\pi i/16) \text{ for } i = 0, 1, 2, \ldots, n_x - 1 \ (n_x = 16)$$

To form the observed signal X_i add Gaussian noise having a specified σ_N to s_i. Determine the value of σ_N for which errors errors begin to occur (an error is observed 10% of the time).

7.5 **Observing signals in noise.** Modify Example 13.40 to implement a matched processor having coefficients matched to input signal

$$s_i = \sin(2\pi i/16) \text{ for } i = 0, 1, 2, \ldots, n_x - 1 \ (n_x = 16)$$

To form the observed signal X_i add Gaussian noise having a specified σ_N to s_i. Determine the value of σ_N for which it is difficult to notice the signal component.

7.6 **Histogram of complementary signals in noise.** Using Example 13.32 as a guide, generate a histogram of one thousand matched processor output values V when pro-

cessing signals in the presence of noise. Choose your favorite complementary signal pair with $n_x = 16$ and a signal-to-noise ratio $\mathcal{E}_s/\sigma_N^2 = 2$. Repeat for $\mathcal{E}_s/\sigma_N^2 = 4$.

7.7 **Estimating probability of error for a matched processor.** Modify Examples 13.41, 13.42, and 13.43 to compute the probability of error as σ_N^2 increases between $0.1\mathcal{E}_s$ and $10\mathcal{E}_s$ by factor $\sqrt{2}$, using signal

$$s_i = \sin(2\pi i/16) \text{ for } i = 0, 1, 2, \ldots, n_x - 1 \ (n_x = 16)$$

7.8 **Designing a signal that has minimum probability of error.** Assume a channel constrains signals to have a limited amplitude

$$|s_i| \leq 1\,\text{V} \text{ for } i - 0, 1, 2, \ldots, n_x - 1$$

Design a signal with $n_x = 16$ that has a maximum $\mathcal{E}_s$. Use two worksheets, with each modeled after Example 13.43 that compares your signal performance with that of the sinusoidal signal

$$s_i = \sin(2\pi i/16) \text{ for } i = 0, 1, 2, \ldots, n_x - 1 \ (n_x = 16)$$

CHAPTER

8

DESIGNING SIGNALS FOR MULTIPLE-ACCESS SYSTEMS

LEARNING OBJECTIVES

After completing this chapter, the reader should be able to:

- Understand the purpose of orthogonal signals for robust data transmission.
- Design signal sets that are orthogonal in time, frequency, and code.
- Investigate the error performance of orthogonal signals in noise.

▶ 8.1 INTRODUCTION

The previous chapter described transmitting a single data bit over a noisy channel by encoding the bit value using a pair of complementary signals. Some data transmission applications need to service multiple users over the same channel simultaneously. An analogous situation arises when one user wants to transmit multiple data bits at one signal transmission time. This chapter considers multiple orthogonal signals that can be transmitted simultaneously over a channel and decoded by separate receivers using matched processors. The concept of orthogonality is described in the time, frequency, and code domains.

The main topics in this chapter are covered in the following sequence.

Multiple simultaneous user systems—Typical data communication systems serve more than one user simultaneously. Wi-Fi systems connect multiple smartphones and laptops to the Internet, and multiple users talk over a cell-phone network simultaneously. This multi-user system also allows a single user to transmit multiple bits simultaneously.

Orthogonal data signals—The conventional processor that operates on received signals forms the sum of weighted signal values. This processing suggests the *orthogonality condition* that prevents signals having the same energy from interfering with each other. Orthogonal signal pairs transmit the binary data produced by multiple users or multiple bits produced by one user over a channel simultaneously.

Orthogonal signal types—Signals can be orthogonal in time, in frequency, and in code word. Each system has its advantages and disadvantages with each application choosing the type that meets its specifications.

Processing orthogonal signals in noise—When a matched processor operates on orthogonal signals that are corrupted with noise, the reliability of detecting the correct (transmitted) signal value is determined solely by the signal-to-noise ratio (SNR).

Probability of error—A simulation procedure is described to estimate the probability of error by performing multiple transmissions in the presence of noise and counting the number of errors.

▶ 8.2 MULTIPLE SIMULTANEOUS USER SYSTEMS

Some digital communication systems need to transmit more than one bit value per signal or to service multiple users, allowing each to transmit and receive data or to *access* the system simultaneously. This additional consideration leads to the need for orthogonal signals, which can be transmitted over a data channel at the same time without interfering with each other. Figure 8.1 shows the extension from a single signal pair to multiple orthogonal signal pairs described in this chapter. Orthogonal signals permit two or more users to communicate over the same channel simultaneously or to allow one user to send multiple data bits in one signal transmission time.

To illustrate the multi-user problem, imagine you are at a party where there is loud music playing and other people are talking while you are trying to carry on a conversation with a friend. There are features in your friend's utterances that allow you to separate his/her voice from the other sounds, even if the other sounds are louder. A similar method for extracting desired signals is being used by another pair of friends in the same party for their conversation. The same approach is employed in simultaneous digital communications between pairs of users (for example, your cell phone and the local transmission tower). A typical cell contains many phones that are communicating with the tower at any one time. Signals having features that allow them to be extracted in the presence of other signals are called *orthogonal*.

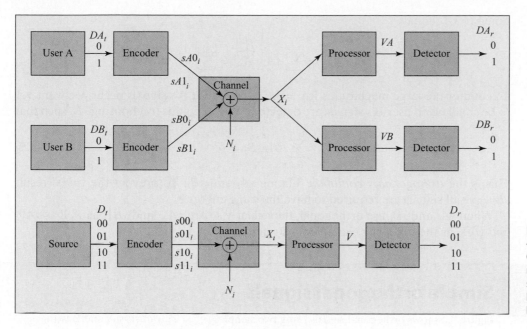

Figure 8.1

Orthogonal signals permit two or more users to communicate over the same channel simultaneously or allow one user to send multiple data bits in one signal transmission time.

8.3 ORTHOGONALITY CONDITION

To illustrate orthogonal signals, we consider the signal transmitted by a single user (user A) who designs data sequence sA_i for $i = 0, 1, 2, \ldots, n_x - 1$ to form complementary signals

$$sA0_i = -sA_i \text{ if } DA_t = 0 \quad \text{and} \quad sA1_i = sA_i \text{ if } DA_t = 1 \tag{8.1}$$

When transmitted over a channel that introduces additive noise N_i for $i = 0, 1, 2, \ldots,$ $n_x - 1$, the signal detected at A's receiver is

$$X_i = sA0_i + N_i = -sA_i + N_i \quad \text{if} \quad sA0_i \text{ is transmitted} \tag{8.2}$$
$$X_i = sA1_i + N_i = sA_i + N_i \quad \text{if} \quad sA1_i \text{ is transmitted}$$

Under ideal (noiseless) conditions, the received signal X_i is simply one of the transmitted signals. This chapter assumes that the amplitudes do not decrease as the signals pass through the channel. The matched processor tuned to sA_i has coefficients matched to $c_i = sA_i$. When $X_i = sA0_i$ for $i = 0, 1, 2, \ldots, n_x - 1$, A's matched processor produces output as

$$VA_{|X=sA0} = \sum_{i=0}^{n_x-1} c_i X_i \tag{8.3}$$

$$= \sum_{i=0}^{n_x-1} sA_i \overbrace{X_i}^{=sA0_i=-sA_i}$$

$$= -\sum_{i=0}^{n_x-1} (sA_i)^2 = -\mathcal{E}_{sA}$$

Similarly, when $X_i = sA1_i$, $VA_{|X=sA1} = \mathcal{E}_{sA}$.

A second user (user B) designs data signal sB_i for $i = 0, 1, 2, \ldots, n_x - 1$ and forms complementary signals $sB0_i$ and $sB1_i$ in a similar manner. When user B transmits a signal, say $sB1_i = sB_i$, over a channel shared by A and B, the output of A's processor tuned to sA_i equals

$$VA_{|X=sB1} = \sum_{i=0}^{n_x-1} sA_i \overbrace{X_i}^{=sB1_i} = \sum_{i=0}^{n_x-1} sA_i \, sB_i \qquad (8.4)$$

To counter the large magnitudes $VA = \pm\mathcal{E}_{sA}$ when user A's signals occur, we want VA to be small when user B's signals are detected. We can do this by choosing sB_i such that

$$\sum_{i=0}^{n_x-1} sA_i \, sB_i = 0 \qquad (8.5)$$

This is the *orthogonality condition* relating sA_i and sB_i. To prevent the trivial result $sB_i = 0$, all signals are required to have the same energy $\mathcal{E}_s$.

When sA_i and sB_i are orthogonal, then clearly so are $sA1_i$ and $sB1_i$ as well as $sA0_i$ and $sB0_i$, as the following example illustrates.

EXAMPLE 8.1 Simple orthogonal signals

Figure 8.2 shows orthogonal signals using two samples ($n_x = 2$) for users A and B with

$$sA_0 = 1 \quad \text{and} \quad sA_1 = 1$$

and

$$sB_0 = 1 \quad \text{and} \quad sB_1 = -1$$

The energy in each signal is

$$\mathcal{E}_{sA} = \sum_{i=0}^{1} sA_i^2 = 1 + 1 = 2$$

$$\mathcal{E}_{sB} = \sum_{i=0}^{1} sB_i^2 = 1 + (-1)^2 = 2$$

Hence, both signals contain the same energy.

Let us verify that sA_i and sB_i are orthogonal signals. The orthogonality condition requires that

$$\sum_{i=0}^{n_x-1} sA_i \, sB_i = 0$$

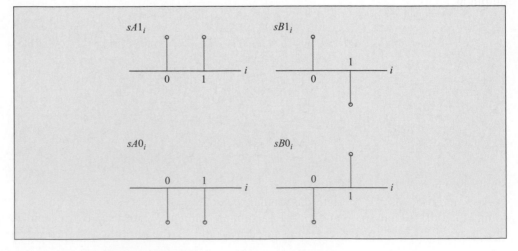

Figure 8.2

Simple orthogonal signal pair. All signal magnitudes are equal to one.

Inserting the values of $n_x = 2$ and these signals, we get

$$\sum_{i=0}^{1} sA_i \, sB_i = (sA_0 \, sB_0) + (sA_1 \, sB_1) = (1 \times 1) + (1 \times -1) = 0$$

Hence, sA_i and sB_i are orthogonal.

We now examine the values produced by user A's matched processor. The VA processor output is given by

$$VA = \sum_{i=0}^{n_x-1} sA_i \, X_i = \overbrace{sA_0}^{=1} X_0 + \overbrace{sA_1}^{=1} X_1 = X_0 + X_1$$

When $X_i = sA1_i = sA_i$,

$$VA_{|X=sA1} = sA_0 + sA_1 = 1 + 1 - 2 \quad (= \mathcal{E}_{sA})$$

When $X_i = sA0_i = -sA_i$,

$$VA_{|X=sA0} = -sA_0 + -sA_1 = -1 - 1 = -2 \quad (= -\mathcal{E}_{sA})$$

When $X_i = sB1_i = sB_i$,

$$VA_{|X=sB1} = sB_0 + sB_1 = 1 + (-1) = 0$$

Similarly, $VA_{|X=sB0} = 0$.

We now examine the values produced by user B's matched processor. The VB processor output is given by

$$VB = \sum_{i=0}^{1} sB_i \, X_i = (\overbrace{1}^{=sB_0} \times X_0) + (\overbrace{-1}^{=sB_1} \times X_1) = X_0 - X_1$$

When $X_i = sA1_i = sA_i$,

$$VB_{|X=sA1} = sA_0 - sA_1 = 1 - 1 = 0$$

Similarly, $VB_{|X=sA0} = 0$.

When $X_i = sB1_i = sB_i$,

$$VB_{|X=sB1} = sB_0 - sB_1 = 1 - (-1) = 2$$

Similarly, $VB_{|X=sB0} = -2$.

Hence, each processor identifies its own signal and ignores the other.

Single-User, Multi-Bit Transmission

The previous discussion considered two users, A and B, transmitting their separate binary data over a single channel. When there is only one user transmitting data over a channel, orthogonal sequences can form complementary signals to transmit *two data bits during one transmission time*. For example, user A designs orthogonal signals $sA0_i$ and $sA1_i$ and forms their complementary pairs. User A combines two data bits into a *two-bit* packet (called a dibit), DA_t, and transmits each packet using the assignment:

$$
\begin{array}{llll}
sA00_i = -sA0_i & \text{if} & DA_t = 00 & \qquad (8.6) \\
sA01_i = sA0_i & \text{if} & DA_t = 01 & \\
sA10_i = -sA1_i & \text{if} & DA_t = 10 & \\
sA11_i = sA1_i & \text{if} & DA_t = 11 &
\end{array}
$$

Clearly, additional orthogonal signals can service additional users or can transmit additional bits simultaneously over a channel. Current data transmission systems do this.

Practical Orthogonal Signals

What makes this problem interesting and challenging is that, unlike complementary signals that form a unique and obvious pair, there are many ways to form orthogonal signals sA_i and sB_i. Consider a channel time interval

$$T_c = n_x T_s \tag{8.7}$$

over which the data sequences occur. There are three common approaches for forming orthogonal signals, with each having its own merits.

Orthogonal in time—in *time-division multiple access (TDMA)* systems, the signals sA_i and sB_i occur during specific, non-overlapping, sub-intervals in T_c. At times other than the specified time slot, a user's signals are zero-valued in order to not interfere with signals of other users.

Orthogonal in frequency—in *frequency-division multiple access (FDMA)* systems, the signals sA_i and sB_i occur during the entire interval T_c but occupy separate frequency bands.

Orthogonal in code—in *code-division multiple access (CDMA)* systems, the signals sA_i and sB_i are *binary* (taking on values equal to either $-V_{max}$ or $+V_{max}$) and occupy the entire interval T_c. CDMA signals have codes that may not be *exactly* orthogonal and that still can be *adequately differentiated* through their *approximately* orthogonal properties.

Theoretically, each technique has the same effectiveness, but each also has practical advantages and disadvantages. The problems and trade-offs encountered are specific to a given application and ultimately determine the appropriate method. We illustrate each technique using a pair of users, A and B, but the methods can be extended to more users with the maximum number of users equal to the maximum number of available orthogonal signals that still operate with sufficient probability of error in noisy channels.

8.4 SIGNALS ORTHOGONAL IN TIME—TDMA

Signals that are orthogonal in time occupy non-overlapping time slots in the channel data interval $0 \le i \le n_x - 1$ using a technique commonly called *time-division multiplexing (TDMA)*. If there are n_u users, each has a slot duration

$$n_d = \frac{n_x}{n_u} \tag{8.8}$$

Typically, n_d is an integer value by choosing n_x to accommodate n_u users. For example, to serve three users, choose $n_x = 15$ to make $n_d = 5$; to serve four, choose $n_x = 16$ to make $n_d = 4$. The system also assigns each user to a specific time slot.

Both signals sA_i and sB_i can have any form—even identical forms—because they occur at different time intervals. Figure 8.3 shows two different signal forms to differentiate the two more clearly.

EXAMPLE 8.2 TDMA signals

Figure 8.3 shows two users, A and B, using signals sA_i and sB_i respectively, that simultaneously access the same channel for transferring data. Each user forms two signals that occupy non-overlapping time intervals within the n_x data interval divided into two sub-intervals. The figure shows signals having duration $n_x = 32$, making

$$n_d = \frac{n_x}{n_u} = \frac{32}{2} = 16$$

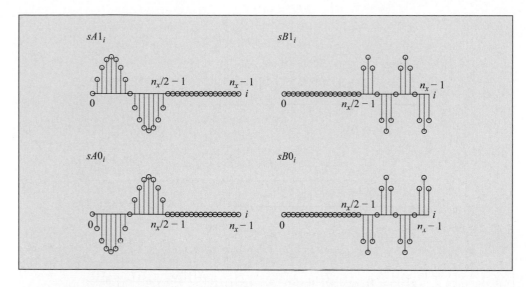

Figure 8.3

TDMA signals.

User A occupies the first half of the interval with signal sA_i for $0 \leq i \leq n_d - 1$. Although user A can use any signal, Figure 8.3 shows a single sinusoidal period

$$sA_i = \sin(2\pi f_A i) \text{ for } 0 \leq i \leq n_d - 1$$

$$= 0 \text{ for } n_d \leq i \leq n_x - 1$$

with $f_A = 1/n_d$ to form a single period. User A uses $sA1_i = sA_i$ for transmitting data $DA_t = 1$ and its complementary signal $sA0_i = -sA_i$ for transmitting $DA_t = 0$.

User B uses sB_i equal to two periods of a sinusoid during the second half of the interval, as

$$sB_i = 0 \text{ for } 0 \leq i \leq n_d - 1$$

$$= \sin(2\pi f_B i) \text{ for } n_d \leq i \leq n_x - 1$$

with $f_B = 2/n_d$ to form two periods. User B uses $sB1_i = sB_i$ for transmitting data $DB_t = 1$ and its complementary signal $sB0_i = -sB_i$ for transmitting $DB_t = 0$.

The energy of the two signals is computed by summing the squares of the non-zero components, producing

$$\mathcal{E}_{sA} = \sum_{i=0}^{n_x-1} sA_i^2 = \sum_{i=0}^{n_d-1} \sin^2(2\pi i/n_d) \quad (= 8 \text{ when } n_d = 16) \tag{8.9}$$

$$\mathcal{E}_{sB} = \sum_{i=0}^{n_x-1} sB_i^2 = \sum_{i=n_d}^{n_x-1} \sin^2(4\pi i/n_d) \quad (= 8 \text{ when } n_d = 16) \tag{8.10}$$

Figure 8.4a shows the signal sequence user A uses to transmit $DA_t = 1$ followed by $DA_t = 0$, and Figure 8.4b shows user B's signals to transmit a 1 followed by a 0. Each n_x interval contains the signals from both users. User A has been assigned the first half of each n_x data interval, and user B has been assigned the second half.

Figure 8.4c shows the composite signal on the channel that contains data transmitted by both users. The channel signal becomes the signal detected by both user A and user B receivers and is equal to

$$X_i = sAT_i + sBT_i \text{ for } i = 0, 1, 2, \ldots, n_x - 1 \tag{8.11}$$

where $sAT_i = sA0_i$ when $DA_t = 0$ is transmitted or $sAT_i = sA1_i$ when $DA_t = 1$ is transmitted. This is similar for sBT_i. Even though X_i looks unfamiliar, the two receivers determine their binary values through their respective matched processor values VA and VB. Each matched processor produces outputs $\mathcal{E}_s$ (for data = 1) or $-\mathcal{E}_s$ (for data = 0) to identify its complementary signals.

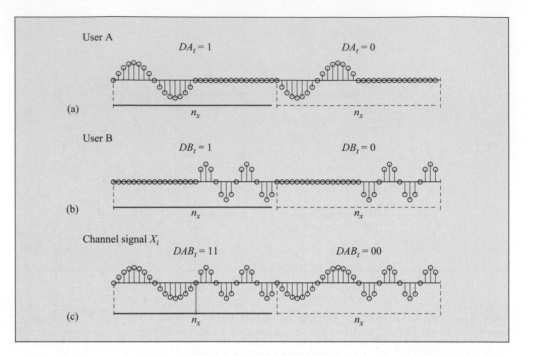

Figure 8.4

TDMA signals: (a) User A transmitting data $DA_t = 1$ followed by $DA_t = 0$. (b) User B transmitting $DB_t = 1$ then $DB_t = 0$. (c) X_i is sum of signals observed on the channel.

VA: Processor evaluates X_i for $i = 0, 1, 2, \ldots, n_x - 1$. For each n_x time interval with VA processor tuned to sA_i we have

$$VA = \sum_{i=0}^{n_x-1} sA_i\, X_i \tag{8.12}$$

When $X_i = sA1_i$,

$$VA_{|X=sA1} = \sum_{i=0}^{n_x-1} sA_i^2 = \mathcal{E}_s \tag{8.13}$$

When $X_i = sBT_i$ (either $sB1_i$ or $sB0_i$), the orthogonality condition produces

$$VA_{|X=sBT} = \sum_{i=0}^{n_x-1} sA_i\, sBT_i = 0 \tag{8.14}$$

VB: Processor evaluates X_i for $i = 0, 1, 2, \ldots, n_x - 1$ and for each n_x time interval assuming $X_i = sB1_i$. When $X_i = sB1_i (= sB_i)$,

$$VB_{|X=sB1} = \sum_{i=0}^{n_x-1} sB_i^2 = \mathcal{E}_s \tag{8.15}$$

When $X_i = sA1_i$, the orthogonality condition produces

$$VB_{|X=sA1} = \sum_{i=0}^{n_x-1} sB_i\, sA1_i = 0 \tag{8.16}$$

Figure 8.5 shows the outputs of the VA and VB processors with X_i equal to the composite channel signal after each of the two n_x data sequences are detected. A large positive processor value indicates that the signal corresponds to the processor's tuned coefficients. In the case shown for sinusoidal signals $\mathcal{E}_s = n_d/2 = n_x/4$.

The VA processor detects the $sA1_i$ TDMA signal in the first $n_x = 32$ data interval by producing a large-magnitude output ($= n_x/4 = 8$) at the end of the interval when the complete signal is within the processor. In user A's TDMA matched processor having 32 coefficients, half the coefficients are zero in order to ignore user B's TDMA signals. The

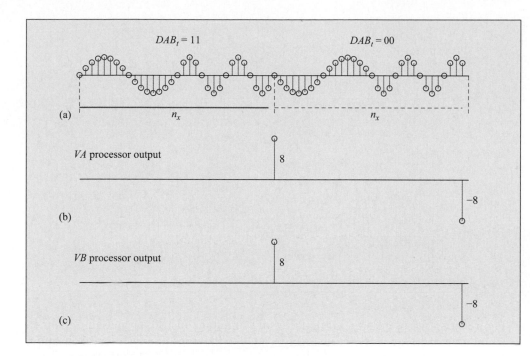

Figure 8.5

Processing TDMA signals. (a) The composite channel signal. (b) The *VA* matched-processor outputs. (c) The *VB* matched-processor outputs. The processor outputs are shown after each of the two n_x data sequences are detected and processed.

composite channel signal also contains $sB0_i$ in the first n_x interval, and the *VB* matched processor detects it by producing a large-magnitude negative value at the end of the interval. The second n_x interval contains $sA1_i$ and $sB1_i$. Both the *VA* and *VB* matched processors detect their respective TDMA signals by producing large output magnitudes at the end of the second interval.

Maximum Number of TDMA Users

The number of users in a TDMA system is limited by the number of orthogonal signals available. Let T_c, or equivalently n_x, be the channel time interval that contains the signal sequences for data transmission that is subdivided among the users. The number of users n_u employing strictly TDMA techniques is limited by the sample time T_s. Thus,

$$n_{u,\max} \le \frac{T_c}{T_s} = \frac{n_x T_s}{T_s} = n_x \tag{8.17}$$

This number can be increased by increasing n_x, but this slows down the entire system.

8.5 SIGNALS ORTHOGONAL IN FREQUENCY—FDMA

In FDMA systems, signals with predetermined frequencies are assigned to specific users. For a practical system, three considerations occur.

1. Each frequency must produce unique signals, which may present a problem in sampled-data systems.

2. To be orthogonal, the frequencies need to be harmonically related.

3. The code length n_x must encode a complete period of the fundamental frequency.

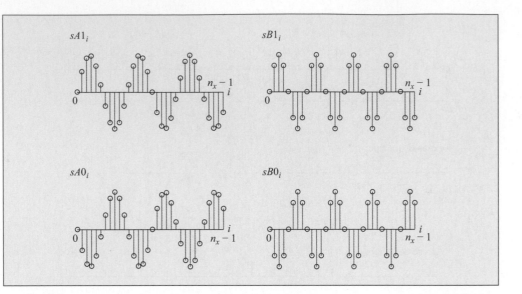

Figure 8.6

FDMA signals with time interval $n_x (= 32)$ and using sampling period T_s. Orthogonal signals for user A is at the third harmonic and for user B at the fourth harmonic.

Figure 8.6 shows users A and B that transmit simultaneously over the same channel. The fundamental frequency f_1 is inversely related to the transmission interval duration, $n_x T_s$, or

$$f_1 = \frac{1}{n_x T_s}$$

where T_s is the sampling period, making $T_c = n_x T_s$ the channel bit signal transmission interval.

EXAMPLE 8.3 — Data interval size and duration with fundamental frequency

With $T_s = 10^{-5}$ s and $n_x = 32$, the signal interval is

$$T_c = n_x T_s = 32 \times 10^{-5} \, \text{s} = 0.32 \, \text{ms}$$

$$f_1 = \frac{1}{n_x T_s} = \frac{1}{32 \times 10^{-5} \, \text{s}} = 3,125 \, \text{Hz}$$

Figure 8.6 shows user A is assigned frequency f_A, which is a harmonic of f_1, to be equal to

$$f_A = 3 f_1 \tag{8.18}$$

and is used to generate

$$sA_i = \sin(2\pi f_A i T_s) \quad \text{for } i = 0, 1, 2, \ldots, n_x - 1$$

User A uses $sA1_i = sA_i$ for transmitting data $DA_t = 1$ and its complementary signal $sA0_i = -sA_i$ for transmitting $DA_t = 0$.

User B is assigned frequency f_B, which is another harmonic of f_1, to be equal to

$$f_B = 4 f_1 \tag{8.19}$$

and is used to generate

$$sB_i = \sin(2\pi f_B i T_s) \quad \text{for } i = 0, 1, 2, \ldots, n_x - 1$$

User B uses $sB1_i = sB_i$ for transmitting data $DB_t = 1$ and its complementary signal $sB0_i = -sB_i$ for transmitting $DB_t = 0$.

When $sA0_i$ is evaluated by the matched processor tuned to sA_i, its output equals

$$VA_{|X=sA0} = \sum_{i=0}^{n_x} sA_i \overbrace{X_i}^{=sA0_i} = -\sum_{i=0}^{n_x} sA_i^2 = -n_x/2 \qquad (8.20)$$

For example, for $sA0_i = -\sin(6\pi i f_1 T_s)$ for $i = 0, 1, 2, \dots, n_x - 1$ ($n_x = 32$). we have

$$VA_{|X=sA0} = -\sum_{i=0}^{15} \sin^2(6\pi i f_1 T_s) = -16 \qquad (8.21)$$

When $X_i = sA1_i$ is processed, the VA output equals

$$VA_{|X=sA1} = \sum_{i=0}^{n_x} sA_i \overbrace{X_i}^{=sA1_i} = \sum_{i=0}^{n_x} sA_i^2 = 16 \qquad (8.22)$$

Similarly, user B uses the same $n_x = 32$ interval but is assigned frequency $f_B - 4f_1$. Thus,

$$sB1_i = \sin(2\pi i f_B T_s) = \sin(8\pi i f_1 T_s) \text{ for } i = 0, 1, 2, \dots, n_x - 1 \qquad (8.23)$$

transmits a 1 data value and complementary signal $sB0_i$ transmits a 0. Similar matched-processor results occur for user B's signals.

To show sA_i and sB_i are orthogonal, we compute the output of the matched processor tuned to sA_i when $X_i = sB1_i (= sB_i)$. Thus,

$$VA_{|X=sB1} = \sum_{i=0}^{n_x} sA_i \overbrace{X_i}^{=sB1_i} = \sum_{i=0}^{n_x} sA_i sB_i = 0 \qquad (8.24)$$

because harmonically related sinusoidal sequences are orthogonal.

We can examine the system operation when both sources transmit their signals over the channel simultaneously. The signal is detected by both matched processors that belong to users A and B. The channel signal is given by

$$X_i = sAT_i + sBT_i \text{ for } i = 0, 1, 2, \dots, n_x - 1 \qquad (8.25)$$

where $sAT_i = sA0_i$ when $DA_t = 0$ is transmitted or $sAT_i = sA1_i$ when $DA_t = 1$ is transmitted. This is similar for sBT_i. Figure 8.7 shows the signals that are produced separately by user A producing two-bit data $DA_t = 10$ and user B producing $DB_t = 10$. The resulting composite signal that appears on the channel is shown. Even though X_i looks unfamiliar, the matched processors can determine their transmitted binary values.

Figure 8.8 shows the outputs of the two matched processors with X_i equal to the composite channel signal after each of the two $n_x = 32$ data sequences are detected. A large positive matched-processor output value ($V = n_x/2 = 16$) indicates the presence of the signal corresponding to the processor's expected value.

The VA processor detects the $sA1_i$ FDMA signal in the first $n_x = 32$ data interval by producing a large positive output ($= n_x/2 = 16$) at the end of the interval (when the complete signal is within the processor). The composite channel signal also contains $sB1_i$ in the first n_x interval, and the VB matched processor detects it by producing a large positive value at the end of the interval. The second n_x interval contains $sA0_i$ and $sB0_i$. Both the VA and VB matched processors detect their respective FDMA signals by producing large negative output values at the end of the second interval.

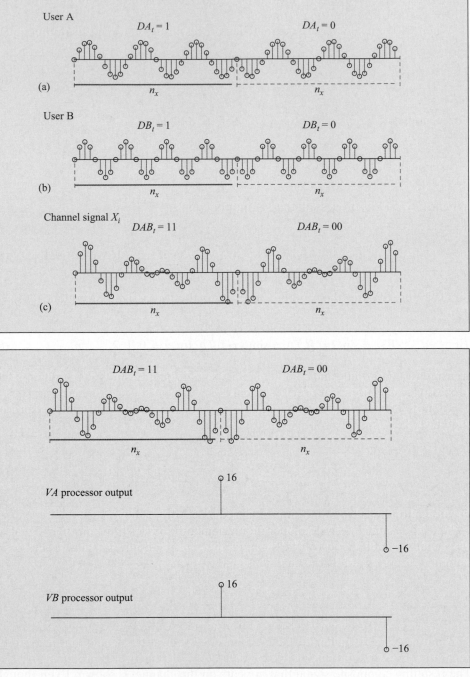

Figure 8.7

FDMA signals from two users A and B and the resulting channel signal. (a) User A transmits DA_t = 10. (b) User B transmits DB_t = 10. The channel data interval n_x = 32. (c) The sum of signals observed on channel (= X_i).

Figure 8.8

Processing FDMA signals. The outputs of the processors with $n_x = 32$ with X_i equal to the summed channel signal (top signal). The four processor outputs are shown after each of the two n_x data sequences are detected. A positive value, typically $n_x/2$, indicates that the signal corresponding to that expected by the processor has been observed.

Maximum Number of FDMA Users

Sampling the waveforms with sampling period T_s restricts the allowable frequencies to a finite number of harmonically related frequencies. To prevent aliasing, the maximum harmonic that can be sampled is

$$f_{\max} < \frac{f_s}{2} = \frac{1}{2T_s} \tag{8.26}$$

The number of valid harmonics n_{harm} equals

$$n_{\text{harm}} < \frac{f_{\max}}{f_1} = \frac{\dfrac{1}{2T_s}}{\dfrac{1}{n_x T_s}} = \frac{n_x}{2} \tag{8.27}$$

That is, only the first $n_x/2-1$ harmonics can be used. Higher harmonics result in aliasing, causing the alias values to equal those produced by one of the original harmonics.

Hence, the maximum number of users equals

$$n_{\text{users,max}} \le \frac{n_x}{2} - 1 \qquad (8.28)$$

Valid and invalid harmonics — EXAMPLE 8.4

Figure 8.9 illustrates the problem of aliasing that limits the number of harmonics that can be used in sampled-data systems. Each sequence contains $n_x = 16$ points.

With $T_s = 10^{-5}$ s, the interval duration is

$$n_x T_s = 16 \times 10^{-5} \text{ s} = 1.6 \times 10^{-4} \text{ s}$$

The fundamental frequency equals

$$f_1 = \frac{1}{n_x T_s} = \frac{1}{1.6 \times 10^{-4}} = 6{,}250 \text{ Hz}$$

The sampling frequency equals

$$f_s = \frac{1}{T_s} = \frac{1}{10^{-5} \text{ s}} = 10^5 \text{ Hz}$$

The maximum frequency that limits the harmonic frequency is

$$f_{\text{max}} < \frac{f_s}{2} = 50{,}000 \text{ Hz}$$

The number of allowed harmonics equals

$$n_{\text{harm}} < \frac{f_{\text{max}}}{f_1} = \frac{50{,}000}{6{,}250} = 8 \quad \left(= \frac{n_x}{2} \right)$$

Hence, only the first seven harmonics

$$f_k = k f_1, \quad \text{for } 1 \le k \le \left(\frac{n_x}{2} \right) - 1$$

can be used.

For example, consider three frequencies: the fundamental frequency, the largest valid harmonic, and a higher non-valid harmonic for $\sin(2\pi f i T_s)$ for $i = 0, 1, 2, \ldots, n_x - 1 (n_x = 16)$.

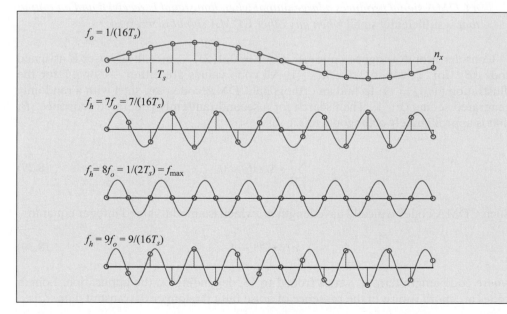

Figure 8.9

Valid and invalid harmonics for FDMA. Valid harmonics are $f_k = k f_1$ for $1 \le k \le (n_x/2) - 1$.

The top curve shows the fundamental frequency

$$f_1 = \frac{1}{16T_s} = 6{,}250\,\text{Hz}$$

The second curve shows the highest valid harmonic $f_k = kf_1$ for $k = (n_x/2) - 1 = 7$ or

$$f_7 = 7f_1 = \frac{7}{16T_s}$$

The third curve shows that the $k = 8$ harmonic, $f_8 = 1/(2T_s)$, is not valid because it produces only zero values. Hence, $\mathcal{E}_s = 0$.

The fourth curve shows a higher non-valid harmonic

$$f_9 = 9f_1 = \frac{9}{16T_s}$$

Note that these values are complements of those already used for f_7. Hence, f_9 is not a valid harmonic.

▶ 8.6 SIGNALS WITH ORTHOGONAL CODES—CDMA

A third common method of transmitting multiple data signals simultaneously over the same channel is to assign a unique *code* to each signal in a way that each signal can be recognized in the presence of the others, which is called *code division multiple assess (CDMA)*.

The CDMA approach has two features that distinguish it from TDMA and FDMA systems.

- CDMA signals are binary valued and consist solely of the channel-limiting values $-V_{\max}$ and $+V_{\max}$. Chapter 7 showed that such binary signals maximize signal energy $\mathcal{E}_s$.

- CDMA signals are formed to be *approximately orthogonal* (that is, they are *sufficiently orthogonal* to result in a practical system). *The matched processor tuned to a CDMA signal produces a large output when that signal is received and an output that is* sufficiently small *when any other CDMA signal is received.*

Consider user A assigned code sA_i for $i = 0, 1, 2, \ldots, n_x - 1$ and user B assigned code sB_i for $i = 0, 1, 2, \ldots, n_x - 1$. All code values are either -1 or $+1$ for the illustration ($V_{\max} = 1$). To find an orthogonal CDMA code pair, start with a randomly generated sequence sA_i. Then search for a second randomly generated sequence sB_i that is approximately orthogonal to sA_i, as

$$\sum_{i=0}^{n_x-1} sA_i \, sB_i \approx 0 \tag{8.29}$$

Such CDMA codes typically have length n_x, which is an odd-valued integer equal to

$$n_x = 2^{n_c} - 1 \tag{8.30}$$

where code length term n_c varies from 2 to 15, depending on the application. Longer codes are more robust in the presence of noise but take longer to transmit data.

The matched processor tuned to sA_i produces the maximum output value

$$VA_{|X=sA} = \sum_{i=0}^{n_x-1} sA_i^2 = \sum_{i=0}^{n_x-1} 1 = \overbrace{1 + 1 + \cdots + 1}^{n_x \text{ values}} = n_x \quad (= \mathcal{E}_s) \qquad \textbf{(8.31)}$$

Note that this $\mathcal{E}_s$ is twice the energy value for sinusoidal signals ($= n_x/2$). When all code samples have magnitude V_{max}, then

$$\mathcal{E}_s = n_x V_{max}^2 \qquad \textbf{(8.32)}$$

The next example indicates how to generate random CDMA codes using a pseudo-random number generator.

Generating random orthogonal codes — EXAMPLE 8.5

Consider a pseudo-random number generator that yields random numbers Y_i that are uniformly distributed over [0,1]. The rule that generates the random CDMA sequence sA_i for $i = 0, 1, 2, \ldots, n_x - 1$ with $n_x = 31$ is

$$\text{If } Y_i < 0.5 \rightarrow sA_i = -1$$

$$\text{If } Y_i \geq 0.5 \rightarrow sA_i = 1$$

We generate the complementary codes

$$sA0_i = -sA_i \quad \text{and} \quad sA1_i = sA_i \text{ for } i = 0, 1, 2, \ldots, n_x - 1$$

Figure 8.10 shows the values of $sA0_i$ and $sA1_i$. The matched-processor outputs for the sA_i signal give

$$VA_{|X=sA0} = \sum_{i=0}^{30} sA_i\, sA0_i = -\sum_{i=0}^{30} sA_i^2 = -31$$

$$VA_{|X=sA1} = \sum_{i=0}^{30} sA_i\, sA1_i = \sum_{i=0}^{30} sA_i^2 = 31$$

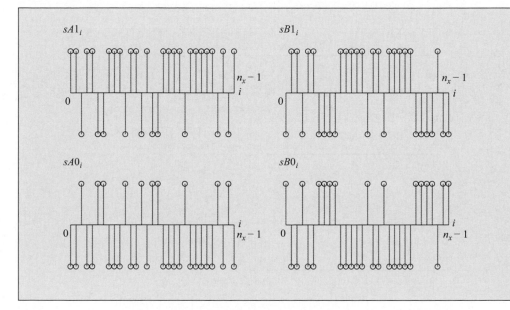

Figure 8.10

CDMA signals for $n_x = 31$. sA_i was chosen randomly to form $sA1_i = sA_i$ and $sA0_i = -sA_i$. sB_i was chosen randomly until $\epsilon \leq 1$ was found. Then, $sB1_i = sB_i$, and $sB0_i = -sB_i$.

We select sB_i in a similar random manner, but after each realization of an sB_i code, we check to determine if the orthogonality condition is met, as

$$\left| VA_{|X=sB} \right| = \left| \sum_{i=1}^{n_x-1} sA_i \, sB_1 \right| \leq \epsilon$$

A successful code has $\epsilon \leq 1$. This is a suitably low value for the matched-processor output when the other signal is detected.

If this approximately orthogonal condition is not satisfied, the search procedure for randomly selecting sB_i is repeated until the condition is met. In practice, this occurs within a reasonable number of repetitions. Figure 8.10 shows the values of $sB0_i$ and its complementary code $sB1_i$, which produced the matched processor output

$$\left| VA_{|X=sB} \right| = \left| \sum_{i=0}^{30} sA_i \, sB_i \right| \leq \epsilon$$

with $\epsilon = 1$.

The signals that have the maximum magnitude V_{max} are obtained by multiplying the ± 1 values by V_{max} to produce the signal energy $\mathcal{E}_s = n_x V_{max}^2 = 31 V_{max}^2$.

We can examine the system operation when both users A and B transmit their signals over the channel simultaneously. The signal detected by the receivers intended for users A and B equals

$$X_i = sAT_i + sBT_i \quad \text{for} \quad i = 0, 1, 2, \ldots, n_x - 1 \tag{8.33}$$

where $sAT_i = sA0_i$ when $DA_t = 0$ is transmitted or $sA_i = sA1_i$ when $DA_t = 1$ is transmitted. This is similar for sBT_i. Figure 8.11 shows the signals that are produced separately by user A transmitting data $DA_t = 10$ and user B transmitting $DB_t = 10$. With sA_i and sB_i containing only -1 and $+1$ values, the composite signal X_i can take on values -2, 0, and $+2$. Even though X_i looks unfamiliar, the two matched processors can determine their binary values.

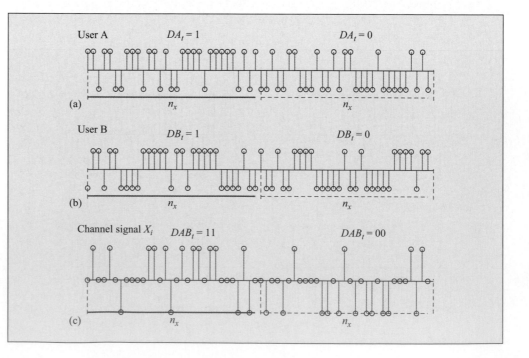

Figure 8.11

CDMA signals from users A and B and the resulting channel signal. (a) User A transmits 10. (b) User B transmits 10. (c) The sum of signals observed on channel (= X_i).

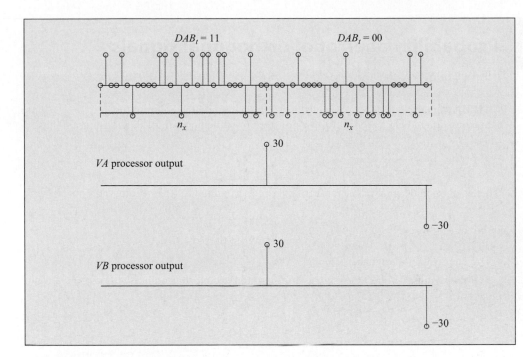

Figure 8.12

Processing CDMA signals. The outputs of the VA and VB processors with $n_x = 31$ with X_i equal to the summed channel signal. The processor outputs are shown after each of the two n_x data sequences are detected. A positive value, typically $V \approx n_x$, indicates that the signal expected by the processor has been observed.

Figure 8.12 shows the outputs of the VA and VB matched processors with X_i equal to the composite channel signal after each of the two n_x data sequences are detected. A large positive processor output indicates that the signal contains the CDMA code corresponding to the processor's tuned value. This output value is typically less than $n_x (= 31)$ because of the interference produced by the other CDMA signal.

The VA processor detects the $sA1_i$ CDMA signal in the first n_x data interval by producing a large positive output at the end of the interval (when the complete signal is within the processor). The composite channel signal also contains $sB1_i$ in the first n_x interval, and the VB matched processor detects it by producing a large positive value at the end of the interval. The second n_x interval contains $sA0_i$ and $sB0_i$. Both the VA and VB matched processors detect their respective CDMA signals by producing large negative output values at the end of the second interval.

Maximum Number of CDMA Users

If additional codes are necessary but those having this optimal quality are no longer available, the non-matching criterion can be relaxed to an acceptable value $|\epsilon| \ll n_x$, which is the signal energy and the value produced by the successful matched processor when $V_{\max} = 1$. Adding additional users requires longer codes, with the trade-off being slower system data transmission rates.

CDMA signals are composed of random binary values. Unlike FDMA systems that occur at specific frequency values, CDMA signals are composed of many frequency components. This feature is important for military applications because this technique, termed *spread-spectrum coding*, is difficult to detect. CDMA signals appear as noise and if detected can be difficult to jam using conventional methods.

⏵ 8.7 DETECTING ORTHOGONAL SIGNALS IN NOISE

This section describes a simulation procedure to estimate the Prob[error] by performing multiple transmissions in the presence of noise and counting the number of errors. The approach follows the methods explained in Section 7.6.

EXAMPLE **8.6**

Probability of error of orthogonal signals

This example compares TDMA, FDMA, and CDMA signals of length $n_x = 16$ shown in Figure 8.13.

TDMA signals: The TDMA signals use the maximum allowed amplitude V_{max} (= 1 V) to maximize signal energy. Signal sA_i is non-zero over the first half of the n_x interval. Thus,

$$sA_i = 1 \text{ for } 0 \le i \le 7$$
$$= 0 \text{ for } 8 \le i \le 15$$

The orthogonal signal sB_i is non-zero over the second half of the n_x interval. Thus,

$$sB_i = 0 \text{ for } 0 \le i \le 7$$
$$= 1 \text{ for } 8 \le i \le 15$$

The energy of both signals is $\mathcal{E}_s = 8$.

FDMA signals: The FDMA signals use first and second harmonics with $f_A = 1/16$ and $f_B = 2/16$. Thus,

$$sA_i = \sin(2\pi f_A i) \text{ for } 0 \le i \le 15$$

and

$$sB_i = \sin(2\pi f_B i) \text{ for } 0 \le i \le 15$$

The energy of both signals is $\mathcal{E}_s = 8$.

CDMA signals: The CDMA signals sA_i and sB_i have maximum V_{max} magnitudes (± 1) that were randomly generated until the orthogonality condition was found with $\epsilon = 0$. Hence,

$$\sum_{i=0}^{15} sA_i \, sB_i = 0$$

with CDMA energy $\mathcal{E}_s = 16$.

A simulation to estimate Prob[error] was performed using the following six steps.

STEP 1. The basic signal orthogonal sequences sA_i and sB_i for $i = 0, 1, 2, \ldots, n_x - 1$, were specified for TDMA, FDMA, and CDMA, and $\mathcal{E}_s$ was computed.

STEP 2. Signals $sA0_i = -sA_i$ and $sA1_i = sA_i$ and $sB0_i = -sB_i$ and $sB1_i = sB_i$ for $i = 0, 1, 2, \ldots, n_x - 1$ were formed.

STEP 3. The number of trials was set equal to 101. Each trial had a specified value for noise SD σ_N with

$$\sigma_N = n\sigma_o \text{ for } 0 \le n \le 100$$

where $\sigma_o = 0.1$ to produce σ_N values ranging from 0 to 10 and σ_N^2 from 0 to 100 ($\gg \mathcal{E}_s$).

STEP 4. For each trial, a large number n_t (= 10^5) of data transmissions were simulated.

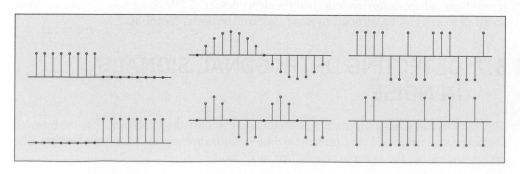

Figure 8.13

TDMA, FDMA, and CDMA orthogonal signals of length $n_x = 16$ and $V_{max} = 1$.

STEP 5. For each transmission, the following operations were performed.

    (a) Transmitted bits DA_t and DB_t were randomly chosen using the uniform PRNG, and the corresponding transmitted signals were formed as

$$sAT_i = sA0_i = -sA_i \ \text{ for } \ DA_t = 0$$
$$= sA1_i = sA_i \ \text{ for } \ DA_t = 1$$

and

$$sBT_i = sB0_i = -sB_i \ \text{ for } \ DB_t = 0$$
$$= sB1_i = sB_i \ \text{ for } \ DB_t = 1$$

    (b) The Gaussian random noise sequence N_i for $i = 0, 1, 2, \ldots, n_x - 1$ having the specified σ_N was generated.

    (c) The received signal sequence was formed as

$$X_i = sAT_i + sBT_i + N_i \ \text{ for } \ i = 0, 1, 2, \ldots, n_x - 1$$

    (d) Matched processors used X_i to produce outputs VA and VB.

    (e) VA and VB were applied to their respective threshold detectors with $\tau = 0$ to determine DA_r and DB_r.

    (f) The error counter N_e was incremented when $DA_r \neq DA_t$ and when $DB_r \neq DB_t$.

STEP 6. At the end of each trial, the probability of error was estimated as

$$\text{Prob[error]} = \frac{N_e}{2\,n_t}$$

The n_t is multiplied by 2, because two bits are transmitted with each X_i for $i = 0, 1, 2, \ldots, n_x - 1$.

This simulation produced the results shown in Figure 8.14. Note the following:

- Prob[error] increases as σ_N increases, as should be expected. The Prob[error] $\to 0.5$ for very large σ_N, because when the noise is large, it is the noise that determines the VA and VB values—no matter what signal was transmitted. Hence, half the time the V values produce data values that agree with those that were transmitted, being correct purely by chance.

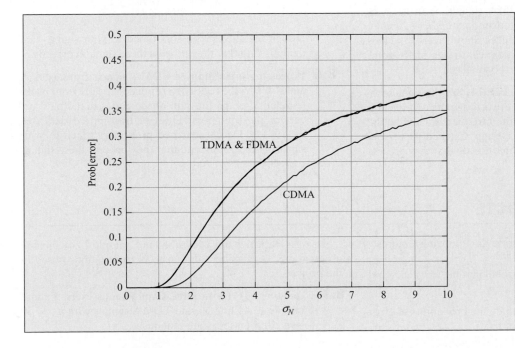

Figure 8.14

Probability of error versus σ_N using TDMA, FDMA, and CDMA orthogonal signals.

- The Prob[error] values for TDMA and FDMA are equal, because both signals have the same $\mathcal{E}_s$. The Prob[error] for CDMA is smaller, because its signals have $\mathcal{E}_s$ that is twice as large. This increased $\mathcal{E}_s$ for CDMA produces the same Prob[error] for a larger σ_N (by a factor of $\sqrt{2}$) than that for TDMA and FDMA.

- While FDMA produces a larger Prob[error] than CDMA because of its smaller $\mathcal{E}_s$, FDMA has the advantage of forming its orthogonal signals through direct computations rather than through a search procedure. Commercial systems use a specified set of CDMA codes, called *Kasami codes*.

8.8 Summary

This chapter described signal processing procedures to detect data signals in multi-access systems. A processor having coefficients tuned to the expected signal was described. Such a matched processor produces maximally different values when the signals encoding binary data 0 and 1 are complementary or negatives of each other. When multiple users access the communication channel simultaneously, each user's signals must be as different as possible (from the matched processor's point of view), and this is accomplished by making signals orthogonal. Three different methods to generate orthogonal signals were described, which included orthogonality in frequency, time, and code. The idea employed for serving several simultaneous users also can be applied to a single user transmitting more than one binary data value per transmission.

8.9 Problems

8.1 **Orthogonal signals.** If $sA1_i$ and $sB1_i$ are orthogonal, show that their complementary signals $sA0_i$ and $sB0_i$ are also orthogonal.

8.2 **Orthogonal signal design in TDMA.** Specify signals sA_i and sB_i for $0 \le i \le 7$ that are orthogonal over time. Design matched processors tuned to your signals and demonstrate that they process these signals correctly.

8.3 **Number of orthogonal FDMA signals.** With $n_x = 8$, compute n_{max}, which is the maximum number of unique orthogonal sinusoidal sequences. Show the $n_{max} + 2$ harmonic has values that are identical to one of the complementary signals in the orthogonal set.

8.4 **Orthogonal signal design in FDMA.** Specify signals sA_i and sB_i for $0 \le i \le 7$, that are orthogonal in frequency. Let $T_s = 10^{-4}$ s and specify their frequency values in the range $0 \le f \le 5{,}000$ Hz. Design matched processors tuned to your signals and demonstrate that they process these signals correctly.

8.5 **Orthogonal signal design in CDMA.** Specify signals sA_i and sB_i for $0 \le i \le 4$ having values equal to either ± 1 that are approximately orthogonal in code. First, flip a coin (tail $\to -1$, and head $\to +1$) to specify the values for sA_i. Then do the same repeatedly to generate sB_i until

$$\epsilon = \left| \sum_{i=0}^{4} sA_i sB_i \right| \le 1$$

Design matched processors tuned to your signals and demonstrate that they process these signals correctly.

8.6 **Processing superimposed CDMA signals from users A and B.** Using the codes determined in the previous problem, let X_i equal the sum of the codes that would occur for the four possibilities of transmitted data (00, 01, 10, 11). Design matched processors tuned to your signals and demonstrate that they process these signals correctly.

8.10 Excel Projects

The following Excel projects illustrate the following features:

- Generate signals that are orthogonal in time, frequency and code.

- Compose a VBA Macro to estimate the probability of error by performing a large number of trials.

Using a Narrative box (described in Example 13.4), include observations, conclusions, and answers to questions posed in the projects.

8.1 **Extending TDMA to three simultaneous users.** Extend Example 13.44 to generate TDMA signals with $n_X = 30$ to serve three users simultaneously.

8.2 Extending FDMA to three simultaneous users. Extend Example 13.44 to generate FDMA signals with $n_X = 32$ to serve three users simultaneously. How many unique FDMA are possible with with $n_X = 32$?

8.3 Extending CDMA to three simultaneous users. Extend Example 13.44 to generate CDMA signals with $n_X = 32$ to serve three users simultaneously. (Hint: Compute three cross-product sums $1 - 2, 1 - 3, 2 - 3$).

8.4 Extending CDMA to three simultaneous users. Extend Example 13.44 to generate CDMA signals with $n_X = 32$ to serve three users simultaneously. (Hint: Compute three cross-product sums $1 - 2, 1 - 3, 2 - 3$).

8.5 Histogram of orthogonal signals in noise. Using Example 13.32 as a guide, generate a histogram of one thousand matched processor output values V when processing signals in the presence of noise. Choose your favorite orthogonal complementary signal pair with $n_X = 16$ and a signal-to-noise ratio $\mathcal{E}_S/\sigma_N^2 = 2$. Repeat for $\mathcal{E}_S/\sigma_N^2 = 4$.

8.6 Estimating probability of orthogonal signals. Modify Examples 13.45 to compute the probability of error as σ_N increases between $0.1\mathcal{E}_s$ and $10\mathcal{E}_s$ by factor $\sqrt{2}$, using one of the orthogonal signal with $n_X = 30$ assigned to one of the users for s_i in the worksheet.

8.7 Comparing probability of orthogonal signals. Modify Examples 13.45 to compare the probability of error as σ_N increases between $0.1\mathcal{E}_s$ and $10\mathcal{E}_s$ by factor $\sqrt{2}$ for the TDMA, FDMA, and CDMA signals with maximum amplitude $|s_i| \leq 10$ with $n_X = 30$ assigned to one of the users for s_i in the worksheet. Which signal type performs best, and which is the worst? Explain why.

SOURCE CODING

► 9.1 INTRODUCTION

This chapter describes source coding techniques that measure the information content of the symbols produced by a source. In many cases, the quantity of data produced by a source can be quite large, often many megabytes or even gigabytes. This chapter illustrates how data compression reduces the quantity of data without any loss of information. A simple version of encryption is described for securing important data.

Previous chapters described how binary data are generated and how the signals encoding them are processed to extract data in the presence of noise. This chapter describes mathematical models of information generation by a source to compute the informational content of data in order to understand how to reduce the data size without affecting the informational content.

The main topics describing source encoding are covered in the following sequence.

Data compression—Data compression reduces the size of data that needs to be transmitted or stored. The quantity of data produced by a source can be quite large—often many megabytes or even gigabytes. *Lossless data compression* occurs without any loss of information, while *lossy compression* causes minor loss in information, which a user would not perceive in practice. This chapter describes lossless compression.

Entropy $\mathcal{E}$—Entropy is a measure of information content. A source produces symbols with each symbol encoded using binary code words. The probabilities of the symbols are used to compute the entropy, having units of bits/symbol. If the entropy value for a source is smaller than the number of bits used in the code words, lossless compression is possible using a Huffman code.

Encryption—Encryption makes information secure by scrambling the binary data in a seemingly random manner. Information is a valuable commodity and should be available only to the intended parties. Data security is a challenge, especially when data are transmitted over a wireless channel, where it can be observed by hackers or sniffed. To keep your important information secure, data are encrypted before they are transmitted or stored. Although practical encryption techniques use advanced mathematical ideas, this chapter presents the fundamental ideas by considering a simple, yet effective, approach.

► 9.2 DATA COMPRESSION

Because channel capacity and digital memory size are always limited, *data compression* techniques were developed to limit the quantity of data that needs to be transmitted, while still maintaining most of its informational content. There are two commonly employed approaches to data compression, depending on whether there is any loss of information in the procedure.

1. **Lossless compression** techniques retain all of the information present in the original data. That is, the original data sequence is recovered *exactly* after the coding/decoding operation. These techniques are used for compressing numerical data, such as monetary account information, computer codes, and for document transmissions. Five-fold data compression allows a 10 MB (megabyte) file to be stored as a compressed 2 MB file for faster transmission over a communication channel.

   The Huffman coding procedure is a lossless compression technique that allows the representation of symbols to take on a variable number of bits to save on the *average number* of bits needed to represent the data. Such a code uses a small number of bits to represent the most commonly occurring symbols and a larger number of bits for those that occur less frequently. This idea was employed to encode telegraph messages using Morse code, composed of dots and dashes. The most common English letter "e" is represented by a single dot and a "t" by a single dash, while the less frequently encountered "q" is represented by a dash-dash-dot-dash sequence.

2. **Lossy compression** techniques allow more drastic reductions (hundred-fold) of the data by permitting some loss in information. Such techniques are typically used for encoding data that are to be *perceived*, such as images or sounds. In such cases, using *sufficiently accurate approximations* to the data—rather than the actual data—can lead to hundred-fold compression with little perceptible distortion.

   An interesting feature of lossy compression techniques is that the amount of compression can be varied by specifying the quality of the result. For example, transmitting a sporting event, such as the Olympics with its fast motion and subtlety of movement, demands a high-fidelity reproduction that requires a large number of bits per second. A technique called MPEG-2 encoding specifies a data rate equal to 6 Mbps. At the other extreme, a video conference call does not involve quick movements and can be encoded with the same MPEG-2 encoding at 64 Kbps. Hence, there is almost a factor of 100 on the amount of data that is required to reproduce video information.

9.2.1 Modeling Information

This section develops a mathematical model of information. Information is more than just data, it contains an element that we did not know previously, otherwise it would not be information. To model the novelty of information we use simple probability theory. We can then compute the entropy to quantify information. A pseudo-random number generator is used to simulate such data sources to generate typical data. Finally, we apply the theory to compute the expected compression of a data file.

Figure 9.1 shows our model of a source having the following elements:

- The source has a vocabulary of m symbols, which are denoted $X_1, X_2, X_3, \ldots, X_m$ or, more concisely, X_i for $1 \leq i \leq m$. By convention, the index starts with 1 rather than 0 used for time sequences.

- Each symbol has an associated probability of occurrence with the probability of the source producing symbol X_i denoted as $P[X_i]$. The set of probabilities are denoted concisely as $P[X_i]$ for $1 \leq i \leq m$. Because the source must produce one of these symbols, these probabilities sum to one:

$$\sum_{i=1}^{m} P[X_i] = 1 \tag{9.1}$$

If the m symbols are equally likely, then the probability of a source producing any particular one equals $1/m$, or

$$P[X_1] = P[X_2] = P[X_3] = \cdots = P[X_m] = \frac{1}{m} \tag{9.2}$$

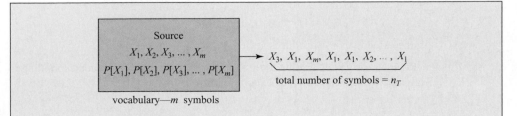

Figure 9.1

A mathematical model of a source.

■ We also consider the case where the source produces a data file that contains a total of n_T symbols that come from the symbol vocabulary containing m unique symbols.

EXAMPLE 9.1 | Binary source

A binary source often models devices that transmit binary data:

- The binary source vocabulary consists of two symbols, making $m = 2$, with symbols

$$X_1 = 0 \quad \text{and} \quad X_2 = 1$$

- Unless there is reason to think otherwise, binary data are typically assumed to be equally probable with

$$P[X_1] = P[X_2] = \frac{1}{2}$$

The sum of these probabilities equals

$$\sum_{i=1}^{2} P[X_i] = P[X_1] + P[X_2] = 1$$

That is, the binary source always produces either a 0 or 1.

Any data file on your computer can be considered to be a sequence of 0's and 1's. Data file sizes vary widely, from a minimum value on many computers being $n_T = 1024$ or a "1 K" file to image and video files containing many megabytes.

EXAMPLE 9.2 | Digital camera source

Consider modeling a digital camera that produces a digital image as a data source:

- The camera vocabulary consists of symbols that represent the 8-bit value of each pixel intensity, composed of RGB colors, yielding $m = 2^8 = 256$ possible symbols, and are denoted by

$$X_i = i - 1 \quad \text{for } 1 \le i \le 256$$

with $X_1 = 0$, up to $X_{256} = 255$, representing all possible values that a byte can have.

- Because the camera can take an almost endless variety of pictures, it is reasonable to assume that all symbols are equally likely to occur. This means their probabilities are equal to

$$P[X_1] = P[X_2] = P[X_3] = \cdots = P[X_{256}] = \frac{1}{256}$$

The sum of these probabilities equals

$$\sum_{i=1}^{m} P[X_i] = \sum_{i=1}^{256} \frac{1}{256} = 1$$

That is, the camera always produces a value between 0 and 255.

A color camera having 10^6 pixels produces three symbols (bytes) per pixel, corresponding to the R, G, B intensities. Hence, an image file contains $n_T = 3 \times 10^6$ symbols.

9.2.2 Source Entropy

We can quantify the average information produced by a source by computing its *source entropy*, which is denoted as $\mathcal{H}_S$. To compare different sources using the same units, we compute $\mathcal{H}_S$ in units of *bits per symbol*.

A source that produces the symbols X_1, X_2, X_3, ..., X_m, with the probability of X_i being $P[X_i]$, has a source entropy computed as

$$\mathcal{H}_S = -\sum_{i=1}^{m} P[X_i] \, \log_2 P[X_i] \text{ bits/symbol} \tag{9.3}$$

The entropy indicates the average number of bits that are required to represent these symbols as a binary code, as illustrated in the following examples.

Entropy of a binary source with equal probabilities

EXAMPLE 9.3

Let us consider a source that generates random bits by transmitting symbols X_1 and X_2 that are equally likely. Then their probabilities equal

$$P[X_1] = P[X_2] = 0.5$$

The entropy of this binary source is computed as

$$\mathcal{H}_S = -\sum_{i=1}^{2} P[X_i] \, \log_2 P[X_i]$$

$$= -P[X_1] \, \log_2 P[X_1] - P[X_2] \, \log_2 P[X_2]$$

$$= -0.5 \log_2 0.5 - 0.5 \log_2 0.5$$

Using the results $\log(1/a) = -\log(a)$ and $\log_2(2) = 1$ gives

$$\mathcal{H}_S = -0.5 \log_2 0.5 - 0.5 \log_2 0.5$$

$$= 0.5 \overbrace{\log_2 2}^{=1} + 0.5 \overbrace{\log_2 2}^{=1}$$

$$= 1 \text{ bit/symbol}$$

It is intuitive that, when a source produces only two symbols (heads/tails, true/false, ...), only one bit is needed to indicate which symbol was produced each time a symbol is generated.

The previous example illustrates that a source producing m symbols that are equally probable has probability values

$$P[X_i] = \frac{1}{m} \quad \text{for} \quad 1 \leq i \leq m \tag{9.4}$$

The source entropy is simple to compute as

$$\mathcal{H}_S = -\sum_{i=1}^{m} P[X_i] \log_2 P[X_i] = -\sum_{i=1}^{m} \left(\frac{1}{m}\right) \log_2 \left(\frac{1}{m}\right) = \log_2 m \tag{9.5}$$

EXAMPLE 9.4 — Entropy of a binary source with unequal probabilities

Consider a source that produces two symbols having unequal probabilities:

$$P[X_1] = 0.25 \quad \text{and} \quad P[X_2] = 0.75$$

Computing the entropy of this source gives

$$\mathcal{H}_S = -\left(\sum_{i=1}^{2} P[X_i] \, \log_2 P[X_i] \right)$$

$$= -\left(P[X_1] \, \log_2 P[X_1] + P[X_2] \, \log_2 P[X_2] \right)$$

$$= -0.25 \log_2 0.25 - 0.75 \log_2 0.75$$

$$= 0.81 \text{ bits/symbol}$$

A result that is expressed in fractional bits can be interpreted using the following logic: A file containing 1,000 symbols can be encoded using 811 bits.

This example illustrates that a binary source producing bits (symbols) having unequal probabilities has a lower entropy than the source producing two symbols having equal probabilities.

> **Factoid:** A source that produces symbols having equal probabilities generates the most information per symbol and thus produces the greatest entropy.

EXAMPLE 9.5 — Source producing one symbol

Consider a source that always produces the same symbol, as in the case of a defective source. If this source is modeled as always producing symbol X_1, then $P[X_1] = 1$. The entropy of this defective source is

$$\mathcal{H}_S = -\sum_{i=1}^{1} P[X_i] \, \log_2 P[X_i] = -(1 \times \overbrace{\log_2 1}^{=0}) = 0 \text{ bits/symbol}$$

That is, a source producing only a single symbol generates zero information.

9.2.3 Effective Probabilities

The previous discussion assumed we know the symbol probabilities. But in many cases these probabilities are not known and must be estimated from the data file.

Consider a data file generated by a source for which we do not know the symbol probabilities. Although a source can produce m symbols, this particular data file may contain only $m_x(\le m)$ different symbols. Let m_x symbols occur in a data file of size n_T, where $n_T \gg m_x$.

- If the file contains decimal digits, $X_1 = 0$, $X_2 = 1, X_3 = 2, \ldots, X_{10} = 9$, and it is likely that all digits occur, then $m_x = m = 10$. For a database file containing telephone numbers or credit card numbers, n_T may be in the millions.

- If the symbols are alphanumeric (ASCII 7-bit) text characters, $X_1, \ X_2, \ X_3, \ldots,$ X_{128}, including all standard punctuation marks. If not all ASCII characters occur $m_x < m = 128$. The data file containing the words in this text has $n_T \approx 10^6$.

- If the symbols are byte values, $X_1 = 00000000$, $X_2 = 0000001$, $X_3 = 00000010$, $\ldots$, $X_{256} = 11111111$. If all symbols occur at least once, then $m_x = m = 256$. The data file of an instruction manual containing images stored on a thumb drive may have n_T in the billions.

Alphanumberic characters in a book · EXAMPLE 9.6

If the symbols are letters of the alphabet, then $m = 26$, although if we count lowercase, uppercase, spaces, and punctuation as separate symbols, $m_x \approx 100$. If the data file of a book contains 600 pages with 200 words per page and an average of five characters per word, the total number of symbols equals

$$n_T = 600 \times 200 \times 5 = 6 \times 10^5$$

A data file contains m_x different symbols with the total number of symbols equal to n_T. If X_i occurs n_i times, the *effective probability* of symbol X_i is determined from its relative frequency, as

$$P_e[X_i] = \frac{n_i}{n_T} \quad \text{for} \quad 1 \leq i \leq m_x \tag{9.6}$$

Valid symbols that could have been produced by the source but do not appear in the file have zero effective probability values by this method, and these are excluded from further consideration.

The book without "e" · EXAMPLE 9.7

The book *Gadsby* contains 50,000 words, yet it does not include the letter "e", which is the most probable letter in English. Hence, when sorting the alphanumeric symbols, symbol $X_5 (= e)$ produces count $n_5 = 0$, making $P_e[X_5] = 0$.

9.2.4 Effective Source Entropy

For data files with unknown symbol probabilities, we use the effective probabilities to compute an estimate of the source entropy. The *effective source entropy*, which is denoted as $\hat{\mathcal{H}}_S$, is computed from the effective probabilities as

$$\hat{\mathcal{H}}_S = -\sum_{i=1}^{m_x} P_e[X_i] \log_2 P_e[X_i] \text{ bits/symbol} \tag{9.7}$$

Effective source entropy · EXAMPLE 9.8

Consider a binary data file containing 1,000 bits of which 600 are 1's and 400 are 0's. There are two ($m_x = 2$) different symbols, and their effective probabilities are computed as

$$P_e[1] = \frac{600}{1,000} = 0.6$$

$$P_e[0] = \frac{400}{1,000} = 0.4$$

The effective source entropy then gives

$$\hat{\mathcal{H}}_S = -\sum_{i=1}^{2} P_e[X_i] \log_2 P_e[X_i]$$

$$= -P_e[1] \log_2 P_e[1] - P_e[0] \log_2 P_e[0]$$

$$= -(0.6 \times -0.74) - (0.4 \times -1.32)$$

$$= 0.97 \text{ bits/symbol}$$

Note that if the counts of 1's and 0's were equal, $\hat{\mathcal{H}}_S$ would equal 1 bit/symbol. The unequal counts of 1's and 0's reduce the $\hat{\mathcal{H}}_S$ value.

9.2.5 Effective File Entropy

The effective file entropy, which is denoted as $\hat{\mathcal{H}}_f$, represents the minimum number of bits that are needed to encode the file. The $\hat{\mathcal{H}}_f$ value determines whether it is worthwhile to compress a file (that is, whether compression will lead to a significant reduction in data size). To compute $\hat{\mathcal{H}}_f$, we multiply the effective source entropy $\hat{\mathcal{H}}_S$ by the total number of symbols n_T, as

$$\hat{\mathcal{H}}_f = n_T \text{ symbols} \times \hat{\mathcal{H}}_S \text{ bits/symbol} \tag{9.8}$$

Note that this determination of whether a file should be compressed can be made before any compression is even attempted.

EXAMPLE 9.9

Effective file entropy

Consider the data file containing only the following ten symbols

$$A\ B\ A\ D\ A\ E\ C\ D\ B\ A$$

There are five ($m_x = 5$) different symbols (A, B, C, D, and E) in the file of ten symbols ($n_T = 10$). The effective probabilities of the symbols listed in alphabetical order are given by

$$P_e[A] = \frac{n_A}{n_T} = \frac{4}{10} = 0.4$$

$$P_e[B] = \frac{n_B}{n_T} = \frac{2}{10} = 0.2$$

$$P_e[C] = \frac{n_C}{n_T} = \frac{1}{10} = 0.1$$

$$P_e[D] = \frac{n_D}{n_T} = \frac{2}{10} = 0.2$$

$$P_e[E] = \frac{n_E}{n_T} = \frac{1}{10} = 0.1$$

The effective source entropy gives

$$\hat{\mathcal{H}}_S = -\sum_{i=1}^{m_x} P_e[X_i] \log_2 P_e[X_i]$$

$$= -(P_e[A] \log_2 P_e[A] + P_e[B] \log_2 P_e[B] + P_e[C] \log_2 P_e[C]$$

$$+ P_e[D] \log_2 P_e[D] + P_e[E] \log_2 P_e[E])$$

$$= -(0.4 \log_2[0.4] + 0.2 \log_2[0.2] + 0.1 \log_2[0.1]$$

$$+ 0.2 \log_2[0.2] + 0.1 \log_2[0.1])$$

$$= 2.12 \text{ bits/symbol}$$

The effective file entropy computes as

$$\hat{\mathcal{H}}_f = n_T \, \hat{\mathcal{H}}_S = 21.1 \text{ bits}$$

That is, this file has an information content equal to 21.2 bits. This means that the ten symbols in the file can be encoded using approximately 22 bits. A file with $\hat{\mathcal{H}}_S = 2.12$ bits/symbol containing 100 symbols can be encoded with 212 bits using the data compression procedure described below.

Without compression, fixed-length code words each having three bits ($2^3 = 8$) are needed to provide a unique code for each of the five symbols. For example, consider the fixed-length code word assignment

$$A : 000, \; B : 001, \; C : 010, \; D : 011, \; E : 100$$

Clearly, not all possible 3-bit code words are used, but 2-bit code words cannot unambiguously encode these five symbols. Thus, 30 bits are required to encode the file using 3-bit code words. We can then decide whether the savings of 8 bits ($= 30 - 22$) justifies the effort to compress the file.

9.2.6 Huffman Code

Lossless data compression is achieved by representing symbols with a variable number of bits. Doing so can save on the *average number* of bits needed to represent the data by using a small number of bits for the most commonly occurring symbols and a larger number of bits for those that occur less frequently. The resulting variable-length code generates an average code word size per symbol that is approximately equal to our two entropy measures.

$\mathcal{H}_S$—The source entropy equals the average number of bits required to encode symbols produced by a source with known symbol probabilities.

$\hat{\mathcal{H}}_S$—The effective source entropy of a data file equals the average number of bits required to encode symbols computed from the effective probabilities.

The method of implementing a variable-length code considered here is the *Huffman code*. This procedure employs a *code tree* that consists of *nodes* connected by *branches* that ultimately terminate in *leaves*. Figure 9.2 shows a code tree that describes the four symbols A, B, C, and D. The node at the top of the tree defines its *root* or starting point. In a *binary tree* the branches originating from each node are designated by the binary values 0 or 1. Hence, only two branches originate from any node. For consistency, the left branch is always designated with a 1 and the right branch with a 0.

Branches terminate in *leaves* with each leaf representing a valid symbol. The sequence of bits in a valid code word specifies a unique path from the root to a leaf. Thus, this sequence of bits (each of which makes a decision at a node) specify the symbol at the

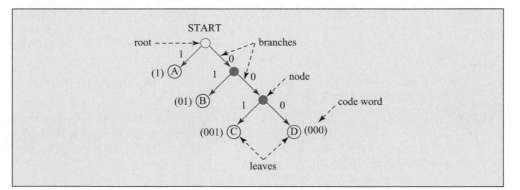

Figure 9.2

A code tree assigns a variable-length code that specifies four symbols.

terminal leaf. The last bit in the code word always terminates at a leaf specifying a symbol.

The tree code in Figure 9.2 indicates that symbol *A* has the shortest code word (a single 1), while symbols *C* and *D* have the 3-bit codes. The corresponding code is called the *Huffman code* and uses the code tree to assign the shortest code words to the most common symbols and longer code words to the ones that occur less often.

EXAMPLE 9.10   Decoding the Huffman code

Using the tree code shown in Figure 9.2, the binary sequence given by

$$00101110001$$

can be decoded into symbols.

Decoding starts at the beginning of the data sequence and the root of the code tree. When a leaf is reached, a symbol has been decoded. The next bit in the data sequence then restarts at the root to search for the next symbol.

Doing this using our binary sequence, we get

$$\underbrace{001}_{C} \quad \underbrace{01}_{B} \quad \underbrace{1}_{A} \quad \underbrace{1}_{A} \quad \underbrace{000}_{D}$$

It is an interesting property of the Huffman code that no valid code word forms the beginning of another code word.

Huffman Coding Procedure

The procedure to form a Huffman code involves the following steps.

STEP 1. *Order the symbols by their probabilities.* We form a list of symbols ordered by decreasing probability, using the measure that is available in the problem: $P[X_i]$ if it is known or $P_e[X_i]$ if it is computed from a data file. If X_3 is the most probable symbol, it appears at the top of the list, and if X_2 is the least probable, it appears at the bottom. This is the *first ordering*.

STEP 2. *Code bit assignment.* Starting from the bottom of the list, we assign a 0 code bit to the least probable symbol and a 1 code bit to next least probable (that is, the one that is second from the bottom). This code bit forms the rightmost (LSB) bit of the code words that differentiate these two symbols.

STEP 3. *Combine symbols by adding their probabilities.* Combine the two least probable symbols into one composite symbol, and the probability of this composite symbol becomes the sum of the constituent probabilities. For example, if the last two symbols on the list are X_2 and X_6, they are combined into the composite symbol denoted X_2-X_6 and

$$P[X_2\text{-}X_6] = P[X_2] + P[X_6] \tag{9.9}$$

when the probabilities are known, otherwise, use the effective probabilities

$$P_e[X_2\text{-}X_6] = P_e[X_2] + P_e[X_6] \tag{9.10}$$

STEP 4. *Re-order symbols according to new probability values.* Generate a new ordering of the symbols using the composite symbol as just another symbol having its summed probability. This is the *second ordering*.

STEP 5. *Repeat Steps 2 to 4.* We repeat the previous three steps until only two symbols or composite symbols remain on the list. This is the *final ordering*. We assign a 0 code bit to the less probable entry and a 1 code bit to the other. This final assignment completes the formation of the code words.

To decode a Huffman code, a *code table* indicating the correspondence of the code words to the symbols must also be stored (or transmitted) along with the the encoded data.

We construct a Huffman code tree from the root by starting with the final ordering, where the 1 in the code word is assigned to the most probable symbol (or composite symbol). This 1 defines the left branch from the root leading to a node in the second row in the tree. A 0 code bit defines the right branch leading to the other node in the second row.

Each node corresponds to either a symbol or a composite symbol. If the node corresponds to a symbol, it is a leaf node, which designates that symbol. No branches originate from a leaf node. If the node corresponds to a composite symbol, two branches from it lead to nodes that represent the constituent symbols determined from the previous ordering. The tree is formed in this manner until all of the branches end in leaf nodes.

The following examples illustrate the Huffman coding procedure.

Huffman coding the sum of two coins — EXAMPLE 9.11

Consider a coin flip with a head indicating a 1 and a tail a 0. We toss two coins and want to encode their sum. The four possible outcomes and their sums are given by

00	(sum=0)
01	(sum=1)
10	(sum=1)
11	(sum=2)

For fair coins, each of the four outcomes is equally probable (with probability $= 0.25$)

Define symbols $X_1 = 0$ (sum $= 0$), $X_2 = 1$ (sum $= 1$), and $X_3 = 2$ (sum $= 2$) with probabilities given by

$P[X_1] = 1/4$	(one of four possible outcomes)
$P[X_2] = 1/2$	(two of four possible outcomes)
$P[X_3] = 1/4$	(one of four possible outcomes)

The entropy of the source that produces the symbols that mimic the two-coin-sum outcomes equals

$$\mathcal{H}_S = -\sum_{i=1}^{3} P[X_i] \log_2 P[X_i]$$

$$= -0.25 \log_2(0.25) - 0.5 \log_2(0.5) - 0.25 \log(0.25)$$

$$= 1.5 \text{ bits/symbol}$$

We toss the two coins ten times and observe the following sequence of sum values.

$$2\ 1\ 1\ 2\ 1\ 0\ 2\ 1\ 1\ 0$$

Fixed-length code: We first encode this sequence using fixed-length code words. We have three symbols (0, 1, and 2), so 2-bit code words are needed to provide a unique code for each symbol. Using the arbitrarily assigned code words $X_1 : 00$, $X_2 : 01$, and $X_3 : 10$, the encoded binary sequence appears as

$$\underbrace{10}_{2}\ \underbrace{01}_{1}\ \underbrace{01}_{1}\ \underbrace{10}_{2}\ \underbrace{01}_{1}\ \underbrace{00}_{0}\ \underbrace{10}_{2}\ \underbrace{01}_{1}\ \underbrace{01}_{1}\ \underbrace{00}_{0}$$

Note that these ten realizations are encoded using twenty bits. Because the code word "11" is not used, this fixed-length code is wasteful.

Huffman code: We can encode this same information with a smaller average number of bits using the Huffman code, as shown in Figure 9.3.

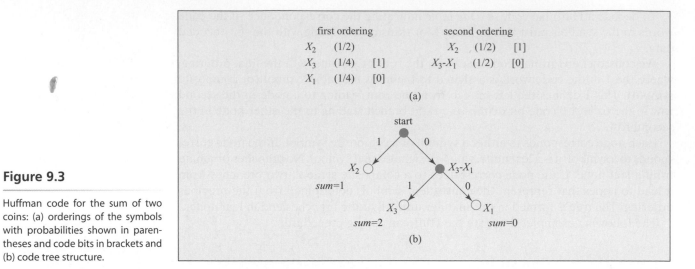

Figure 9.3

Huffman code for the sum of two coins: (a) orderings of the symbols with probabilities shown in parentheses and code bits in brackets and (b) code tree structure.

In the first ordering, X_1 is assigned the code bit 0 and X_3 a 1. Then these two symbols are combined to form the composite symbol X_1-X_3, with the probability of this composite symbol equal to

$$P[X_1\text{-}X_3] = P[X_1] + P[X_3] = 0.25 + 0.25 + 0.5$$

In the second ordering, because the remaining probabilities are equal, X_1-X_3 is arbitrarily assigned a 0 and X_1 a 1. There were only two symbols in this ordering, so the process is finished.

The resulting Huffman code is given by the code table

sum=0: $X_1 \rightarrow 00$
sum=1: $X_2 \rightarrow 1$
sum=2: $X_3 \rightarrow 01$

The Huffman-encoded version of the ten observations is given by

Note that the same ten symbols are encoded in 15 bits, for an average of 1.5 bits/symbol, which equals the source entropy $\mathcal{H}_S$ as computed here.

EXAMPLE 9.12   Theoretical Huffman coding of text files

Consider a text file consisting of all capital letters and six punctuation marks (space, comma, period, question mark, semicolon, and colon). These 32 symbols can be encoded with 5-bit code words ($2^5 = 32$). If these symbols were equally probable (with $P[X_i] = 1/32$), the entropy of the source producing them would equal

$$\mathcal{H}_S = \log_2 32 = 5 \text{ bits/symbol}$$

Studies by linguists indicate that the actual probabilities result in the entropy of English text being equal to 4.2 bits/symbol. Hence, compressing text files with a Huffman code using effective probabilities would result in a 16% reduction in the size of a data file.

Huffman coding of data files EXAMPLE 9.13

The histograms of data in three files types are shown in Figure 9.4.

The first file is the text file of this chapter with the numbers corresponding to ASCII value of each letter. Note that the numbers lie in the range [0,127], because the most significant bit of an ASCII code is always 0 ($m_x = 128$). The most common symbol value is 32 (the ASCII code for a *space*).

The second file is the postscript file sent to the printer to produce this chapter. Note that the numbers lie in the range [0,255] ($m_x = 256$). Control characters lie outside the normal ASCII range [0,127].

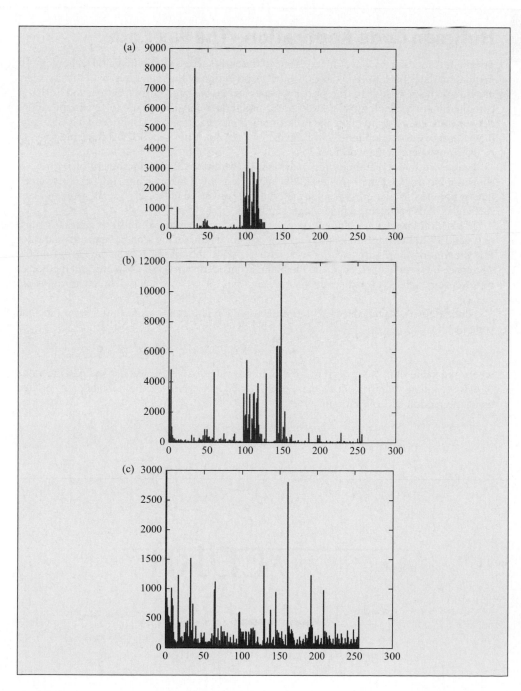

Figure 9.4

Histograms of byte values in three data files. (a) The text file of this chapter with the numbers corresponding to ASCII value of each letter. (b) The Postscript data file that printed this chapter. (c) The binary version of the assembly code of a typical computer program.

The third file includes binary instructions in a typical computer program. Note that these data are distributed almost uniformly over the entire range [0,255], indicating that almost all of the 8-bit codes appear ($m_x = 256$).

For compressing these data, Huffman coding is more effective when the histogram is more *peaked*, as in the first two data files. These peaks indicate symbols having large effective probability values, which would be assigned short code words. A large number of short code words would result in an efficient data compression. Histograms that are approximately uniformly distributed, as in the third file, result in minimal data reduction.

EXAMPLE 9.14 Huffman Code Application—The Fax Code

The digital facsimile machine (or *fax*) is one of the early success stories in digital technology. Its first patent dates back to 1843 (shortly after the invention of the telegraph) with the first commercial use being in 1865 to transmit newspaper images between Paris, London, and Berlin. A combination of technology, standardization, and the deregulation of the telecommunications network was necessary to make the fax a commercial success and a part of everyday life. Anyone with access to a telephone could connect two fax machines to send or receive messages to or from anywhere in the world.

The resolution of a standard fax machine is a width of 1/8 mm and height of 1/4 mm, as shown in Figure 9.5. When a page of text is inserted into a fax machine, the page is scanned line by line. Text is treated as graphic material and converted into a set of small white or black rectangles, thus producing a binary data sequence.

The typical binary data sequence produced by a fax machine exhibits groupings (or runs) of 0's and 1's to indicate white and black regions. For example, the scan in Figure 9.5 indicates that the tops of the E's are encoded with sixteen 1's and the space between the E's as four 0's. The space between the left edge and the first E is encoded with 203 0's. An efficient method of representing such data is to encode the *run lengths* of 1's or 0's, which is denoted as *run-length coding*.

For example, consider the case of n consecutive 1's. This run is represented in fax machine code as

$$n = 64m + r$$

where the value of m is between 0 and 27 and the value of r is between 0 and 63. The value of m is the integer number of 64's that are contained in n, while r is the remainder, computed using the modulo-64 computation as

$$r = [n]_{\mathrm{mod}(64)}$$

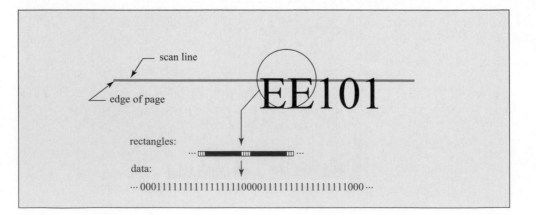

Figure 9.5

Fax sampling operation.

Make-up code words (64m)			Terminating code words (r)		
64m	white	black	r	white	black
64	11011	0000001111	0	00110101	0000110111
128	10010	000011001000	1	000111	010
192	010111	000011001001	2	0111	11
256	0110111	000001011011	3	1000	10
			4	1011	011
1600	010011010	0000001011011	5	1100	0011
1664	011000	0000001100100	6	1110	0010
1728	010011011	0000001100101	7	1111	00011
			8	10011	000101
			9	10100	000100
			10	00111	0000100
			EOL	000000000001	000000000001

Figure 9.6

Part of the standard Huffman code table used in a fax machine.

In the standard fax code, the term 64m is encoded separately as the *make-up* code word and r is encoded as the *terminating* code word. While all code words contain r, most codes contain a 64m component followed by an r component. For small values of n, corresponding to short white or black segments, the 64m component is zero and only the r component is transmitted.

For example, the make-up (64m) and terminating (r) values for the following n values are computed as:

$n = 100$: $100 = 64 + 36$, so we have $64m = 64$ and $r = 36$.

$n = 8$: Because $8 = 0 + 8$, we have $64m = 0$ and $r = 8$.

$n = 2,000$: Dividing by 64 gives 31.25. Taking m to be the integer part gives $64m = 64 \times 31 = 1,984$ and r is the remainder computed as $r = 2,000 - 1,984 = 16$.

To reduce the number of bits that are transmitted, the values of 64m and r are applied to a Huffman code to encode the most common runs with the shortest code words. To obtain the effective probability values for the various runs of 0's and 1's, a wide variety and large number of text documents and line drawings were scanned by manufacturers.

Figure 9.6 shows selected elements of the resulting standard fax code. The end-of-line (*EOL*) is a separate code word that terminates a scan line. The remaining white space on a line does not need to be transmitted, thus saving time. It has been observed that the use of the Huffman code allows a ten-fold reduction, explaining the commercial success of the fax.

9.3 ENCRYPTION

With wireless networks being accessible to anyone with a Wi-Fi receiver, and with remote data storage provided by backup services and cloud computing, data security has become a major concern. The current method of providing adequate (although not perfect) security is through *encryption*. Although commercial encryption methods (such as the *Data Encryption Standard (DES)*) involve complex mathematical algorithms, the basic idea is not very complicated and is explained here.

The easiest way to implement an encryption scheme is with an *Exclusive-OR (ExOR)* gate applied in a bit-wise manner to the binary data and a random binary sequence (RBS). The ExOR gate produces the encrypted data sequence. The RBS is typically produced with a *pseudo-random number generator (PRNG)* that produces an identical set of random bits at both the source and destination. The clever feature of this scheme is that a RBS is applied to the data for encryption at the source and the same RBS is applied to the encrypted data at the destination to yield the original data. Meanwhile, the transmitted data on the channel appears to be a meaningless random collection of bits.

In the usual encryption operation, binary data are encoded at the source to conceal the values, the encrypted version is transmitted, and the received data are decoded at the destination to reveal the original data. Because large quantities of data are usually transmitted, the coding and decoding operations must be done quickly and efficiently. The ExOR gate is ideal for this purpose, because performing the ExOR operation encrypts the data at the source and then decrypts it at the destination to recover the original data.

To explain the encryption process, we use the symbol $\oplus$ to indicate the bit-by-bit ExOR operation on the data sequence $\mathcal{D}_i$ and random binary sequence RBS_i to produce the encrypted binary sequence $\mathcal{E}_i$ for $1 \le i \le n_x$. Note the change of index initial value from 0 to 1 for notational convenience.

For example, if data $\mathcal{D}_i = 0\ 1\ 0\ 1$ and $\text{RBS}_i = 1\ 1\ 0\ 0$, the bit-wise ExOR output produces the encrypted data $\mathcal{E}_i$ as

$$\mathcal{E}_i = \mathcal{D}_i \oplus \text{RBS}_i$$

$$= 0 \oplus 1 \quad 1 \oplus 1 \quad 0 \oplus 0 \quad 1 \oplus 0$$

$$= 1\ 0\ 0\ 1$$

In practice, the implementation may be with a single ExOR gate to which each bit in $\mathcal{D}_i$ and RBS_i are introduced sequentially or using multiple ExOR gates operating in parallel. For example, eight gates can operate simultaneously on each bit in a data byte.

To accomplish encryption successfully requires the same RBS at both the transmitting and receiving ends. A PRNG uses a seed (the *key*) as the initial value to compute the RBS. The encrypted message $\mathcal{E}$ is then transmitted to the destination, where the receiving party can compute the same RBS employed at the transmitter by using the same key value. To retrieve the original data $\mathcal{D}_i$, the ExOR operation is again performed, but this time on $\mathcal{E}_i$ and RBS_i, as

$$\mathcal{D}_i = \mathcal{E}_i \oplus \text{RBS}_i \quad \text{for} \quad 1 \le i \le n_x \tag{9.11}$$

By the magic of the ExOR operation, this operation reproduces the original data at the destination.

The following example illustrates this simple encryption scheme. We then describe how the random bits are generated with a pseudo-random number generator (PRNG) algorithm and, finally, how the secret value of the key can be (reasonably) securely transmitted over a spy-ridden communication channel.

EXAMPLE 9.15   Encrypting data

Let us consider data that consists of 16 hexadecimal numbers. As a simple illustration, let these be in order, from 0 to F, with each encoded using its equivalent 4-bit value, as shown in Figure 9.7.

There are then $4 \times 16 = 64$ bits in this message, forming the binary data sequence $\mathcal{D}_i$ for $1 \le i \le 64$. To encrypt this binary sequence, the PRNG produced the RBS_i for $1 \le i \le 64$, as shown in Figure 9.7. To form the encrypted binary sequence $\mathcal{E}_i$, use the ExOR operation as

$$\mathcal{E}_i = \mathcal{D}_i \oplus \text{RBS}_i \quad \text{for} \quad 1 \le i \le 64$$

The figure shows the encrypted values. The last column gives the hexadecimal equivalents of the encrypted binary values. Note that the hexadecimal values of $\mathcal{E}_i$ do not resemble those corresponding to the original message encoded in $\mathcal{D}_i$.

Encryption at Source

D	Binary Code	RN (Hex)	RBS	$\varepsilon = D \oplus \text{RBS}$	Encrypted Value
0	0000	E	1110	1110	E
1	0001	5	0101	0100	4
2	0010	D	1101	1111	F
3	0011	0	0000	0011	3
4	0100	1	0001	0101	5
5	0101	9	1001	1100	C
6	0110	8	1000	1110	E
7	0111	6	0110	0001	1
8	1000	5	0101	1101	D
9	1001	9	1001	0000	0
A	1010	7	0111	1101	D
B	1011	B	1011	0000	0
C	1100	8	1000	0100	4
D	1101	9	1001	0100	4
E	1110	A	1010	0100	4
F	1111	F	1111	0000	0

Decryption at Destination

ε	Binary Code	RN (Hex)	RBS	$D = \varepsilon \oplus \text{RBS}$	Decrypted Value
E	1110	E	1110	0000	0
4	0100	5	0101	001	1
F	1111	D	1101	0010	2
3	0011	0	0000	0011	3
5	0101	1	0001	0100	4
C	0110	9	1001	0101	5
E	1110	8	1000	0110	6
1	0001	6	0110	0111	7
D	1101	5	0101	1000	8
0	0000	9	1001	1001	9
D	1101	7	0111	1010	A
0	0000	B	1011	1011	B
4	0100	8	1000	1100	C
4	0100	9	1001	1101	D
4	0100	A	1010	1110	E
0	0000	F	1111	1111	F

Figure 9.7

Example of encrypting hexadecimal values at the sources and decryption at the destination. RN is a random hexadecimal number that forms the RBS.

After $\mathcal{E}_i$ is received at the destination, the received data is decrypted by again applying it to the ExOR gate and the same RBS_i, as

$$\mathcal{D}_i = \mathcal{E}_i \oplus \text{RBS}_i \quad \text{for} \quad 1 \le i \le 64$$

with results shown in the figure. Note that the hexadecimal values of the decrypted values are identical to the original data.

Encrypting waveforms EXAMPLE 9.16

Figure 9.8 shows the encryption of a quantized sinusoidal waveform. Figure 9.8a shows the integer values in the range [0,15] of a sinusoidal waveform generated as

$$st_i = \text{round}[7.5(1 + \sin(2\pi \, i/32)] \quad \text{for} \quad i = 0, 1, 2, \ldots, n_x - 1 \; (n_x = 33)$$

The corresponding 4-bit code words generated the binary data sequence D_i containing $4 \times 33 = 132$ bits. A RBS sequence of 132 random bits was generated with a PRNG. The source

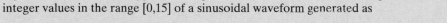

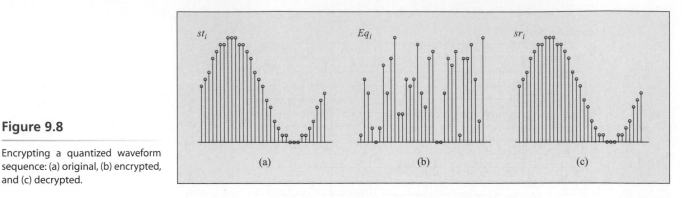

Figure 9.8

Encrypting a quantized waveform sequence: (a) original, (b) encrypted, and (c) decrypted.

encrypted the data with the ExOR operation as

$$\mathcal{E}_i = \text{RBS}_i \oplus \mathcal{D}_i \text{ for } 0 \leq i \leq 131$$

This $\mathcal{E}_i$ was applied to a 4-bit DAC to construct the waveform Eq_i. Figure 9.8b shows that there is no resemblance between Eq_i and st_i, demonstrating the effectiveness of encryption.

At the destination the encrypted binary data is decrypted by forming $D_{r,i}$ by the ExOR operation using the same RBS_i as in the encryption as

$$\mathcal{D}_i = \text{RBS}_i \oplus \mathcal{E}_i \text{ for } 0 \leq i \leq 131$$

This decrypted sequence is applied to a 4-bit DAC to construct the waveform sr_i. Figure 9.8c shows that sr_i is identical to st_i, verifying that the data has been retrieved exactly.

EXAMPLE 9.17 Encrypting speech waveforms

Figure 9.9 shows the encryption of a speech waveform generated by a laptop microphone containing 200 samples. Each sample was expressed as an 8-bit sequence, forming $\mathcal{D}_i$ for $i = 0, 1, 2, \ldots, n_x - 1$ with $n_x = 1,600$. A PRNG generated 1,600 random bits to form RBS_i. The encrypted sequence was formed by the ExOR operation as

$$\mathcal{E}_i = \text{RBS}_i \oplus \mathcal{D}_i \text{ for } 0 \leq i \leq 1,599$$

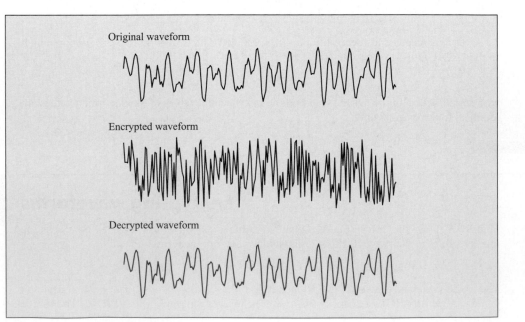

Figure 9.9

Encrypting a speech waveform.

The encrypted values were applied to a DAC that formed the encrypted waveform displayed in Figure 9.9.

The $\mathcal{E}_i$ was decrypted by forming D_i with the ExOR operation using the same RBS_i as in the encryption, as

$$\mathcal{D}_i = \mathrm{RBS}_i \oplus \mathcal{E}_i \text{ for } 0 \le i \le 1{,}599$$

The decrypted values were then converted into the audio samples and displayed in Figure 9.9 to show that the decrypted waveform matches the original.

Encrypting images   EXAMPLE 9.18

Figure 9.10 shows the encryption of a color image of a cat. A jpeg image taken with a digital camera was decoded to form a matrix containing 500 rows and 500 columns (250,000 pixels) with a third dimension that encoded 8 bits (256 levels) for red, 8 bits for green, and 8 bits for blue. The total number of data bits was equal to $3 \times 8 \times 2.5 \times 10^5 = 6 \times 10^6$ data bits, forming D_i, for $1 \le i \le 6 \times 10^6$. A PRNG generated six million random bits, forming RBS_i. The encrypted binary sequence was formed by the ExOR operation as

$$\mathcal{E}_i = \mathrm{RBS}_i \oplus \mathcal{D}_i \text{ for } 1 \le i \le 6 \times 10^6$$

The encrypted values are displayed in Figure 9.10 to indicate there was no hint of a cat.

The encrypted image was decrypted by the ExOR operation using the same RBS_i as in the encryption as

$$\mathcal{D}_i = \mathrm{RBS}_i \oplus \mathcal{E}_i \text{ for } 1 \le i \le 6 \times 10^6$$

The decrypted values are displayed in Figure 9.10 to verify that it appears exactly as the original.

9.3.1 Simulating Randomness by Computer

The computer simulates randomness with an *algorithm* called a *pseudo-random number generator* (PRNG). A good PRNG for encryption needs to produce numbers that have the following three properties.

1. They should not be obviously *predictable* from past values, even though a deterministic formula is used to generate the numbers.

2. They should be *uniformly distributed* over the interval $[0, n_{\max})$.

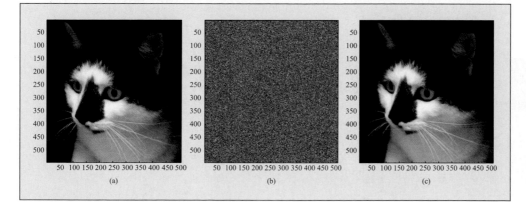

Figure 9.10

Encrypting a color image: (a) original image, (b) encrypted image, and (c) decrypted image.

3. They should be *repeatable*, so that the same sequence of random numbers can be produced at a different time or in a different place.

We first consider generating random integers and then describe a method to generate random bits.

Generating Random Integers

Consider a sequence of n random integers, which is denoted

$$X_1, \ X_2, \ X_3, \ \ldots, \ X_n$$

A common formula used to compute random integers is given by

$$X_i = \text{mod}[(\alpha \ X_{i-1} + \beta), \ h] \tag{9.12}$$

where

α is a factor that multiplies the previous integer in the sequence (X_{i-1}),

β is a constant that prevents the sequence from degenerating into a set of zero values and

h is the divisor in the modulus operation.

All values are typically large positive integers. The parenthesis within the *mod* operator helps prevent confusion when the result of the multiplication and addition is a large number that contains commas. (Seeing too many commas makes it difficult to identify the value of h.)

The sequence of random numbers starts with X_1. To compute X_1 we need a starting value X_0, which is called the *seed* (the *encryption key*). This seed starts the sequence and different seeds produce different random sequences. Using the same number for the seed repeats the same random integer sequence. This is a useful property for encryption that needs the same random sequence at the source and destination.

The random integers produced by our PRNG using the modulus operator range from 0 (when the result within the parentheses is a multiple of h) to the largest remainder value ($h - 1$). Hence, the random numbers lie in the range $0 \leq X_i \leq h - 1$.

Why does the formula in Eq. (9.12) produce random-looking numbers? The reason is that the modulo-h operation discards the quotient (the predictable part of the division) and retains the remainder (which is difficult to predict). It is not a perfect (but reasonable) compromise between true randomness and a repeatable computation.

A successful PRNG should generate billions of different random numbers. This requires α, β, and h to be large integers. The *mod* operator produces integers between 0 and $h - 1$. Hence, the value of h limits the number of different random numbers. For example, $h = 1000$ can only produce (at most) one thousand different random numbers in the range [0,999]. Once a random integer X_i is repeated, the PRNG formula begins to repeat the entire sequence.

EXAMPLE 9.19   A pseudo-random sequence of 4-digit integers

Consider generating a sequence of 4-digit random integers. A 4-digit integer falls in the interval $[0, 9{,}999]$ by setting $h = 10{,}000$. Making h a power of 10 (say $h = 10^n$) is particularly convenient for computing Eq. (9.12) because the modulus operation result is simply the n least significant digits in the computation. With $h = 10^4$, the modulus operation produces the four least significant digits of the multiplication.

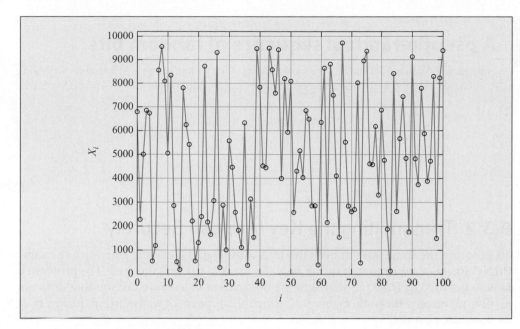

Figure 9.11

Pseudo-random sequence X_I for $1 \le i \le 100$ generated with $\alpha = 97{,}531$, $\beta = 54{,}321$, and $h = 10{,}000$ with $X_0 = 6{,}789$.

There are no optimum values for α, β, and X_0 other than being large. We arbitrarily set $\alpha = 97{,}531$, $\beta = 54{,}321$, and the seed $X_0 = 6{,}789$. These values generate the sequence of random numbers as

$$X_1 = \mathrm{mod}[(97{,}531 \times 6{,}789 + 54{,}321), 10{,}000]$$

$$= \mathrm{mod}[(662{,}192{,}280), 10{,}000] = 2{,}280$$

$$X_2 = \mathrm{mod}[(97{,}531 \times 2{,}280 + 54{,}321), 10{,}000]$$

$$= \mathrm{mod}[(222{,}425{,}001), 10{,}000] = 5{,}001$$

$$X_3 = \mathrm{mod}[(97{,}531 \times 5{,}001 + 54{,}321), 10{,}000]$$

$$= \mathrm{mod}[(487{,}806{,}852), 10{,}000] = 6{,}852$$

Figure 9.11 shows X_i for $1 \le i \le 100$. Note that it is difficult to predict the value X_i from X_{i-1}, and all of the values fall uniformly in the range $[0, h-1] = [0, 9{,}999]$, which is often expressed as $[0, h) = [0, 10{,}000)$.

Generating Random Bits

The random integers generated by Eq. (9.12) are approximately uniformly distributed over the interval $[0, h)$. We choose an even value for h to form two equal sub-intervals defined as

$$[0, h/2) \quad \text{and} \quad [h/2, h) \tag{9.13}$$

The random binary sequence value RBS_i is determined by the sub-interval X_i falls into, as

$$\text{If } 0 \le X_i < h/2 \;\rightarrow\; \mathrm{RBS}_i = 0 \tag{9.14}$$

$$\text{If } h/2 \le X_i < h \;\rightarrow\; \mathrm{RBS}_i = 1$$

EXAMPLE 9.20

A pseudo-random sequence of random bits

A sequence of three random bits is generated from the random integers in Example 9.19 using $h = 10{,}000$. Repeating the sequence of random integers and testing if $0 \le X_i < h/2(= 5000)$, we get

$$X_1 = 2{,}280 \;\rightarrow\; \text{RBS}_1 = 0$$

$$X_2 = 5{,}001 \;\rightarrow\; \text{RBS}_2 = 1$$

$$X_3 = 6{,}852 \;\rightarrow\; \text{RBS}_3 = 1$$

9.3.2 Transmitting the Key (Almost) Securely

To generate the same RBS at both the transmitting and receiving ends with the same PRNG formula, we need to transmit only the key and use it as the seed. The problem is how to get the key from the source to the destination in a secure fashion that does not involve a personal meeting. Figure 9.12 shows the process of transmitting the key over an unsecure channel.

One clever solution uses an interesting property of modulus arithmetic. Let a and N be integers. In practice, both a and N are typically large, currently 128-bit or 256-bit numbers, with the larger numbers being more secure (that is, tougher to crack) but at the expense of additional computations. The two parties (transmitter T and receiver R) agree on the values of a and N and suspiciously assume that everyone else knows these values as well. T secretly chooses integer x and computes $\mathcal{X}$, the modulus-N value of a^x, denoted as

$$\mathcal{X} = [a^x]_{\text{mod}(N)} \tag{9.15}$$

T transmits $\mathcal{X}$ to R over a channel that everyone sees. R also secretly chooses integer y and computes

$$\mathcal{Y} = [a^y]_{\text{mod}(N)} \tag{9.16}$$

and R transmits $\mathcal{Y}$ to T. Both keep secret x (known only to T) and y (known only to R).

Having received $\mathcal{Y}$ from R, T computes

$$K_T = [\mathcal{Y}^x]_{\text{mod}(N)} \tag{9.17}$$

Having received $\mathcal{X}$ from T, R computes

$$K_R = [\mathcal{X}^y]_{\text{mod}(N)} \tag{9.18}$$

With modulus arithmetic, it turns out that these last two computations produce the same result:

$$K_T = K_R \tag{9.19}$$

Figure 9.12

Encrypting information can be done over an non-secure channel by using modulo arithmetic to transmit the value of the *key*.

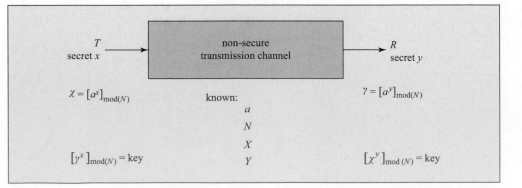

Because both T and R arrive at the same number, it can be the seed in the PRNG for generating the RBS for encryption for both parties.

The reason that this simple scheme works is because (even if the values of a, N, $\mathcal{X}$, and $\mathcal{Y}$ are known to all) the problem of determining the value of the key from these quantities is a very difficult and time-consuming computational process, especially if x and y are large integers.

Computing the encryption key   EXAMPLE 9.21

For a simple example, let $N = 23$ and $a = 13$. Secretly, T chooses $x = 7$ and R chooses $y = 10$.

- T computes $\mathcal{X}$ as

$$\mathcal{X} = [a^x]_{\text{mod}(N)} = [13^7]_{\text{mod}(23)} = [62,748,517]_{\text{mod}(23)} = 9$$

  and sends it to R.

- R computes $\mathcal{Y}$ as

$$\mathcal{Y} = [a^y]_{\text{mod}(N)} = [13^{10}]_{\text{mod}(23)} = [137,858,491,849]_{\text{mod}(23)} = 16$$

  and sends it to T.

- T computes the key as

$$K_T = [\mathcal{Y}^x]_{\text{mod}(N)} = [16^7]_{\text{mod}(23)} = [268,435,456]_{\text{mod}(23)} = 18$$

- R computes the key as

$$K_R = [\mathcal{X}^y]_{\text{mod}(N)} = [9^{10}]_{\text{mod}(23)} = [3,486,784,401]_{\text{mod}(23)} = 18$$

This key is used by both T and R as the seed in their identical PRNGs to generate the same RBS for encryption (by T) and decryption (by R).

In the previous example, even if we know that $N = 23$, $a = 13$, $\mathcal{X} = 9$, and $\mathcal{Y} = 16$ by intercepting data transmission, to find the key we need to solve for the values of x and y that obey the formula

$$[16^x]_{\text{mod}(23)} = [9^y]_{\text{mod}(23)} \tag{9.20}$$

The solution is not easy to find, because we cannot simply plug in a value for one to find the other. Finding the value of the key from the known quantities requires trying all possible pairs of integers (x, y) to check which pair satisfies the equation

$$[\mathcal{X}^y]_{\text{mod}(N)} = [\mathcal{Y}^x]_{\text{mod}(N)} \tag{9.21}$$

If the pair of integers x and y are each in the range $[1,100]$, then an exhaustive search needs to examine $10^2 \times 10^2 = 10^4$ combinations. On average, an answer would be found by performing half of these computations, because some (x,y) pairs occur sooner in the search than others. However, even the large range $[1, 10^9]$ is small by encryption standards.

Number of calculations for finding key   EXAMPLE 9.22

Assuming you know $\mathcal{X}$ and $\mathcal{Y}$ and N, let us compute the average number of computations needed to find the key value if x and y are 128-bit numbers.

A 128-bit integer spans the range from 0 to

$$2^{128} = 2^8 \times \left(2^{10}\right)^{12} = 256 \times \left(10^3\right)^{12} = 2.56 \times 10^{38}$$

The number of pairs of two 128-bit numbers equals

$$(2.56 \times 10^{38})^2 = 6.55 \times 10^{76}$$

On average, the key could be found by performing half this number or 3.3×10^{76}. If it takes 1 μs to perform the computation in Eq. (9.21), to do the calculation this many times would take

$$3.3 \times 10^{76} \times 10^{-6} \, \text{s} = 3.3 \times 10^{70} \, \text{s}$$

With 32 million seconds per year, the computation would take 10^{62} years!

In addition, evaluating the modulo-h value for 128-bit numbers is an involved computation requiring about 40 digits of precision, which is well beyond the integer range employed by most current computers.

9.4 Summary

This chapter introduced source coding concepts of data compression and encryption. The source entropy was used as a measure of information. The source entropy equals the average number of bits that are needed to encode the symbols produced by a source, while the effective file entropy measures the number of bits required to encode a file. The Huffman coding procedure generates variable-length code words as a means for reducing the quantity of data that needs to be stored or transmitted without any information loss. The idea is to assign short code words to the most common symbols and the longest code words to the least common. A code word translation table also needs to be transmitted. Some applications (fax machines) can generate a code and incorporate the table within the system, thus eliminating the need to transmit it. Data security is reasonably maintained through encryption with a pseudo-random number generator that is common to both the transmitter and receiver. Transmitting the seed is performed by employing modular arithmetic, which requires exhaustive computations to crack.

9.5 Problems

9.1 **Modeling a source with a die toss.** A die has six sides and produces a symbol corresponding to the number of dots showing on the top face. A fair die produces symbols that are equiprobable. What are the symbols and their probabilities?

9.2 **Source entropy of the die-toss source.** Compute $\mathcal{H}_S$ for the source that produces symbols X_1 through X_6 with probabilities modeled by a fair die toss.

9.3 **Source entropy of the unfair die-toss source.** Compute $\mathcal{H}_S$ for the source that produces symbols with probabilities modeled by an unfair die toss in which a 1 appears one quarter of the time and 2 through 6 have equal probabilities.

9.4 **Source entropy of the decimal digit source.** Compute $\mathcal{H}_S$ for the source that produces ten symbols that have equal probabilities.

9.5 **Data sequences having specified $\mathcal{H}_S$.** Compose a representative sequence containing ten or more symbols that would be typical of a source having the following source entropy value.

(a) $\mathcal{H}_S = 0$ bits/symbol

(b) $\mathcal{H}_S = 1$ bit/symbol

(c) $\mathcal{H}_S = 2$ bits/symbol

9.6 **Effective probabilities of symbols in a text file.** A file contains the following text:

```
i need a vacation!!!
```

Include the space as a separate symbol. What are the values for the vocabulary size m_x, total file size n_T, and the effective probabilities of the symbols in this file?

9.7 **Effective source entropy of a text file.** Use the values in Problem 9.6 to compute $\hat{\mathcal{H}}_S$.

9.8 **Huffman code a text file.** Generate a Huffman code for the text file in Problem 9.6. Encode the file with the variable-length code and compare the total bit count to the $\hat{\mathcal{H}}_f$ value.

9.9 **Valid or invalid Huffman code.** Does the following code represent a valid Huffman code? (Hint: draw the tree.)

$$X_1 : 0 \quad X_2 : 01 \quad X_3 : 10 \quad X_4 : 11$$

9.10 Huffman code for digits. Generate the Huffman code for the ten digits 0 through 9 that have equal probabilities.

9.11 Data compression. Consider the following data file.

AAACAAABAAACAAADAAAE

Implement a Huffman code to compress this file. What is the average number of bits per symbol?

9.12 Huffman tree encoding. Find the binary sequence generated by the symbol sequence

$$X_1 \ X_3 \ X_2 \ X_5 \ X_4 \ X_6$$

using the Huffman code tree shown in Figure 9.13.

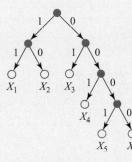

Figure 9.13

Huffman code tree for Problem 9.12.

9.13 Huffman tree decoding. What symbols are represented by the binary sequence

01110010001000010

using the Huffman code tree shown in Figure 9.14.

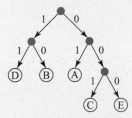

Figure 9.14

Huffman code tree for Problem 9.13.

9.14 Huffman code tree for symbols having unequal probabilities. A coin flip produces a 0 if a tail appears and a 1 for a head. A source produces symbols that have the same probabilities as the sums that are observed when three fair coins are flipped. For example, the outcome HTH produces 101, which sums to 2. Determine the Huffman code tree for this source.

9.15 Source encoding. A source generates six symbols with $P[X_1] = 0.05$, $P[X_2] = 0.47$, $P[X_3] = 0.07$, $P[X_4] = 0.20$, $P[X_5] = 0.11$, and $P[X_6] = 0.10$.

(a) Compute the source entropy $\mathcal{H}_S$.

(b) Generate the Huffman code tree.

(c) Encode the ten-symbol sequence

$$X_3 \ X_1 \ X_3 \ X_5 \ X_6 \ X_1 \ X_6 \ X_3 \ X_1 \ X_3$$

(d) Compute the average number of bits per symbol used for encoding the sequence.

(e) Compare the values of the source entropy and the average bits/symbol computed.

(f) Why does the $\mathcal{H}_S$ value differ significantly from the average bits/symbol?

9.16 Huffman code for a symbol file. Consider the data file containing the ten symbols

$$X_3 \ X_3 \ X_5 \ X_1 \ X_2 \ X_5 \ X_1 \ X_1 \ X_2 \ X_4$$

(a) Specify a fixed-length code to encode the symbols and compute the total number of bits that encode the file.

(b) Compute the effective source entropy of the file $\hat{\mathcal{H}}_S$.

(c) Generate a Huffman code to encode the file.

(d) Encode the file using this code and compute the total number of bits that encode the file.

9.17 Encryption. Let

$$\text{RBS}_i = 10011001$$

Use RBS_i to encrypt the data

$$\mathcal{D}_i = 11001011$$

to form the encrypted binary sequence $\mathcal{E}_i$ and to decrypt it.

9.18 Pseudo-random number algorithm. Letting $X_0 = 123$, generate X_1 and X_2 using the PRNG formula

$$X_i = \text{mod}[\ (\alpha \ X_{i-1} + \beta), \ h]$$

with $\alpha = 766, \beta = 369, h = 1,000$.

9.19 Random numbers to random bits. Assume the pseudo-random number algorithm in Problem 9.18 generated the eight random numbers

$$979, 122, 475, 986, 730, 286, 66, 899$$

Describe how you would generate eight random bits from this sequence and generate the random bit sequence.

9.20 Encryption key. Compute the value of the encryption key using the values: $n = 10, a = 13, x = 5, y = 6$.

9.6 Excel Projects

The following Excel projects illustrate the following features:

- Generate random source symbols with specified probabilities.

- Compute the source entropy and the effective entropy of a data file.

Using a Narrative box (described in Example 13.4), include observations, conclusions, and answers to questions posed in the projects.

9.1 **Generating equi-probable source symbols.** Using Example 13.46 as a guide, consider a source producing $m = 9$ symbols $X1, X2, \ldots, X8$ that have equal probabilities. Produce $n_T = 30$ random symbols.

9.2 **Computing effective entropy.** Using Example 13.47 as a guide, generate $n_T = 30$ realizations of the $m = 9$ symbols in Project 9.1 and compute effective probabilities $P_e[X_i]$ and effective entropy $\mathcal{H}_e$.

9.3 **Huffman Code.** Using Example 13.48 as a guide, generate a Huffman Code of the symbols in Project 9.2

9.4 **Pseudo-Random Number Generator.** Using Example 13.49 as a guide, use your own values for A, B, h, and seed X_0 to generate 16 pseudo-random numbers that generate 16 pseudo-random bits. What limits your values of A and B for a successful PRNG in Excel?

9.5 **Encryption of sinusoidal data.** Using Example 13.50 as a guide, quantize $n_x = 17$ samples (period $= 16$) of sinusoidal waveform to 4-bit data and your PRNG to generate a random bit sequence having the same number of bits to encrypt the data. Compare plots of the original waveform and encrypted data. At the receiver decrypt the data and plot the decrypted data.

9.6 **Secure Key Transmission.** Using Example 13.51 as a guide, illustrate transmission of the key. If x and y are both ≤ 15, what is the maximum value of N that still produces successful key transmission?

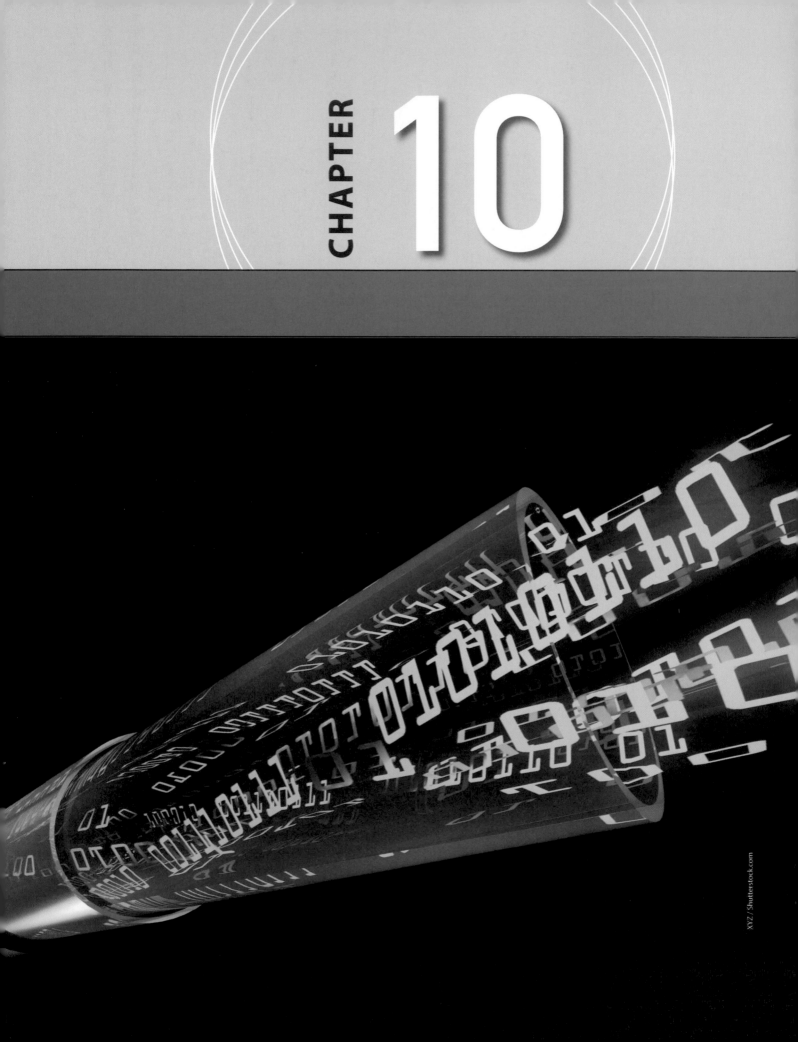

CHAPTER

10

CHANNEL CODING

LEARNING OBJECTIVES

After completing this chapter, the reader should be able to:

- Understand how to model errors in digital data.
- Detect and correct single errors in digital data under a variety of error conditions.
- Understand the trade-off between error-correction capability and data quantity.
- Compute the data carrying capacity of a data transmission channel having a specified bandwidth and signal-to-noise ratio.

10.1 INTRODUCTION

This chapter describes channel coding techniques for reliable storage or transmission over a communication channel. Several schemes are suggested to detect and even correct errors in data. The simplifying *single error assumption (SEA)* is employed to illustrate the basic ideas. Commercial systems correct more than one error but are mathematically more complicated and are topics for future study. All error-correction methods use some form of redundancy by adding bits whose values are computed from the original data. This redundancy results in an increased data size that is measured with the *data increase factor (DIF)*. These data are transmitted over a channel having a limited *channel capacity*, which measures the rate data can be reliably transmitted over a noisy channel. For reliable data transmission, the channel capacity must be greater than the data rate produced by the source.

This chapter introduces the problem of transmitting data from a source to destination accurately and reliably. To insure successful operation, a data transmission system must confront the following issues.

Errors—Data errors are inevitable because the Prob[error] > 0 for practical data transmission systems. Because the magnitude of Gaussian noise is theoretically unlimited, there is always the possibility of error occurrence. Error-detection codes determine when errors occur, and error-correction codes restore the correct data. But these codes come with the cost of increased data size and computation times. Errors in most properly operating systems are rare. A practical system usually begins by working flawlessly. As the system degrades, errors begin to occur. It is at this time that error correction is useful: Not only does it allow the system to continue to operate as intended, but detecting the occurrence of errors indicates that the system is beginning to fail so that the system can be repaired or replaced.

Error correction—Although practical error correction techniques use advanced mathematical ideas, this chapter presents the fundamental ideas by considering these operations under the simplifying constraint of single errors.

Data rate $\mathcal{D}_S$—For data transmission, it is important to measure how many bits a source produces per second. The source data rate $\mathcal{D}_S$ is expressed in bits per second and is convenient for comparing data sources.

Channel capacity $\mathcal{C}$—Binary data are transmitted over a channel having a capacity measured in bits per second that is limited by noise. For reliable data transmission, the channel capacity $\mathcal{C}$ must be greater than the data rate $\mathcal{D}_S$ of the source.

10.2 DATA STRUCTURES

Figure 10.1 shows the different data structures discussed in this chapter. These include

Data: Binary data are produced by a source, which can be single-bit or multiple-bit units.

Figure 10.1

Data are produced by a source, form code words for error detection and correction, and are packed into data packets for transmission. The packet interval is the total count of bits in a packet.

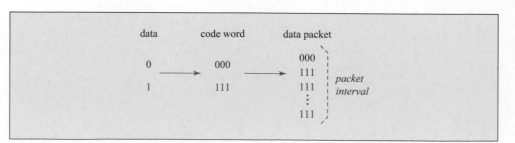

Code word: Code words add redundant bits (that is, bits computed from the data values) to form a fixed-length data structure to facilitate error detection and/or error correction.

Data packet: Data packets are blocks of code words that contain additional redundant code words for error correction. Data packets form the basic data structures for transmission over a communication channel or for storage in a digital memory. Data packet sizes range from a single code word to thousands of code words.

Typical communication systems employ standard data packets with code words that are bytes (containing 8 bits) plus an appended parity bit. Data packets terminate in a checksum byte with its own parity bit. Such data packets conform to computer data structures to simplify processing and storage tasks. For sources that produce different fixed-length or variable-length code words, the transmitter *packs* these binary data into standard data packets and the receiver *unpacks* these standard data packets to form the original code words.

10.3 CODING FOR ERROR CORRECTION

Error correction is the process of adding redundant bits to *data* produced by a source to form *code words*. The code words are grouped into *data packets* in order to detect and correct errors during transmission over a channel or storage in a digital memory. Other terms for this topic are *channel encoding* and *forward error correction*. While this field is mathematically rich, this chapter presents the basic ideas by constraining the number of errors that can occur.

The basic simplifying assumption used in this book is the *single error assumption (SEA)*: *No more than one error can occur during a data packet interval*, as shown in Figure 10.2. Data packet intervals range in size from the smallest data packet that occurs for error correction (a 3-bit code word) to large data packets that can be formed by grouping a large number of code words into a single block.

The validity of the SEA for specific channel determines the data packet interval: If errors are common, the packet interval must be small (3 bits). If errors are extremely rare, data packets can be large, with the maximum size being determined by transmission factors other than errors. Channel conditions can change, and if multiple errors cause the SEA to be no longer valid for a particular packet interval, the packet interval must be reduced. The shortest packet interval is a 3-bit code word that accomplishes error correction (shown next).

Unlike analog waveforms, digital data can be expanded to include additional bits that help the system detect and even correct errors when they occur. To understand

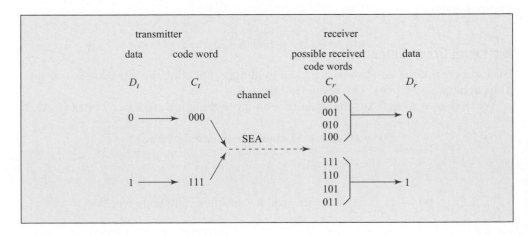

Figure 10.2

Signal error assumption (SEA) where no more than one error can occur during a data packet interval. The simplest data packet is shown, consisting of a single 3-bit code word.

the trade-offs in providing error detection and correction, we compute a *data increase factor* (DIF) that measures the cost of achieving these capabilities.

10.3.1 Redundancy and Data Increase Factor

Errors occasionally occur when data are transmitted. To make the system robust, it is desirable to detect when an error has occurred, and if the system requirements demand it, to correct the error. Codes can be designed to detect errors in the received data by adding *redundant* bits to data to form *transmission code words*. Such bits add nothing to the information but serve to confirm it. Redundant bits have values that can be computed from the original data. This computation is also performed by the receiver upon reception of the code word and the result is compared to the observed values of the transmitted redundant bits included in the code word. If the two computed and transmitted values of the redundant bits agree, there is no error. Otherwise, error detection or correction can be performed, depending on the system design.

We often use redundancy in our conversations: A *yes* answer is often accompanied by a nod of the head, and a verbal instruction to turn right is often clarified by a hand gesture. When the visible gesture is inconsistent with the verbal information, the recipient detects that an error has occurred. In a similar manner, transmitted code words should contain a mechanism by which the receiver is able to detect that an error has occurred. This is the purpose of error-detection codes.

The basic error recovery mechanism at the detector is *redundancy*—either by repetition, by including additional bits, and/or by designing specific bit patterns. A little redundancy provides *error detection*, while additional redundancy provides *error correction*. However, these additional data take longer to transmit and require more memory to store.

More robust error-correction codes produce more data. We define the *data increase factor (DIF)* as

$$DIF = \frac{\text{final size of data packet}}{\text{original data size}} \tag{10.1}$$

We now describe error detection and error correction methods.

10.3.2 Bit-Level Error Detection

Different coding systems employ special forms of error detection and correction. Each form is designed to treat the types of errors each system encounters. This section starts with the simplest systems and increases the system complexity to show how the considerations change.

Bit-Level Error Detection

Bit-level error detection can be achieved by simply repeating the bits (that is, transmitting each data bit twice). This results in DIF = 2.

When does it work? As long as there is no more than one error per 2 bits (SEA), so

$$00 \rightarrow 01 \text{ or } 10 \quad \text{(single errors are detected)}$$

and

$$00 \rightarrow 11 \quad \text{(two errors produce another valid code word)}$$

Error detection by repetition   **EXAMPLE 10.1**

Errors can be detected by repeating bits. Let the original data sequence be

1 0 1 0

A transmission data code word is formed by repeating each data bit. The corresponding 2-bit code words would appear as

11 00 11 00

Repeating the data, of course, does not add any new information, so these second bits in each 2-bit code word are redundant. These redundant bits allow an error in the transmission of a code word to be detected. If the received data appeared as

11 01 11 00

it would be readily apparent to the receiver that an error had occurred in the second code word. However, the receiver could not determine if the true value of the code word is either 11 or 00. That determination is the function of error-correcting codes.

The trade-off for this error-detection capability is that the size of the original data has been doubled in forming the transmission code words. Thus, DIF = 2.

10.3.3 Bit-Level Error Correction

Single errors can be corrected by adding additional redundant bits to the code words. Upon reception, these redundant values are computed from the data and compared against the values that were received. With the assistance of error detection, which indicates the code word containing the error, error correction corrects the error.

Bit-level error correction is achieved by repeating the bits three times. This results in DIF = 3. The error-correction mechanism uses the majority count to determine the correct bit value.

When does it work? As long as there is only one error per three bits, the error can be corrected. If there are two errors, the error can be *detected*, but the *correction* will be wrong. Figure 10.3 shows an interpretation of an error correction code that illustrates an idea proposed by a Bell Labs engineer Richard Hamming in the 1950s. Representing an error correction code in a three-dimensional space, we can see how far (called a *Hamming distance*) from a valid code a single error will take us. A single error-correction technique has a Hamming distance equal to one and still keeps the code within the vicinity of a valid code word.

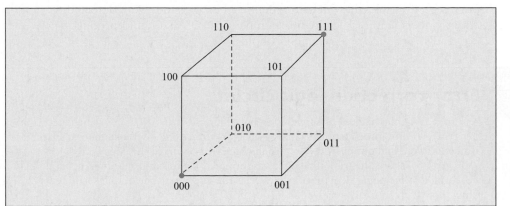

Figure 10.3

Representing an error correction code in a three-dimensional space. A single error keeps the code within the vicinity of a valid code word (either 000 or 111).

EXAMPLE 10.2

Error correction by repetition

Errors can be corrected by repeating bits. A simple error-correction method is to form a 3-bit code word from each original data bit. Let the original data sequence be represented by

$$1\ 0\ 1\ 0$$

A 3-bit code word representation of these data would appear as

$$111\ 000\ 111\ 000$$

Clearly, repeating the data does not add any new information, so these additional two bits are redundant. These redundant bits allow a single error in the transmission code word to be corrected. If the received data appeared as

$$111\ 010\ 111\ 000$$

a majority count performed on each 3-bit code word involves noting the values of most of the bits in the received code word. It would be readily apparent to the receiver that an error had occurred in the second code word and that the true value should be 000. Correcting binary data is especially simple, requiring only flipping the erroneous bit value from a 1 to a 0 in this case to produce the corrected sequence. Hence,

$$111\ 000\ 111\ 000$$

The trade-off for this error-correction capability is that the size of the data has been tripled in forming the transmission code words. Thus, DIF = 3.

Error Correction Logic Circuit

A repeated-bit transmitted triplet permits error correction. Let each binary value D_t be transmitted as a triplet of identical values DA_t, DB_t, and DC_t with

$$DA_t = DB_t = DC_t = D_t \tag{10.2}$$

That is, if $D_t = 0$, the transmitted values will be $DA_t\ DB_t\ DC_t = 000$, and if $D_t = 1$, the transmitted values will be $DA_t\ DB_t\ DC_t = 111$.

These transmitted values travel over a communication channel that introduces noise that can cause an error at the receiver. Under SEA, only one error at most can occur in the transmitted data triple. If $DA_t\ DB_t\ DC_t$ are transmitted, the receiver detects the triplet, forming the received data as $DA_r\ DB_r\ DC_r$, which can only have one of the following eight values

$$000\quad 001\quad 010\quad 011\quad 100\quad 101\quad 110\quad 111$$

Since $DA_t\ DB_t\ DC_t$ had identical values, transmission errors occurred in the second through seventh triples, in which non-identical binary values were detected.

An error-correction logic circuit examines $DA_r\ DB_r\ DC_r$ and produces $D_r = 0$ if $D_t = 0$ and $D_r = 1$ if $D_t = 1$ when SEA holds.

EXAMPLE 10.3

Error-correction logic circuit

Assume SEA holds and only a single error can occur in $DA_r\ DB_r\ DC_r$. Let D_t be transmitted as a code word of three identical values. An error-correction circuit examines each received 3-bit code word and outputs $D_r = D_t$ even if one error occurs in the code word.

Figure 10.4 shows the truth table that was formed using this information. The input section was formed by including all of the possible received 3-bit values. The output section sets $D_r = 0$ when there were more 0's received than 1's, and $D_r = 1$ otherwise.

DA_r	DB_r	DC_r	D_r
0	0	0	0
0	0	1	0
0	1	0	0
0	1	1	1
1	0	0	0
1	0	1	1
1	1	0	1
1	1	1	1

Figure 10.4

Truth table for error correction with received data DA_r, DB_r, DC_r as inputs when transmitted data $DA_t = DB_t = DC_t = D_t$. Output D_r equals D_t under SEA.

10.3.4 Data Packet Error Correction

If errors are extremely rare, the SEA extends to larger data sizes called *data packets* that contain data in 8-bit units (bytes). To implement a data packet error-correction system takes two steps:

STEP 1. Provide error detection for each code word.
STEP 2. Provide error detection for each *bit position* in a code word.

This combination identifies the erroneous code word and the bit position of the error. This information is all that is needed to correct single errors in the data packet merely by flipping the erroneous bit. The allowable data packet size is related to the validity of SEA: If SEA is valid over a large number of bits, the data packet can be large. As a channel degrades, however, multiple errors begin to be encountered in each data packet and is sensed by the error detection logic. However, with our scheme, two errors cannot be corrected, although more sophisticated error-correction techniques can correct such multiple errors. When error-detection logic shows more than one error per packet, the SEA is no longer valid, necessitating the reduction in the data packet size. These smaller packets pay the cost of an increase in the DIF, as shown next.

Code Word-Level Error Detection

Code word-level error detection is achieved by adding an extra bit, called a *parity bit*, to each byte. The parity bit value is such that the count of 1's in the byte and the parity bit is always either an even or odd number for all code words. When the count is even, it is called *even parity*, and when odd, it is *odd parity*. For interpreting the count, zero is considered to be an even number. The choice of which parity to use is predetermined—known to both the transmitter and receiver—and depends on the failure mechanism that causes errors, as explained next. This error detection scheme is successful as long as the SEA holds. If the data size is n bits, the code word size is $n + 1$, producing

$$\text{DIF} = \frac{n + 1}{n} \tag{10.3}$$

For the common byte-sized data $n = 8$ and the DIF $= 9/8 = 1.125$.

EXAMPLE 10.4

Parity bit for code-word error detection

Let the original data sequence consist of 2-bit data values (dibits). An example of four dibits is

$$\underbrace{00}_{\text{dibit}} \quad \underbrace{01}_{\text{called dibit}} \quad \underbrace{10}_{\text{dibit}} \quad \underbrace{11}_{\text{dibit}}$$

An even-parity system forms code words by appending a parity bit to each dibit such that each code word contains an even number of 1's. Forming 3-bit even-parity code words from the dibits produces 3-bit code words, as

$$\underbrace{\underbrace{00}_{\text{dibit}} \ \mathbf{0}}_{\text{code word}} \quad \underbrace{\underbrace{01}_{\text{dibit}} \ \mathbf{1}}_{\text{code word}} \quad \underbrace{\underbrace{10}_{\text{dibit}} \ \mathbf{1}}_{\text{code word}} \quad \underbrace{\underbrace{11}_{\text{dibit}} \ \mathbf{0}}_{\text{code word}}$$

Since the parity bits do not add any new information, they are redundant. These parity bits allow an error in the transmission of each 3-bit code word to be detected.

To illustrate the error-detection procedure, consider receiving code words equal to

$$010 \quad 011 \quad 101 \quad 110$$

It is readily apparent to the receiver that an error occurred in the first received code word because its parity is not even. However, the receiver could not determine the bit position of the error. If the error occurred in the first bit position, the correct code word is 110, and if the error occurred in the third bit position, the correct word is 011.

This error-detection capability is achieved at the expense of data size, which has been increased by one bit in forming the 3-bit code words. Thus,

$$\text{DIF} = \frac{2+1}{2} = 1.5$$

EXAMPLE 10.5

Choice of parity

The choice of even or odd parity is determined by the failure mode of the system. For example, the standard computer memory is structured to contain byte-sized data, each with an additional ninth bit for parity to form 9-bit code words. Assume the particular technology used to implement the computer memory fails by having a faulty memory location produces all 0's, as

$$\underbrace{\underbrace{00000000}_{\text{data byte}} \ 0}_{\text{9-bit code word}}$$

These nine 0's produce a valid even-parity code word, so the bad memory location would not be detected when using even parity. In this case, the memory should employ odd parity so the all-0's value would be detected as a failure.

Now consider a memory technology that fails by producing all 1's, as

$$\underbrace{\underbrace{11111111}_{\text{data byte}} \ 1}_{\text{9-bit code word}}$$

These nine 1's produce a valid odd-parity code word. In this case, the memory should employ even parity so the all-1's value would be detected as a failure.

Correcting Data Packet Errors

If adding parity bits to bytes detects code-word errors, adding a parity bit to detect errors in each *bit position* would provide sufficient information to correct errors in a data packet. This bit-position parity is implemented by appending a *longitudinal redundancy check code word (LRC)* to the end of a data packet. The LRC value is computed from the data in the packet, so it is redundant. Because each data byte appends a parity bit (for code word error detection), providing a parity bit for each bit position in the code word would locate the position within the code word that contains the error. For data packets, SEA is interpreted as *no more than one error in the entire data packet, including the LRC*. Once the erroneous bit is detected, it can be corrected by merely flipping its value.

Forming a data packet with a LRC code word

EXAMPLE 10.6

Consider a system that performs code word error correction on data bytes with even-parity bits that form 9-bit code words. A data packet with three data bytes, even parity, and an LRC is formed using the following four steps, as shown in Figure 10.5.

STEP 1. Append even-parity bits to the three data bytes to form code words.

STEP 2. Stack the code words so the bit positions align in columns.

STEP 3. Add a LRC byte whose bit values form even parity in each data-bit column.

STEP 4. Form an LRC code word by appending an even parity bit to the LRC byte. (The parity bit value of the LRC also forms the parity for the entire data packet.)

The original data consisted of three bytes (24 bits). Code-word parity bits increased the data packet to 27 bits and adding the 9-bit LRC produced a data packet containing 36 bits. The data increase factor is

$$\text{DIF} = \frac{36}{24} = 1.5$$

The DIF decreases when the data packet contains more data. A data packet that contains m 8-bit bytes has m 9-bit code words. The LRC code word increases the packet

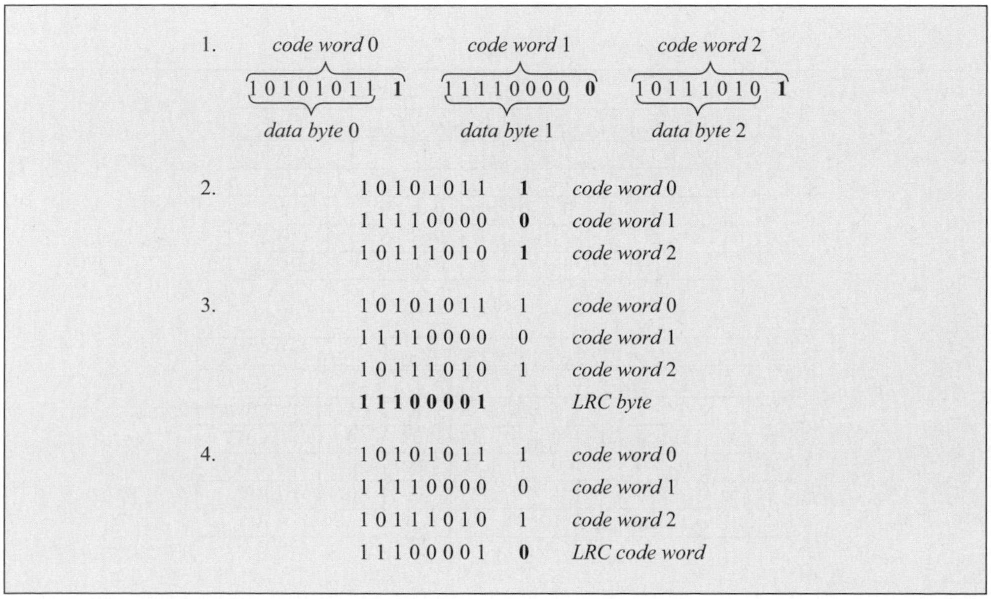

Figure 10.5

Steps in forming a data packet with even parity and an LRC code word.

size by 9 bits. Thus,

$$DIF = \frac{9(m+1)}{8m} \tag{10.4}$$

The previous example had $m = 3$, giving DIF $= 36/24 = 1.5$.

As the data packet size m increases, the additional LRC code word becomes less significant and the DIF approaches $9/8 = 1.125$, which is the value for parity-bit error detection. This small DIF makes the addition of an LRC code word an efficient and preferred method for transmitting data packets over cellular systems and the Internet.

EXAMPLE 10.7 Correcting a single data packet error

A data packet contains four data bytes within four 9-bit code words CW0, CW1, CW2, and CW3 and a LRC_{odd} that forms an odd parity in each column. Each code word in the data packet has its own odd-parity bit denoted as P_{odd}. The data packet was transmitted over a noisy channel. The received data packet shown in Figure 10.6a contains one error.

Find and correct the error using the following five steps.

STEP 1. The receiver computes its own parity values P'_{odd} using *only the data bytes*. That is, the parity bits in the received data packet (P_{odd}) are not included in the receiver's parity calculations.

STEP 2. The receiver compares its computed P'_{odd} to the received P_{odd}. Differing values indicate the code word that contains an error. Figure 10.6b shows an error occurred in CW1.

STEP 3. The receiver computes its own odd-parity LRC code word, which is denoted as LRC'_{odd}, to form odd parity using *only the data columns in the data packet*. The bits in the received data packet LRC_{odd} are not included in the receiver's LRC'_{odd} calculations.

STEP 4. The receiver compares LRC_{odd} to LRC'_{odd}. Differing bit values indicate that a transmission error occurred in that bit position. The difference in data column b_4 indicates an error in that code word column.

STEP 5. The differing P'_{odd} and LRC'_{odd} values point to b_4 in CW1 as being the error. It is a simple matter for the receiver to change the received $b_4 = 1$ in CW1 in Figure 10.6b to the correct value $b_4 = 0$.

Figure 10.6

Data-packet error correction in Example 10.7. (a) Received data packet contains one error. (b) Processed data packet detects error in b_4 of CW1. Error clues shown in bold text indicate b_4 in CW1 is an error.

b_7	b_6	b_5	b_4	b_3	b_2	b_1	b_0	P_{odd}	CW#
1	0	1	0	1	1	1	1	1	CW0
1	1	0	1	1	1	1	0	0	CW1
0	0	1	1	0	0	1	0	0	CW2
0	1	0	0	1	0	1	0	0	CW3
1	1	1	0	0	1	1	0	0	LRC_{odd}

(a)

b_7	b_6	b_5	b_4	b_3	b_2	b_1	b_0	P_{odd}	P'_{odd}	CW#
1	0	1	0	1	1	1	1	1	1	CW0
1	1	0	**1**	1	1	1	0	**0**	**1**	CW1
0	0	1	1	0	0	1	0	0	0	CW2
0	1	0	0	1	0	1	0	0	0	CW3
1	1	1	**0**	0	1	1	0	0	0	LRC_{odd}
1	1	1	**1**	0	1	1	0			LRC'_{odd}

(b)

b_7	b_6	b_5	b_4	b_3	b_2	b_1	b_0	P_e		CW#
1	0	1	0	1	1	1	1	0		CW0
1	1	0	1	1	1	1	0	0		CW1
0	0	1	0	1	0	1	0	1		CW2
0	1	0	1	1	0	1	1	1		LRC_e

(a) Error-free data packet

b_7	b_6	b_5	b_4	b_3	b_2	b_1	b_0	P_e	P'_e	CW#
1	0	1	0	**0**	1	1	1	**1**	1	CW0
1	1	0	1	1	1	1	0	0	0	CW1
0	0	1	0	1	0	1	0	1	1	CW2
0	1	0	1	**1**	0	1	1	1	1	LRC_e
0	1	0	1	**0**	0	1	1			LRC'_e

(b) Processed data packet containing two errors

Figure 10.7

Data-packet error correction in Example 10.8. (a) Error-free data packet. (b) Processed data packet containing errors in b_3 and P_e of CW0 prevents error correction although LRC difference indicates error detection.

Error correction not possible with two errors

EXAMPLE 10.8

Figure 10.7a shows an error-free data packet that contains three code words CW0, CW1, and CW2 using even parity and an LRC that uses even-parity in the column bits.

Two errors are introduced in the data packet (a SEA violation) at b_3 and P_e of CW0, as shown Figure 10.7b. The figure shows the processing performed in the receiver that corrects single errors. The following steps illustrate that two errors prevent error correction.

STEP 1. The receiver computes P'_e, compares it to P_e. and finds no disagreements because CW0 contains errors in b_3 and P_e.

STEP 2. The receiver computes LRC'_e and compares it to LRC_e. The differing values in the b_3 column indicate an error in that code word position.

STEP 3. The receiver determines that an error occurred in the b_3 column but cannot determine the code word that contains the error. Thus, *error correction* is not possible, but *error detection* is still possible.

Different systems treat uncorrectable errors in data packets in different ways. If the data must be correct (such as in banking transactions) the receiver sends a message back to the transmitter that the data packet contains an error and to re-transmit the data packet. In time-critical (real-time) systems, such as streaming video, re-transmitted data packets would arrive too late to be useful. In this case, these systems often substitute the previous correct data packet for the erroneous one. This causes a stuttering video playback if there are many erroneous data packets.

10.4 DATA RATE

The *data rate* measures the amount of data a source generates in one second. Let $\mathcal{R}$ denote the *source symbol rate* equal to the number of symbols a source produces every second. For example, if a source transmits ten symbols per second, $\mathcal{R} = 10$ symbols/s.

A source that has a vocabulary of m symbols produces the equivalent of $\log_2 m$ bits of data when it transmits a symbol (or $\log_2 m$ bits/symbol). For example, the source that has $m = 4$ (symbols X_1, X_2, X_3, and X_4) can encode two bits ($\log_2 4 = 2$) of data with each symbol transmission.

The data rates of different sources are compared using common units of *bits/second* (*bps*). Multiplying the symbol rate $\mathcal{R}$ (symbols/second) and $\log_2 m$ (bits/symbol) we obtain the *source data rate*, which is denoted as $\mathcal{D}_S$ with units bps, as

$$\mathcal{D}_S = \mathcal{R} \log_2 m \text{ bps} \tag{10.5}$$

EXAMPLE 10.9 — Smartphone in a Wi-Fi network

A smartphone transmits binary data using two signals ($s0$ and $s1$) that are considered symbols ($m = 2$) at a source symbol rate equal to

$$\mathcal{R} = 10^6 \text{ symbols/second}$$

Its data rate is computed as

$$\mathcal{D}_S = 10^6 \overbrace{\log_2 2}^{=1} = 10^6 \text{ bps}$$

EXAMPLE 10.10 — Digital audio source

The microphone converts your speech into a waveform that is limited to have a maximum frequency $f_{max} = 3.5$ kHz. The ADC in your smartphone samples this waveform using sampling rate $f_s = 8$ kHz. Each sample is a symbol, and the source symbol rate equals

$$\mathcal{R} = 8{,}000 \text{ symbols/second}$$

The ADC quantizes each sample with 8-bit code words. Thus, the ADC has a vocabulary of different symbols having a size equal to

$$m = 2^8 = 256$$

The data rate your smartphone produces when you talk on the phone is equal to

$$\mathcal{D}_S = 8{,}000 \overbrace{\log_2 256}^{=8} = 64{,}000 \text{ bps}$$

▶ 10.5 CHANNEL CAPACITY

The *channel capacity* measures the number of bits per second that can be transmitted reliably over a channel. The quality of a channel for transmitting data depends on two parameters.

Bandwidth $\mathcal{B}$ is the frequency range from the low frequency limit f_L to the high frequency limit f_H (measured in Hz) that the data transmission signals can occupy with $\mathcal{B} = f_H - f_L$. As $\mathcal{B}$ increases, the duration of the data signals decreases. Thus, it takes less time to transmit data.

Signal-to-noise ratio (SNR) is the ratio of the energy of signals propagating on a channel $\mathcal{E}_s$ to noise on the channel measured by its variance σ_N^2. Chapters 7 and 8 describe the significance of $\mathcal{E}_s/\sigma_N^2$ in extracting data values from detected signals. Noise present on the channel limits the amount of data that can be reliably transmitted over a channel.

A formula developed by Claude Shannon (of entropy fame) and Ralph Hartley is called the *Shannon-Hartley theorem* and computes the channel capacity, which is

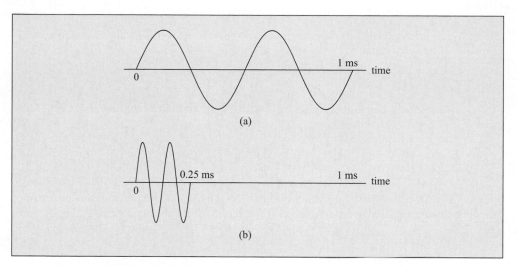

Figure 10.8

Data signals at two bandwidth values: (a) $B = f_H = 2$ kHz and (b) $B = f_H = 8$ kHz.

denoted as C in units of *bits per second (bps)* as

$$C = B \log_2 (1 + \mathrm{SNR}) \text{ bps} \tag{10.6}$$

Let us examine the effects of bandwidth B and SNR separately.

This formula states that the channel capacity C varies linearly with bandwidth. Thus, doubling B doubles C. Reducing B increases the duration of the data signals that can propagate on the channel and thereby reduces C.

Effect of B on the channel capacity EXAMPLE 10.11

To simplify, let the lower frequency limit $f_L = 0$, which makes $B = f_H$. Assume a data signal consists of two periods of a sinusoidal waveform and has a frequency equal to $B = f_H$, as shown in Figure 10.8.

When $B = f_H = 2$ kHz, the two-period data signal has a duration equal to 1 ms.

When $B = f_H = 8$ kHz, the data signal has a duration equal to 0.25 ms. Thus, four data signals can be transmitted when $B = 8$ kHz in the time that one data signal can be transmitted when $B = 2$ kHz. Thus, C varies linearly with B.

The C value also varies as the logarithm of the signal-to-noise ratio. For data signals the $\mathrm{SNR} = \mathcal{E}_s / \sigma_N^2$ where $\mathcal{E}_s$ is the energy in the signal and σ_N^2 is the noise variance. Substituting these values into the capacity equation gives

$$C = B \log_2 \left(1 + \frac{\mathcal{E}_s}{\sigma_N^2} \right) \text{ bps} \tag{10.7}$$

Effect of noise on the channel capacity EXAMPLE 10.12

To interpret the effect of noise on channel capacity, consider a channel having a fixed bandwidth B and vary the noise variance σ_N^2.

When $\sigma_N^2 = 0$ (true only for an ideal channel) the capacity equation gives

$$C = B \log_2 \left(1 + \frac{\mathcal{E}_s}{0} \right) \rightarrow B \log_2 (\infty) = \infty$$

The infinite capacity occurs because, when there is no noise, we can differentiate *any two signals* no matter how close they are to each other. For example, we can make a quantizer step size Δ extremely small and still differentiate two adjacent levels. Recall that the number of quantizer bits are related to Δ as

$$\Delta = \frac{V_{\max}}{2^b} \quad \rightarrow \quad b = \log_2\left(\frac{V_{\max}}{\Delta}\right)$$

If Δ is very small, the number of bits b that can be transmitted becomes very large. Thus, a large capacity is achieved.

At the other extreme, $\sigma_N^2 = \infty$ gives

$$C = B\log_2\left(1 + \frac{\mathcal{E}_s}{\infty}\right) \rightarrow B\log_2(1) = 0$$

In this case, the noise is so large that it overwhelms any signal and prevents any information from being transmitted over the channel. Thus, the channel capacity is zero.

EXAMPLE 10.13 Telephone channel capacity

The telephone channel has a bandwidth equal to $B = 3 \times 10^3$ Hz. A telephone channel with SNR $= 1,000$ has capacity equal to

$$C = B\log_2(1 + \text{SNR})$$
$$= 3 \times 10^3 \times \log_2(1 + 1,000)$$
$$= 3 \times 10^3 \times 9.97 = 29,900 \text{ bps}$$

An approximate value for C is obtained without a calculator by recalling $1,000 \approx 2^{10}$. Then $\log_2(1,001) \approx 10$, making $C = 30$ kbps.

For a practical data transmission system, the channel must reliably transmit all the data produced by the source. Thus, the *channel capacity must be greater than or equal to the source data rate*, as

$$C \geq \mathcal{D}_S \tag{10.8}$$

The capacity equation is useful for computing the effect of bandwidth B and SNR on channel performance. However, its values are slightly larger than achievable in practice. As a practical measure of system performance, we substitute for the C value calculated in Eq. (10.6) with the *data transmission capability* routinely achieved by commercial data channels. Figure 10.9 lists some common data transmission systems, the technology they use, and their data transmission capability.

To accommodate the two common data rates, units used for we use the convention that the rate expressed in *bits per second* is written as *bps* and the rate in *bytes per second* as *Bps*.

Figure 10.9

Data transmission systems, their technology, and their data transmission capabilities.

System	Technology	Data Transmission Capability
WiFi	Short-range radio wave	1 Mbps (10^6 bps)
Cellular 4G LTE	Long-range radio wave	100 Mbps (10^8 bps)
Ethernet	Long copper cable	100 Mbps (10^8 bps)
USB 2.0	Short copper cable	480 Mbps (0.48×10^9 bps)
USB 3.0	Short copper cable	5 Gbps (5×10^9 bps)
Fiber-optic	Glass fiber	1 Gbps (10^9 bps)

EXAMPLE 10.14

Watching a video download

A video source produces data at the rate equal to $D_V = 10^4$ bytes per second (Bps) and your smartphone must display it at that rate. Let the size of the data file that stores the video be denoted by n_f.

Assume a video segment is stored in a video file of size $n_f = 1$ MB (10^6 bytes). Thus, the video segment lasts a time denoted by T_V and equal to

$$T_V = \frac{n_f}{D_V} = \frac{10^6 \, B}{10^4 \, Bps} = 100 \, s$$

Let the channel capacity (data transmission capability) of your smartphone connection equal $C = 7.5 \times 10^3$ Bps. Because $C < D_V$, some buffering is required on the smartphone to produce continuous playback without pauses. The data transfer time over the channel is denoted as T_D and equals

$$T_D = \frac{n_f}{C} = \frac{10^6 \, B}{7.5 \times 10^5 \, Bps} = 133 \, s$$

There are two approaches that can be used for your smartphone to display a continuous video.

One approach is to first download the entire video file (in 133 seconds) to your smartphone and play the 100-second video.

The alternative approach is to download a sufficient amount of data to start the video and transmit data while the video is playing. Ideally, the video ends just after the last byte in the file is transmitted. With the video lasting T_V seconds, the video can start at time denoted as T_{start} and equal to

$$T_{start} = T_D - T_V = 133 - 100 = 33 \ \text{seconds}$$

During this time the quantity of data transferred to your smartphone equals

$$33 \text{ s} \times 7.5 \times 10^3 \, \text{Bps} = 0.25 \, \text{MB}$$

During the one hundred seconds the video is playing on the smartphone, the remaining data is transferred and is equal to

$$100 \text{ s} \times 7.5 \times 10^3 \, \text{Bps} = 0.75 \, \text{MB}$$

thus completing the 1 MB data transfer.

10.5.1 Capacity Loss with Transmission Range

As your smartphone moves away from the cell-phone company transmitting antenna the signal strength $\mathcal{E}_s$ detected at your smartphone drops (your bars become smaller). This section examines the effect of the signal energy ($\mathcal{E}_s$) attenuation on the channel capacity C.

Signal Energy Decay with Transmission Range

Imagine shining a flashlight onto a wall, as shown in Figure 10.10. As the flashlight (transmitter) range from the wall increases, the beam spot becomes larger and dimmer, because the transmitter power (which is fixed) spreads over a larger area. The intensity of the transmitted signal at the receiver (your smartphone) equals the transmitter power divided by the area. The smartphone antenna converts the intensity into the detected signal energy $\mathcal{E}_s$.

The decay of intensity with range is governed by the *inverse square law*, which states that the intensity is reduced by the square of the range to the transmitter. To show this, let R_o denote the radius of the beam at range r_o and $A(r_o)$ denote the beam area range $r = r_o$. Thus,

$$A(r_o) = \pi R_o^2 \tag{10.9}$$

When the range doubles to $r = 2r_o$, the radius doubles and the area $A(2r_o)$ equals

$$A(2r_o) = \pi (2R_o)^2 = 4\pi R_o^2 = 4A(r_o) \tag{10.10}$$

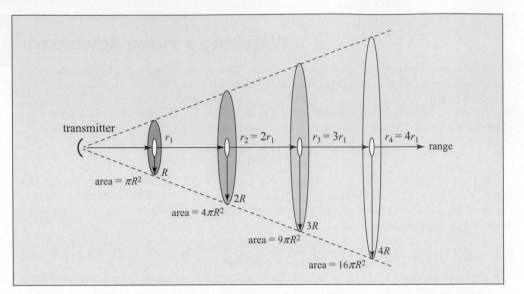

Figure 10.10

Transmission intensity loss with range. The smartphone antenna converts intensity into signal energy and is indicated as a small circle along the transmitter beam axis.

In general, when the range increases by factor α, the area increases by α^2. The area at range $r = \alpha r_o$ equals

$$A(\alpha r_o) = \alpha^2 A(r_o) \qquad (10.11)$$

Because the signal intensity at your smartphone decreases as $A(\alpha r_o)$ increases, the signal energy detected by your smartphone $\mathcal{E}_s$ also decreases as

$$\mathcal{E}_s(\alpha r_o) = \frac{\mathcal{E}_s(r_o)}{\alpha^2} \qquad (10.12)$$

A calibration procedure is required to employ this formula. When the receiver is located at a known range $r = r_o$ from the transmitter, the system is calibrated by measuring the signal energy detected by the smartphone $\mathcal{E}_s(r_o)$.

Figure 10.11 shows the consequences of the decay of $\mathcal{E}_s(\alpha r_o)$ with range $r = \alpha r_o$. At close range (small α), $\mathcal{E}_s(\alpha r_o)$ is large. As range increases (α increases), $\mathcal{E}_s(\alpha r_o)$ reduces, as in Eq. (10.12). Problems arise when the signal energy falls below a minimum value, which is denoted as $\mathcal{E}_{min}$, because errors become too common.

The maximum range of satisfactory service (that is, the Prob[error] is sufficiently small) is denoted as r_{max} and equals

$$r_{max} = \alpha_{max} r_o \qquad (10.13)$$

The minimum signal energy value equals

$$\mathcal{E}_{min} = \frac{\mathcal{E}_s(r_o)}{\alpha_{max}^2} \qquad (10.14)$$

For range $r = \alpha r_o < r_{max}$ (that is, $\alpha < \alpha_{max}$) satisfactory service is provided to the users, because $\mathcal{E}_s(\alpha r_o) > \mathcal{E}_{min}$ produces signals well above the noise level, as shown in Figure 10.11a. When $r = \alpha r_o > r_{max}$ (that is, $\alpha > \alpha_{max}$), $\mathcal{E}_s(\alpha r_o) < \mathcal{E}_{min}$ and the signals are difficult to detect in the presence of noise, as shown in Figure 10.11b. In this latter case, the Prob[error] becomes sufficiently large to make the system unreliable.

To maintain satisfactory service, the cellular system must locate the network of transmitting antennas so that a smartphone can always detect sufficiently strong signals.

Channel Capacity Reduction with Range

The variation of signal energy $\mathcal{E}_s(\alpha r_o)$ with range expressed in Eq. (10.12) can be included in the capacity equation in Eq. (10.7) to show the variation in channel capacity with

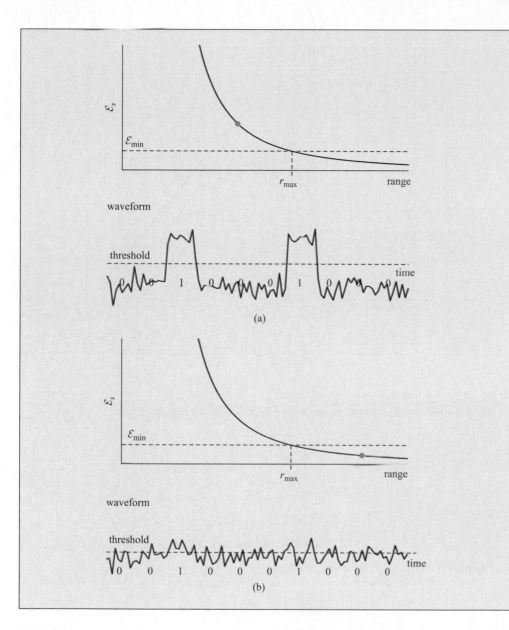

Figure 10.11

Signal energy decay through the inverse square law. (a) Digital signals at $r < r_{max}$ are detected reliably. (b) Digital signals at $r > r_{max}$ are unreliably detected in the presence of noise.

range, as

$$C(\alpha r_o) = \mathcal{B} \log_2 \left(1 + \frac{\mathcal{E}_s(\alpha r_o)}{\sigma_N^2} \right) = \mathcal{B} \log_2 \left(1 + \frac{\mathcal{E}_s(r_o)}{\alpha^2 \sigma_N^2} \right) \tag{10.15}$$

Assume the bandwidth $\mathcal{B}$ and noise variance σ_N^2 do not change with range. Figure 10.12 shows the expected decrease in capacity as range increases when $\mathcal{B} = 10^4$ Hz and $\mathcal{E}_s(r_o)/\sigma_N^2 = 10^4$ at $r_o = 1$ m. At $r = r_o = 1$ m, the capacity equals $C(r_o) = 1.3 \times 10^5$ bps and decreases to 10 bps at $r = 4{,}000$ m.

Capacity of cellular telephone system   EXAMPLE 10.15

Consider a cell-phone system operating with constant bandwidth $\mathcal{B} = 10^5$ Hz and noise variance σ_N^2.

At range $r_o = 1$ m, $\mathcal{E}_s(r_o)$ is measured to be one million times greater than σ_N^2, or

$$\mathcal{E}_s(r_o) = 10^6 \, \sigma_N^2$$

making the SNR $= 10^6$.

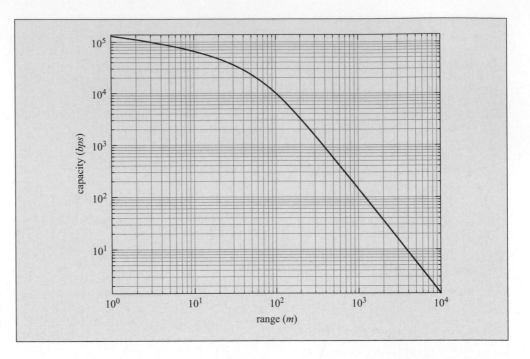

Figure 10.12

Capacity as a function of range. $\mathcal{B} = 10^4$ Hz, and $\mathcal{E}_s(r_o) = 10^4 \sigma_N^2$ with $r_o = 1$ m.

The capacity of this short-range transmission channel is computed as

$$C(r_o) = \mathcal{B}\log_2\left(1 + \frac{\mathcal{E}_s(r_o)}{\sigma_N^2}\right)$$

$$= 10^5 \times \log_2\left(1 + \frac{10^6 \times \sigma_N^2}{\sigma_N^2}\right)$$

$$= 10^5 \times \log_2(1{,}000{,}001)$$

$$= 10^5 \times 19.93 \approx 20 \times 10^5 \text{ bps}$$

The approximation also occurs by noting that $10^6 = 2^{20}$.

When the smartphone moves to $r = 1$ km, $\alpha = 10^3$, the detected signal energy equals

$$\mathcal{E}_s(10^3\, r_o) = \frac{\mathcal{E}_s(r_o)}{(10^3)^2} = \frac{\mathcal{E}_s(r_o)}{10^6}$$

Substituting $\mathcal{E}_s(r_o) = 10^6\,\sigma_N^2$ gives

$$\mathcal{E}_s(10^3\, r_o) = \frac{10^6\,\sigma_N^2}{10^6} = \sigma_N^2$$

Thus, at a 1-km range, the signal energy equals the noise variance. The channel capacity at this range equals

$$C(10^3\, r_o) = \mathcal{B}\log_2\left(1 + \frac{\mathcal{E}_s(10^3 \text{ m})}{\sigma_N^2}\right)$$

$$= 10^5 \times \log_2\left(1 + \frac{\sigma_N^2}{\sigma_N^2}\right)$$

$$= 10^5 \times \log_2(2) = 10^5 \text{ bps}$$

Note that, even though the signal energy decreased by one million, the channel capacity has decreased by a factor of twenty. This is a consequence of the logarithmic relationship of the SNR in computing the channel capacity.

Inverse-square law while driving
EXAMPLE 10.16

This example examines the reduction of signal-strength bars on your smartphone as your journey starts close to a cell-phone transmission tower and you drive away.

The signal energy at your smartphone is measured to be $\mathcal{E}_s(r_o) = 10^4 \sigma_N^2$ at $r_o = 0.5$ km from the tower, providing the calibration value and a maximum signal-strength indication.

At $r = 25$ km, your smartphone begins to lose the signal (the number of bars decreases to zero). Thus, the maximum reliable operating range is

$$r_{max} = 25\,\text{km} = 50 r_o \quad \rightarrow \quad \alpha_{max} = 50$$

The minimum signal energy equals

$$\mathcal{E}_{min} = \mathcal{E}_s(\alpha_{max} r_o) = \frac{\mathcal{E}_s(r_o)}{\alpha_{max}^2} = \frac{10^4 \sigma_N^2}{2500} = 4\sigma_N^2$$

Increasing antenna power
EXAMPLE 10.17

A more powerful antenna is added to the system described in Example 10.16 and increases $\mathcal{E}_s(r_o)$ by a factor of 100 from $10^4 \sigma_N^2$ to $10^6 \sigma_N^2$. The new maximum operating range is

$$r'_{max} = \alpha'_{max} r_o$$

The minimum energy required for proper operation is

$$\mathcal{E}_s(\alpha'_{max} r_o) = \mathcal{E}_{min}$$

Substituting

$$\mathcal{E}_s(\alpha'_{max} r_o) = \frac{\mathcal{E}_s(r_o)}{(\alpha'_{max})^2} = \frac{10^6 \sigma_N^2}{(\alpha'_{max})^2}$$

and $\mathcal{E}_{min} = 4\sigma_N^2$ from the previous example gives

$$\frac{10^6 \sigma_N^2}{(\alpha'_{max})^2} = 4\sigma_N^2$$

Solving for α' yields

$$\alpha'_{max} = \sqrt{250{,}000} = 500$$

The new maximum range is

$$r'_{max} = \alpha'_{max} r_o = 250\,\text{km}$$

Thus, increasing the antenna power by a factor of 100 increases the operating range from 25 km to 250 km (a factor of 10).

This result can be generalized: Increasing the transmitter power by a factor increases the operating range by the square root of the factor.

Space probe channel capacity
EXAMPLE 10.18

A space probe has the bandwidth $\mathcal{B} = 10^4$ Hz that does not vary with range. The channel capacity measured at range r_o is $C(r_o) = 10^6$ bps.

The SNR of the detected signals at r_o is computed from

$$\frac{C(r_o)}{\mathcal{B}} = \frac{10^6\,\text{bps}}{10^4\,\text{Hz}} = 100 = \log_2\left(1 + \frac{\mathcal{E}_s(r_o)}{\sigma_N^2}\right)$$

Because $\mathcal{E}_s(r_o)/\sigma_N^2 \gg 1$, we use the approximation $\log_2(1 + x) \approx \log_2(x)$ to give

$$\log_2\left(\frac{\mathcal{E}_s(r_o)}{\sigma_N^2}\right) = 100$$

Raising both sides to the power of 2 yields the SNR at r_o as

$$\frac{\mathcal{E}_s(r_o)}{\sigma_N^2} = 2^{100}$$

The probe ventures out to range $r = 1024r_o$ (making $\alpha = 2^{10}$). If σ_N^2 remains the same, the new SNR of the signals detected on Earth is found by computing the decay in the signal energy, as

$$\mathcal{E}_s(\alpha r_o) = \frac{\mathcal{E}_s(r_o)}{\alpha^2} = \frac{\mathcal{E}_s(r_o)}{2^{20}}$$

The new SNR equals

$$\frac{\mathcal{E}_s(1024r_o)}{\sigma_N^2} = \frac{\mathcal{E}_s(r_o)/\sigma_N^2}{2^{20}} = \frac{2^{100}}{2^{20}} = 2^{80}$$

If $\mathcal{B}$ remains the same, the capacity at $r = 1024r_o$ is

$$\mathcal{C}(1024r_o) = \mathcal{B}\log_2(\text{SNR}) = 10^4\log_2(2^{80}) = 10^4 \times 80 = 8 \times 10^5 \text{ bps}$$

Note: Even though the range increased by a factor of more than 1,000, the capacity dropped by only 20%.

10.6 Summary

This chapter described several schemes to detect and correct errors in data. The simplifying single error assumption (SEA) was employed to illustrate the basic ideas. All error-correction methods use some form of redundancy by adding bits whose values are computed from the original data. This redundancy results in an increased data size that is measured with the data increase factor (DIF).

Sources that produce audio or video data exhibit a specific data rate. Data are transmitted over a channel having a capacity that measures the rate at which data can be reliably transmitted. For real-time operation the channel capacity must exceed the source data rate. The channel capacity is determined by its bandwidth and the signal-to-noise ratio of the data signals. The inverse square law described the decrease in signal energy with range. The decrease of the channel capacity with range allows cell-phone systems to position antennas in a network.

10.7 Problems

In the problems below, the rate expressed in bits per second is written as bps and the rate bytes per second as Bps.

10.1 Error correction by repetition. Design a bit-level code that corrects up to 2 errors per code word. What is the data increase factor for your code?

10.2 Data-packet error correction. A data packet contains the 5-digit number 24680. Each code word contains a single digit encoded using binary-coded-decimal (BCD) codes that form quadbit (4-bit) data plus an even-parity bit. Generate a data packet that includes a LRC code word. What is the data increase factor for your data packet?

10.3 Data rate of texting. You enter text on a smartphone keypad at a rate of ten characters per second and an 8-bit ASCII code word is generated for each character. What is your data rate in bps and Bps?

10.4 Data rate of a video source. A digital video is a sequence of digital frames with each forming a 480×640 pixel image and with each pixel encoded with 24-bit color. The video displays 15 frames/sec. What is the data rate produced by this video source?

10.5 Digital camera source. My 10 Megapixel camera has three (R, G, and B) sensors in each pixel and each sensor outputs 256 levels of intensity. The camera allows me to take multiple pictures at a rate of four per second when I keep the shutter button depressed.

(a) What is the maximum data rate from the camera into the camera memory card when I keep the shutter button depressed?

(b) If I want to store 300 images, what size memory card should I purchase for my new camera? (Specify size as an integer number of gigabytes, e.g., X Gb memory.)

10.6 Data rate in the telephone system. If the audio waveform sampling period in the smartphone is $T_s = 0.1$ ms and each sample is quantized to 256 levels, what is the data rate produced by your smartphone?

10.7 Time to transfer 1 TB disk data. A external hard disk drive has a 1 terabyte (TB) capacity. How long does it take to transfer 1 TB through a USB 2.0 port having a 480 Mbps transmission rate? How long does it take if you upgrade to a USB 3.0 port having a 5 Gbps transmission rate?

10.8 Video download to your smartphone. A 10 MB video file requires a data rate $D_V = 6$ Mbps for continuous playback. The file is transmitted over a channel having transmission data rate equal to 1 Mbps. At what point in the data transfer should your smartphone begin to play the video to produce continuous playback and with minimal delay?

10.9 Data transmission between your smartphone and antenna tower. A powerful antenna tower transmits data *down* to your smartphone at a rate equal to 300 Mbps, while your less-powerful smartphone transmits data back *up* to the tower at 75 Mbps. Assume the bandwidths and noise levels are the same in each transmission direction. How are the signal energies at the tower and cellphone related? Evaluate for $\mathcal{B} = 1$ MHz and 10 MHz. (Hint: Use the channel capacity equation and ignore the +1 in the $\log_2$ term.)

10.10 Minimum signal-to-noise ratio. The data rate produced by your smartphone is $\mathcal{D} = 2 \times 10^5$ bps. Your smartphone transmits data over the frequency band from 100 kHz to 200 kHz. What is the minimum SNR required to transmit your data in real time?

10.11 Channel capacity reduction with range. The signal energy detected by your smartphone at range $r_o = 0.5$ km equals $\mathcal{E}_s(r_o) = 10^6 \sigma_N^2$. The bandwidth $\mathcal{B} = 100$ kHz and is constant with range.

(a) What is the value of $\mathcal{C}(r_o)$?

(b) As the smartphone distance from the antenna increases, at what range does $\mathcal{E}_s(\alpha r_o)$ become less than $100\sigma_N^2$?

(c) What is the channel capacity at that range?

10.12 Channel capacity of space probe. A space probe transmitter operates with $\mathcal{B} = 10^5$ Hz and $\mathcal{E}_s(r_o)/\sigma_N^2 = 10^6$ at $r_o = 1$ km. The probe transmits images that each contain 1 Mb of data. If an image takes approximately one hour to transmit, what is the range of the probe from the antenna on Earth?

10.8 Excel Projects

The following Excel projects illustrate the following features:

- Perform error correction.
- Compute the channel capacity as a function of the signal-to-noise ratio.

Using a Narrative box (described in Example 13.4), include observations, conclusions, and answers to questions posed in the projects.

10.1 Bit-level error correction. In work sheet *Repeated bits*, form a truth table that has input section formed by logic values DA_r, DB_r and DC_r, and write an Excel formula that determines D_r the correct value under the single error assumption.

10.2 Data-packet error correction. Following Example 13.53 generate a data packet with random binary data having five rows and five columns. Introduce a single-bit error to show that the erroneous bit can be located. Introduce two errors to illustrate the inability to correct them using parity bits and LRC.

10.3 Channel capacity. Using Example 13.54 as a guide, compute the channel capacity as a function of range when $\mathcal{E}_s/\sigma_N^2 = 2 \times 10^6$ at range $r_o = 1$ km and $B = 20$ kHz.

(a) At what range does $\mathcal{E}_s < 10\sigma_N^2$?

(b) What is the channel capacity at that range?

CHAPTER

11

DATA NETWORKS

LEARNING OBJECTIVES

After completing this chapter, the reader should be able to:

- Understand how data packets are transmitted asynchronously over a data channel.
- Recognize the structure of data packets for robust communication over the Internet.
- Appreciate the differences between circuit switching and packet switching.
- Probe the topology of the Internet.
- Describe packet collisions that occur in wired and wireless networks.
- Comprehend the possible utility of cloud computing.

⊙ 11.1 INTRODUCTION

The *Internet* is a network of interconnected computers called *router nodes* that direct data from sources to specified destinations. Many sources generate data at irregular and random time instants, and such data transfer upon demand is called *asynchronous data transmission*. Data of various sizes form data packets that are transmitted over a channel. We describe how data packets are structured to ensure that data are received reliably and without error. The network interconnections provide robust operation by automatically by-passing individual routers that fail. Problems associated with packet collisions are described. Cloud computing resources are discussed with examples of the flexibility they provide.

The Internet copes with data communication that varies widely in timing, quantity, and delays, as illustrated in Figure 11.1 This chapter describes how a networks manages to operate successfully by considering the following issues.

Internet packets and protocols—Data packets that travel between different transmitters to different receiver experience variable delays. Data packets contain source and destination addresses for proper routing, a data size value for smaller packets, and a data segment number to re-sort data that get delayed.

Asynchronous data transmission—Data packets occurring at random times contain more than just data in order to avoid confusing the network by incorporating features that indicate their presence on the communication channel. The simplest packets using these techniques travel from a specific transmitter to a specific receiver and have a specific data size, such as the remote control (transmitter) that switches channels on your TV (receiver).

Probing Internet topology—To prevent erroneous data packets from endlessly circulating within the network, data packets contain a timer code word that expires if a problem occurs. This timer feature is used to probe the Internet to find the number of computers between a source and destination, as well as the travel time of the data packet.

Detecting collisions—Practical limitations cause data packets to collide. Techniques are described to detect collisions in wired and wireless data communication systems.

Figure 11.1

Data within a network travel from various sources to one or more destinations with data packets being transmitted and received at random times and with different delays. C are components (sensors) and D are devices (smartphones).

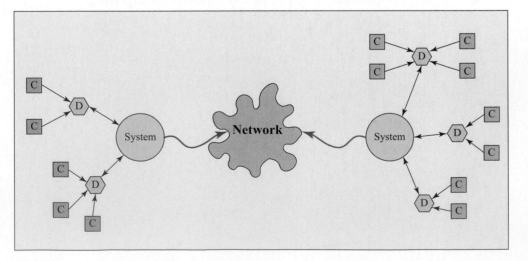

11.2 EVOLUTION OF DATA NETWORKS

So, where are digital networks heading, and why? We have seen computers grow smaller and faster, consume less energy, and communicate information using graphic displays on tablet or smartphone screens in place of computer monitors. Connections to the Internet, which used to be solely by cable, are now increasingly wireless. It all started with email, then cell phones that merged with Blackberries, and currently with ubiquitous tablets and smartphones.

Connectivity describes the study and practice of how to interconnect these devices so that they work together seamlessly. While the programming issues are beyond the scope of this book and usually in the domain of computer science, the electrical engineering issues include models and protocols. The topics covered in the previous chapters permit a meaningful overview of these connectivity issues.

Protocols are rules governing operation. Internet protocol (IP) was established to allow computers to communicate with each other. In data transmission, solving data communication problems between users in complex networks usually means that additional information is attached to the data packet and is sufficient for the network to solve problems. In data transmission, this additional information is stripped away and never seen by the application requesting the data, such as a video player.

32-bit to 64-bit addresses EXAMPLE 11.1

The Internet started getting crowded, with many additional Web sites appearing daily, and IP addresses were running out. The solution was to expand the address size from 4 bytes (= 32 bits) to 16 bytes (= 128 bits). With 32-bit addressing, under Internet Protocol Version 4 (IPv4), the number of unique addresses is

$$2^{32} = 2^2 \times \overbrace{2^{10}}^{=10^3} \times \overbrace{2^{10}}^{=10^3} \times \overbrace{2^{10}}^{=10^3} = 4 \times 10^9$$

or nine billion separate addresses. The 128-bit addressing under IPv6 provides

$$2^{128} = \left(2^{32}\right)^4 = (4 \times 2^{30})^4 \approx (4 \times 10^9)^4 = 256 \times 10^{36}$$

separate addresses, enough to have one for each conceivable device: computers, smartphones, televisions, GPS, automobiles, tablets, and appliances. In the envisioned Internet of things (IoT), each device incorporates a small Web-Interface-computer (WIC), each having its own address that has the ability to receive and send data.

11.2.1 Packet Switching and Circuit Switching

Historically, there have been two main methods for transmitting data from a source to a destination on a system level or, equivalently, from a transmitter to a receiver when describing operations on a circuit level.

- *Packet switching (the postal model).* In this model, the complete data file is broken up and packed into envelopes (data packets) that can be accommodated by the postal system. In today's electronic transfers, the address of the destination and the return address of the source are *Internet protocol (IP) addresses*. When you type a Web name (such as www.yale.edu), which is technically called a *Universal*

Resource Locator (URL) this URL is looked up by a *Domain Name Server (DNS)* in a local computer, called a *router* to find the corresponding IP address of the destination, in this case (130.132.35.53). This IP address was assigned by an Internet administration group when Yale registered both the name and a particular computer on the Yale campus, which is identified by its unique *Media Access Control (MAC)* number specified by the computer manufacturer when the computer was built. This process produces a unique IP address assigned to www.yale.edu.

- *Circuit switching (the telephone operator model).* In early telephone days, a telephone connection was made by an operator physically connecting a wire from your phone to that of the person being called. More recent telephone systems made the connection with electrically-operated switches. In both cases, the result is a dedicated wired connection between the source and destination. Circuit switching is used for special (and costly) applications where data transmission time is critical.

The rest of the chapter describes the more common packet-switching method. Sufficiently fast Internet speeds make these two methods less distinguishable. Special services demanding high-speed data transfer in the future will likely provide a hybrid solution.

11.2.2 Internet Transmission Protocols

While data on the Internet are transmitted in the form of data packets, different applications have different requirements. This section describes the two most common *protocols* or methods for data transfer that represent a trade-off between reliability and speed.

1. *Transmission control protocol/Internet protocol (TCP/IP)* guarantees that the transmitted data are correct—no matter how long it takes. The term *TCP* indicates the structure of data packets, while *IP* specifies the addressing and data-packet sizes to enable communication between two computers. Data integrity at the data receiver is ensured by including error detection elements (parity bits and a checksum character) within a data packet. If no error is detected, the destination transmits a short *acknowledgment data packet (ACK)* back to the source, as Figure 11.2 illustrates. If the source receives no ACK within a specified time delay, it assumes the data packet was lost and re-transmits the data packet. If a data packet error is detected at the destination—or along the path to the destination—a short *not-acknowledge data packet (NAK)* is sent back to the source requesting re-transmission.

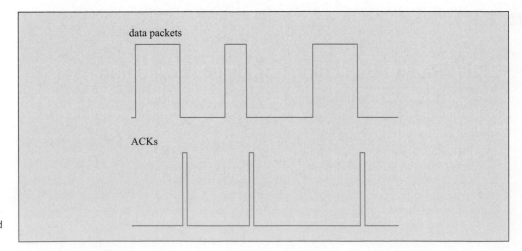

Figure 11.2

Data packets are transmitted and acknowledged.

Sending ACKs and NAKs (with re-transmissions) slows the data transfer rate, but data are transferred successfully. TCP/IP is the current method used for Web surfing and downloading documents on the Internet. While most data packet transfers are handled in a timely manner, transmission delays become noticeable when the data traffic becomes congested. Under severe congestion, Web pages fail to load at first request.

2. *Universal datagram protocol/Internet protocol (UDP/IP)* transmits data from source to destination with minimal delay by eliminating the re-transmission of erroneous packets. UDP/IP is employed for applications requiring real-time operation, such as telephone calls over the Internet called *voice over Internet protocol (VoIP)*, or for video streaming. If a received data packet contains an error under TCP/IP, the time delay involved in the destination sending a NAK and receiving a re-transmitted correct packet would cause annoying pauses in the audio or video playback. To eliminate such pauses, UDP/IP uses its *best effort* to transmit the data. This means not transmitting acknowledgment (ACK or NAK) data packets and no data packet re-transmissions.

In this case, the destination needs to accommodate received data packets that contain errors. Different systems have different approaches: some replace an erroneous data packet with the previous good data packet (common in video streaming), while others use empty data packets or *average* data packets (common in VoIP). When you make a call using your smartphone, the data are usually transmitted using UDP/IP. If you listen carefully, you will hear short pauses caused by missing or erroneous data packets. As long as there is less than a 5% error rate, the audio does not sound too objectionable. However, in times of high Internet congestion, delays and errors cause the audio to have an annoying choppy quality, sometimes making a telephone conversation frustratingly unintelligible.

As an alternative to UDP/IP, if you are willing to accept a significant delay in order to get good sound quality, you could transfer an audio or video file using standard TCP/IP, wait until all (or most of) the data are received, and then start playing it on your computer. However, this approach would have too long of a delay to allow for normal telephone conversations using VoIP.

▶ 11.3 ASYNCHRONOUS DATA TRANSMISSION

Transmitting data between two digital systems requires special considerations. For example, data packets typically occur at random times: when a mouse is clicked, when a key on the keyboard is pressed, or when the next frame of video is required. In *asynchronous data transmission*, data packets are transmitted when they become available.

To illustrate transmission of randomly occurring data, we describe the transmission of a single data byte that is enclosed within a time waveform called a *data character* that contains additional timing information sent by a transmitting circuit (transmitter) to a receiving circuit (receiver). The transmitter and receiver both use the same timing interval T_B that is set when specifying the transmission rate (*setting the baud rate*). Other systems use *self-clocking codes* (described in Chapter 12) that are less efficient.

The transmitter controls the signals on the channel and the receiver monitors the channel for the arrival of new data. In asynchronous data transmission, the transmission channel is idle for much of the time and has an *idle-channel* value supplied by the transmitter that indicates the transmitter is connected. For example, the idle-channel value being 5 V indicates that the transmitter is working, while a 0 V level does not.

The asynchronous data character contains the following four components shown in Figure 11.3.

1. **Start bit.** The beginning of each data character sent by the transmitter is indicated with a start bit whose value produces a change from the idle-channel value (for

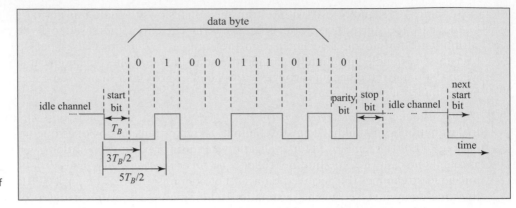

Figure 11.3

Asynchronous character format of data byte 01001101.

example, from 5 V to 0 V). The receiver uses this start-bit transition time to start timing the data character. The start-bit duration is T_B seconds (s) and is used by the transmitter and receiver.

2. **Data.** The data bits are transmitted one at a time with each lasting T_B s. The receiver determines the value of each data bit by examining the data character waveform value in the middle of its T_B interval. That is, the middle of the first data bit interval occurs at time $3T_B/2$ as measured from the start-bit transition time as shown in Figure 11.3. The middle of the second data bit occurs at time $5T_B/2$, and so on.

3. **Parity bit.** The data character appends a parity bit to the data for error detection. The parity bit is optional and not included for reliable transmitter-receiver connections to increase the data transmission rate. For systems using even parity, the value of the parity bit is such that the number of 1's in the data plus the parity bit is even.

4. **Stop bit(s).** The data character is terminated with one or more stop bits—each have the idle-channel value and last for T_B seconds. Stop bits provide a delay between character transmissions. The transmitter inserts a sufficient number of stop bits to allow the receiver to process the received data before it transmits the next character.

EXAMPLE 11.2   ## Asynchronous transmission

A data source transmits data characters asynchronously over a transmission channel having a capacity equal to $C = 10^4$ bps. The transmission time for each bit over the channel is computed as

$$\frac{1}{C} = \frac{1}{10^4 \text{ bps}} = 0.10 \text{ ms/bit}$$

which makes $T_B = 0.1$ ms.

A data character contains a start bit, eight data bits, a parity bit, and one stop bit, which is eleven bits for each byte of data. The data character duration is denoted T_{char} and equals

$$T_{char} = 11 \ T_B = 1.1 \text{ ms}$$

The source transmits eight bits of data during each T_{char}. The source data rate is computed as

$$\mathcal{D}_S = \frac{8 \text{ bits}}{T_{char}} = \frac{8 \text{ bits}}{1.1 \text{ ms}} = 7,270 \text{ bps}$$

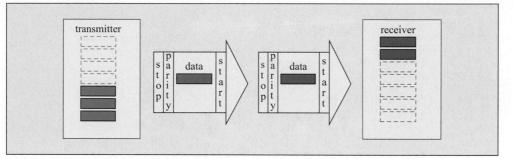

Figure 11.4

Asynchronous transmission divides large data files into smaller units that are transmitted as data characters.

Data characters typically have short durations as longer durations increase the probability of incurring errors. Large data files are broken up into smaller and more manageable units that are transmitted as a series of manageable data characters, as shown in Figure 11.4.

 # 11.4 INTERNET DATA PACKETS

Communicating over the Internet often involves the transmission of large quantities of data, such as downloads or videos. In such cases, the data is broken up into small segments and transmitted as a sequence of separate data packets. To illustrate the data exchange process, consider using the Internet to request funds from your bank account at an automated teller machine or to validate a credit card purchase. Information is sent in the form of these data packets into a network using a data format that allows the packet to travel through the network and arrive at the desired destination. Figure 11.5 shows a simplified data packet that contains six *fields* of useful information.

1. The *address* field contains the routing information about the destination and the source. If a response is required, such as an acknowledgement or error condition, the destination knows the return address.

2. The *time-to-live (TTL)* field contains a byte value that limits the number of times the packet can be relayed through intermediate nodes. The TTL value equals 255 in the packet transmitted by the source. Each time the packet is relayed by a computer node or *router*, the TTL is decremented. If the packet TTL count reaches zero at a particular node, that node deletes the packet and sends a message back to the source indicating the packet has been deleted. The TTL prevents packets from circulating endlessly through the Internet due to some malfunction. Otherwise, errant data packets would clog the Internet and even bring it to a halt. This feature is also used for determining the structure of the Internet in the next section.

3. The *data length specifier* field indicates the number of bytes in the data field. Data size is an important consideration for the operation of the network. Text data could be transmitted one byte at a time, but this would be inefficient because the overhead in the packet structure must be transmitted with each byte. Data sizes between 500 and 1,500 bytes are found to be transmitted most efficiently.

address	TTL	data length	tag	data	CRC

Figure 11.5

Simplified structure of an Internet data packet.

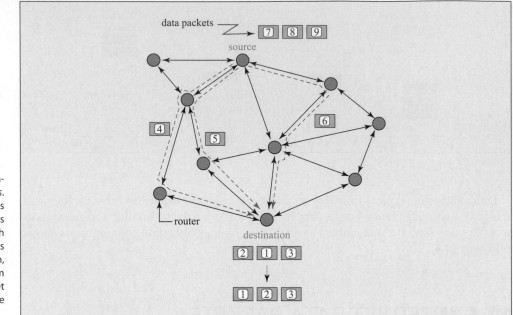

Figure 11.6

A network consists of many switching computer nodes or *routers*. These routers form multiple links with their neighbors. Data packets can travel over any available path (indicated in dashed line) that links the source with the destination, causing packets to arrive at random times. The tag in the data packet allows the destination to arrange the packets in the proper order.

4. The *tag* field is a byte that indicates the packet number in a sequence. The Internet is a dynamic, changing environment consisting of many switching routers between the source and the destination, as shown in Figure 11.6. Transmitting a large data file is usually done by partitioning it into multiple smaller data packets. Each individual packet can travel over different paths, using a different set of routers to travel from source to destination. Figure 15.6 shows alternate paths as dashed lines. Such flexibility maintains the integrity and reliability of the network. With such a flexible network, however, one of the packets may encounter a slow path, as when passing through a router that is handling a large quantity of data. This packet can arrive at the destination after a packet that was transmitted at a later time. In such a case, the packets must be put into the correct order at the destination, and this is the function of the *tag* field. The tag is a single byte with values from 0 to 255. The tag forms a modulo-256 counter, because having more than 256 packets in the network at one time is rare and the destination keeps track of tag overflow values.

5. The *data* field contains the bytes that are being transmitted. Parity bits are not appended to the data bytes to conserve data packet size.

6. The *cyclic redundancy check (CRC)* field performs the error-detection function in the data packet. It is a one-byte value that is computed at the source by summing the bits that are 1's in the data field and computing its modulo-256 value. This value detects a change in as few as one bit in the data and fails if the data field changes by a multiple of 256 (a rare occurrence). As the packet passes through each node, the router computes this sum and compares this value to the CRC value stored in the packet. If the two are equal, the probability is high that the data packet contains no errors.

   Data packets transmitted over a network occasionally experience *burst errors* that effect a large number of bits in the data packet. Burst errors occur when disturbances are comparable in duration to that of the data packet, such as lighting strikes or power surges produced by heavy equipment. The CRC is better suited than parity bits for detecting burst errors.

Multiple paths from source to destination allow the system to continue to function—even though one or more routers or links stop operating. The exact path to be taken by a particular packet is often chosen randomly, because deterministic routing schemes often result in bottlenecks. More sophisticated networks monitor the traffic congestion and select the fastest paths available. However, such monitoring requires additional circuits and complicates the network.

11.4.1 Probing the Internet

We can probe the structure of the Internet and determine data transfer times using a program called *Tracert*[1] (pronounced *trace route*). The *Tracert* (or Traceroute in Macs) command is followed by entering a Web address. For example, entering the command *tracert www.yale.edu* (on a PC) displays the number of nodes and data-packet travel times between your computer and the destination (in this case) www.yale.edu.

Tracert operates in the following manner.

1. It repeatedly transmits data packets to the URL until the destination is reached.

2. Tracert instructs your computer to set the first data packet TTL value to one and start a timer. Then the first node that this data packet encounters decrements it, making TTL = 0. That zero-valued TTL causes the first node to transmit a "NAK" that includes the node IP address back to the source (your computer). When your computer receives the "NAK", it stops the timer to determine the round-trip travel time to the first node. The minimum time recorded is 3 ms and time resolution is 1 ms on a PC.

3. Each transmission is repeated, and the travel times are reported three times to display any variations in travel times caused by congestion. Reported times to the same node can vary by a factor of two or more.

4. Tracert increments the TTL value in each subsequent data-packet transmission, causing the NAK to be sent by successive nodes. On a PC tracert limits the maximum TTL value to access 30 nodes.

5. The process stops when the destination is reached and tracert receives an ACK from the destination.

Tracert results   EXAMPLE 11.3

Running the tracert program while at Yale and specifying the destination as Illinois Institute of Technology (www.iit.edu—my alma mater) produces the results shown in Figure 11.7. The Yale computer system has seven internal nodes from my computer's Wi-Fi connection to node 8, which is the router that connects Yale's network to the *Internet Service Provider (ISP)*. The node IP addresses indicate that the route enters the IIT computer system (with first-byte IP address 216) at node 11. The last entry (node 13) is the computer that hosts the IIT Web home page and tracert indicates 31 ms lapsed before the round-trip acknowledgements from the that page were received.

A distance calculator on the Web gives the distance between New Haven, CT, and Chicago, IL, to be 1223.6 km. The data speed over the Internet is denoted c_d and can be computed from the round trip travel time as

$$c_d = \frac{2 \times 1223.6 \text{ km}}{31 \text{ ms}} = 8 \times 10^7 \text{ m/s}$$

Recalling that the speed of light $c = 3 \times 10^8$ m/s, the data packet speed over the Internet is approximately 1/4 the speed of light!

Executing tracert from Yale to Oxford (www.oxford.ac.uk) shows 17 nodes and a 100 ms delay caused by crossing the Atlantic Ocean.

---

[1] In Windows, *tracert* is accessed in the *cmd.exe window* (entered by typing "cmd" in the start text box). Macs access *Traceroute* in the *Terminal* window found the *Applications/Utilities* folder.

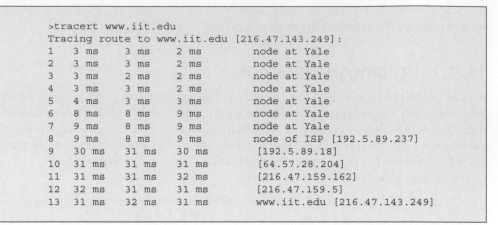

Figure 11.7

Results produced by "tracert www.iit.edu" for the data-packet path from Yale to Illinois Institute of Technology.

11.5 DETECTING PACKET COLLISIONS

Computers that are connected to a network either through a cable called an *ethernet cable* or (increasingly more often) through a wireless connection. Fundamental and practical limitations lead to data-transmission interference called *data packet collisions*.

Wired systems: Because signals travel over a pair of wires at approximately the speed of light (equal to one foot per nanosecond), data transmitted by one system is observed by a distant system after a significant time delay. The distant system sensing an idle channel begins to transmit its own packet, which collides with the arriving data packet transmitted the first system.

Wireless systems: Because wireless routers are positioned in the center of an accessible area, two systems on opposite sides of the area can sense the router signals but not those produced by each other. This *hidden terminal* problem occurs when both systems, sensing an idle channel, begin to transmit wireless data packets that collide at the router.

The following sections describe the solutions to packet collisions in wired and wireless systems.

11.5.1 Packet Collisions in Wired Systems

This section discusses the technique that allows two or more computers to exchange information reliably over a pair of wires that implements the communication channel. We use the simple example of two computers to illustrate the basic idea, but the approach extends to many computers.

Figure 11.8 shows two computers (C1 and C2) connected to a pair of wires over which data transmission occurs. The computers use the *data line wire* (labeled *DL*) for signalling and the other wire (GND) to establish a common ground voltage reference. DL is connected to supply voltage V_s through resistor R (the *pull-up resistor*) to set the idle-channel level of DL at V_s and to limit currents in the communication process. Each computer reads the value of DL with a *sensing gate* (*G*) that outputs a logic level based upon the DL voltage as

$$G = 0 \quad \text{if} \quad DL = V_s \tag{11.1}$$
$$G = 1 \quad \text{if} \quad DL = 0$$

C1 uses sensing gate G1 and C2 uses G2.

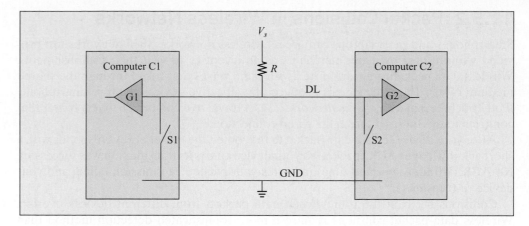

Figure 11.8

Two computers (C1 and C2) transmit data packets over the DL line using switches (S1 and S2) while reading DL with sensing gates (G1 and G2).

Each computer contains a switch that is normally open between DL and GND. C1 has switch S1 and C2 has switch S2. If both switches are open, $DL = V_s$. When either computer closes its switch, $DL = 0$. A data value D is communicated on DL using *negative logic* as

$$D = 0 \quad \text{if} \quad DL = V_s \qquad (11.2)$$
$$D = 1 \quad \text{if} \quad DL = 0$$

With negative logic, a computer transmits $D = 1$ by closing its switch to connect the data line to ground, thus making $DL = 0$.

C1 transmits $D = 1$ by closing S1 to make $DL = 0$ and verifies the DL value by reading the G1 output: If $G1 = 1$ (because it reads $DL = 0$) the operation was successful. C1 transmits a data packet on DL by using the asynchronous method described previously.

Now let's observe what C2 is doing during C1's transmission. When $DL = 0$, C2 reads $G2 = 1$. Now this is the clever part: *Because C2 did not close S2, $G2 = 1$ indicates to C2 that another computer (C1 in this simple case) is transmitting data.* C2 acquires the data packet by reading G2. If C1 is transmitting data to C2, C1 transmits the address of C2 in the data packet to inform C2 the data is directed to it.

A packet collision occurs if C2 closes S2 at the same time as C1 closes S1. This is not a problem initially because both C1 and C2 are transmitting $D = 1$ and read their gates to verify they are doing so correctly. A packet collision is noted when the values transmitted by C1 and C2 are different. For example, C1 transmits $D = 1$ (by closing S1 to make $DL = 0$) and C2 transmits $D = 0$ at the same time (by leaving S2 open). When C2 reads G2 to verify the transmitted value it gets the wrong value, because $DL = 0$ makes $G2 = 1$. Thus, C2 senses a packet collision, stops transmitting immediately, and continues to read G2 to determine if the data packet contains its address. Note that C1's data transmission is still valid. When C2 senses an idle channel ($DL = 0$ for an extended time), it can begin to transmit its data packet.

Because transmission delays may prevent immediate collision detection, both C1 and C2 may sense packet collisions. In this case, both stop transmitting immediately. Then each computer uses its uniform PRNG to generate a random time delay before starting to transmit again. The computer having the smaller delay starts transmitting its data packet first. The other computer has a larger delay, senses the channel is active, and delays its transmission until it senses an idle channel. This random-delay assignment allows priority to be given to premium customers by adjusting their PRNG to generate smaller random delays.

11.5.2 Packet Collisions in Wireless Networks

Smartphones and most laptops can access wireless networks. Such networks are provided without cost by many merchants as an incentive to visit their establishments. Wireless data packets are transmitted with radio waves that travel through the shared medium of air. The radio signals produced by all active devices (laptop, smartphone, iPod Touch, ...) in a *local area network (LAN)* travel to a common *wireless router* that connects to the Internet through a wired connection.

After your device sends a data packet to the router, the router always responds with a short acknowledge (ACK) packet, so your device knows its transmission was successful. No ACK (within a specified time) indicates your packet transmission failed, and your device re-transmits it.

Collisions occur at the router when data packets from different devices overlap. Wireless data-packet collisions require a more sophisticated detection method than that described previously for wired systems because of the *hidden terminal problem*. In this problem, two devices in the same LAN do not detect each other's packets (although both communicate with a central router). Thus, they cannot sense that the other device is transmitting its packet before transmitting its own packet. To detect collisions, the router uses constraints in the packet construction, which include the following.

- A known start-bit sequence called a *preamble* occurs at the beginning of each data packet. It is analogous to a long start bit. If a packet begins with a bit sequence that is different than the expected preamble, the router assumes a collision has occurred.

- Parity or checksum errors computed by the router indicate interfering packets, and no ACK(s) are sent.

- A maximum size is imposed on data packets. Let my device start sending a data packet, and the router detects its correct preamble. But before the data-packet transmission finishes, another (hidden) device starts transmitting a data packet that interferes with my packet. This collision would not have occurred if my data packet were shorter and had finished before the other device started transmitting. Thus, shorter data packets reduce the probability of collisions.

EXAMPLE 11.4 **Wireless packet collision detection**

Figure 11.9 shows a properly operating two-user system with data packets and ACKs transmitted at separate times. The Wi-Fi router assigns a unique IP address to each device and acknowledges each data packet it receives by broadcasting an ACK. Every wireless device in the LAN receives the ACKs and decodes only packets addressed to its IP address.

Each device checks for the presence of radio transmissions to ensure the channel is idle before transmitting data, which does not work when one device is hidden from the other. When the router senses simultaneous packet transmissions, it determines a collision has occurred through error detection and broadcasts a NAK to all users, as shown in Figure 11.9 as a negative-going signal. Devices that recently transmitted data packets (and have not received ACKs) wait a random amount of time before they re-transmit their data packet.

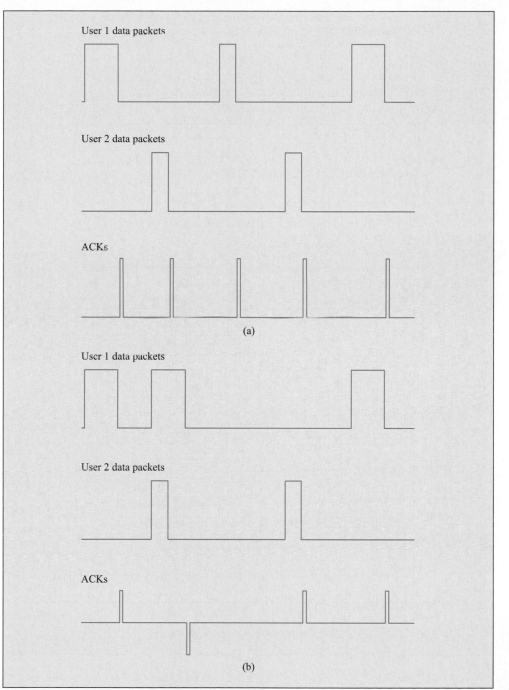

User 1 data packets

User 2 data packets

ACKs

(a)

User 1 data packets

User 2 data packets

ACKs

(b)

Figure 11.9

Data packet collisions: (a) properly operating two-user system with data and ACKs transmitted at separate times and (b) when the router senses a data packet collision, it broadcasts a NAK (shown as a negative pulse).

Modeling wireless packet collisions

EXAMPLE 11.5

The probability of data-packet collisions is reduced by using shorter data packets. Figure 11.10 shows two data packets having the same duration that start at random times during a constant data transmission time interval. A collision occurs if the data packets overlap. The collision probability increases as the data packet becomes a significant fraction of the data transmission interval. Shorter packets have lower collision probabilities, but each packet transmits less data. Engineers implement systems that maximize the data throughput from a set of sources to a set of destinations.

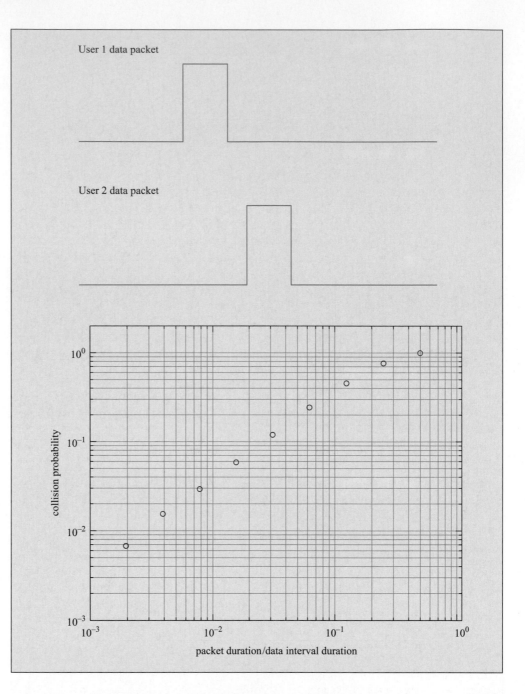

Figure 11.10

Packets of same size and duration starting at random times during a constant data interval duration. Collision probability is reduced by using shorter data packets.

11.6 CLOUD COMPUTING

A computer-cloud installation is a collection of hardware (CPUs and memories) and software (operating systems and applications) resources that takes advantage of the inter-connectivity and communication speed provided by the Internet. In a manner similar to the electrical service to your home, cloud computing is viewed as a utility that can be used when it is needed and in the quantity required to get the job done according to your schedule. When you buy a PC or laptop, you specify the computing resources, CPU speed, and memory size you need. A computer cloud offers more flexibility to accommodate changing computational needs.

The driving force behind cloud computing is the efficiency in costs that can be achieved.

Acquisition: You do not use your computer at all times. Cloud computing can use "your computer" for another job when you don't need it. The most eager customers for cloud computing are new companies that can easily expand their computer power as their business grows: They can start small (cheaply) to get the business going and acquire additional cloud computing resources as and when they are needed.

Power/cooling: Because a typical cloud installation can have as many as 10^4 servers, power can be more effectively managed by turning off entire sections if they are not used. Similar to multiple-core CPUs, server sections can be cycled to reduce cooling costs.

Maintenance: A cloud facility typically will purchase or lease all of its servers from a single manufacturer. Identical devices are easy to maintain and swap out if a problem occurs. Backups and software updates are similarly simplified for the IT staff.

So what is needed? The current thinking is that there are at least four consumer needs that can be served by cloud computing.

Shared peripherals: Specialized hardware is used rarely by a particular user, but when many users are present, the specialized hardware is used more often. Hence, by charging a small usage fee, even the most specialized hardware can be made available. This is an extension of devices like network printers, which can now be color (rather than black-and-white) or network hard drives that can handle temporary needs of terabyte storage.

Shared computers: Fast computers (super-computers) are available on an hourly use basis, but require complicated sign-up procedures. With tablet computers having flexible GUI's but lacking computing horsepower, sharing powerful computers on the cloud using tablet interfaces is an obvious extension to the super-computer model.

Shared software: There are many shrink-wrapped applications that are very specialized and expensive. Rather than each user buying a license for installation on one's computer, a less expensive fee can be charged for using and sharing the application on the cloud.

Flexible procurement: Buying a computer is a bigger task than simply going on-line and making a purchase. A careful analysis of computer power, memory requirements, and availability is needed and is prone to error by either over- or under-specification. Miscalculations can be costly. The cloud offers the possibility to lease the computer that satisfies current needs without large equipment costs. As the needs change, more or less computing power and memory can be easily obtained without actually getting new equipment that requires additional space, cooling, and maintenance.

The needs are typically generated by application programs that require large computation resources, larger storage capacity, or both. A further advantage of the cloud is the capability of easily sharing information or designs for interactive collaborations. For example, an architect and client can view a design with each using their own computer at their own location.

Figure 11.11 shows two extremes for hosting guests in a computer cloud. At one extreme, the cloud facility contains many physical computers (on the order of 10,000) that are pre-configured with various desirable CPU and memory properties. A guest subscriber can lease a suitable computer having the closest-available desired computing power and memory size along with the applications (CAD, Simulation package) and operating system (OS) for the time that is needed and in a timely fashion. For example, one guest may require a fast computer with a small memory tomorrow for data analysis and sorting, while another guest may require a less powerful computer and a large memory next week for searching a large database when it becomes available. The access to the machine is through a high-speed data network using a *network interface card*

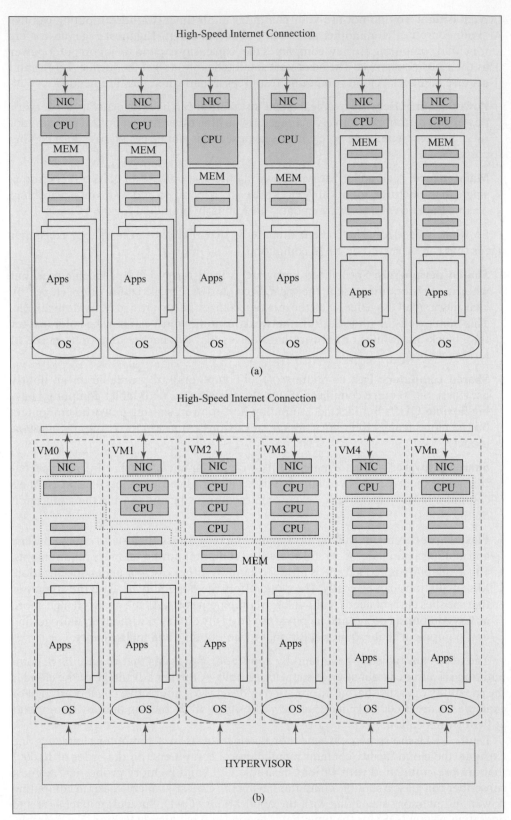

Figure 11.11

Two versions of cloud computing: (a) dedicated physical computers (solid rectangles) having specified computing power and memory components that are dedicated to guests and (b) virtual computers (dashed rectangles) configured by a hyperviser for use by guests.

(NIC). When the guest is done, the same machine can be leased to another guest. Assigning and maintaining such a facility is similar to a rental car model.

The other extreme is a share-hardware model that allows a finer resolution in meeting the needs of particular guests. A *hypervisor* is the cloud supervisor that controls massive hardware consisting of CPU and memory elements. The CPU needs of a particular guest are met by combining the requisite number of elemental CPUs. The memory needs are providing by allocating a segment of memory. Operating systems and applications that are commonly used can be accessed from an application database with appropriate licensing. The challenge is to maintain the security of such a complex system by preventing guests from viewing the data of others.

The cloud side has the following benefits.

- Computer resources can be used 24 hours a day by shifting from one guest to another.

- Computer resources can be allocated in server banks, so if a bank is not being used, it can be turned off, conserving power and providing cooling needs.

- Computer resources from the same vendor are easier to maintain in terms of support personnel and spare parts.

- A facility that is designed to accommodate banks of computers can be powered and cooled more efficiently.

- Common applications can be maintained and updated, so the newest versions arc available.

Cloud use by a start-up company guest EXAMPLE 11.6

When a company is initially formed, venture capitalists prefer to provide funds for salaries and specialized equipment rather than computers. The start-up company can lease a small-sized computer from the cloud for initial designs. As the business grows, additional computing resources can be leased accordingly.

A number of technologies are needed to mature simultaneously to make the cloud feasible.

1. An accessible high-speed network provided by the Internet.

2. Standardized interfaces provided by several popular Web browsers, including Internet Explorer, Firefox, Chrome, and Safari.

3. Security and privacy provided by encryption.

It currently appears that cloud computing is a permanent feature in the computing landscape.

11.7 Summary

This chapter described data packet transmission over a network of interconnected computers. Asynchronous data transmission was described to accommodate data packets generated at random time instants. Data of various sizes form data packets that are transmitted over a channel. The chapter described how data packets are structured to ensure that data are received reliably and without error. Problems associated with packet collisions were described. Cloud computing resources were described with examples of the flexibility they provide.

11.8 Problems

11.1 **Asynchronous byte transmission.** A 1MB data file must be transmitted to another computer using asynchronous data transmission described in Example 11.2. How many bits are transmitted over the channel?

11.2 **Asynchronous data characters.** Construct two serial data characters to transmit bytes 10101010 and 11001100. Assume each character is transmitted in a separate character waveforms each using a start bit, odd parity, and two stop bits. Sketch the two character waveforms of the data and label the character elements. Assume a transmission rate of 1,000 baud (bps), the idle channel level is 5 V, and a logic 1 is 5 V and logic 0 is 0 V.

11.3 **Probing the Internet with Tracert.** This problem asks you to estimate the speed of data transmission over the Internet from data produced tracert.

    (a) Find four Web addresses, each one from a different continent. University web pages are good addresses to use for this problem.

    (b) Find a distance calculator on the Web that determines the distance (in km) from your present location to the city hosting the Web address.

    (c) Tracert reports three round-trip travel times to each node on the way to the destination. Use the minimum of these three values for only the destination as the time that the data packet took to travel the round trip distance between your location and the destination.

    (d) Compute the data travel speed by dividing the round-trip distance by the round-trip travel time.

    (e) Compare your results to the speed of light c.

11.4 **Maximum delay for a wired-line telephone call.** Assume an electrical signal travels along a wire at the speed of light. What is the minimum travel delay that occurs for a telephone waveform from NYC to LA (distance = 4,000 km)? What is the minimum delay between any two cities on Earth assuming signals travel along the Earth's surface?

11.5 **Delay for a geostationary satellite telephone call.** Assume a radio signal travels at the speed of light. What is the minimum travel delay that occurs for a telephone waveform travelling to and from a geostationary satellite that is orbiting at an altitude of 36,000 km?

11.9 Excel Projects

The following Excel projects illustrate the following features:

- Design an asynchronous transmission character.
- Determine data packet travel times and speeds to remote destinations.
- Simulate data packet collisions.

Using a Narrative box (described in Example 13.4), include observations, conclusions, and answers to questions posed in the projects.

11.1 **Asynchronous data packet character design.** Let sampling period $T_s = T_b/4$. Design the a data packet that transmits the ASCII number 5. Show T_b idle channel durations before and after packet. The packet contains a start bit, eight data bits, one even parity bit, and two stop bits.

11.2 **Tracert to remote destinations.** Estimate the speed of data transmission over the Internet from the tracert data to two URLs, one in the U.S. and the second outside North America.

11.3 **Packet packet collision simulation.** Using Example 13.55 as a guide, determine the Prob[error] for $T_p = 1$ ms with data interval equal to $T_d = 10$ ms for three users.

11.4 **Packet packet collision simulation.** Using Example 13.55 as a guide, determine the packet duration that provides the maximum data throughput.

CHAPTER 12

SYMBOLOGY

LEARNING OBJECTIVES

After completing this chapter, the reader should be able to:

- Appreciate the issues regarding machine-readable codes.
- Recognize the utility and limitations of bar codes.
- Understand how codes are embedded in data files.

12.1 INTRODUCTION

This chapter describes methods to encode digital data in machine-readable form. Magnetic sensing decodes magnetic stripes on ATM and credit cards. Techniques that allow robust scanning of various bar codes are described. Current bar-code symbols contain not only data but also features that determine code location and orientation. Optical sensing is used in the most common scanners and is described for one-dimensional and two-dimensional symbols. Steganography is a technique for embedding secret codes into data files that are difficult to detect. Early scanning systems were expensive, situated in controlled settings, and used by trained people. Current scanning of digital data is performed by smartphone cameras under a variety of conditions and is used by nearly everyone.

This chapter describes systems that facilitate the transfer of data from products and users to a computer system. To insure successful operation, machine-readable systems must confront the following issues.

Machine-readable codes—Digital data are encoded in bar codes. Such codes are designed to allow reliable recognition of the data under a variety of scanning conditions. Early scanning systems were expensive, situated in controlled settings, and used by trained people. Current scanning of 2D symbols is performed by smartphone cameras, under a variety of conditions, and by nearly everyone.

Machine-unreadable codes—Symbols that humans read easily (using human reasoning) help distinguish human users from malicious computer programs.

Steganography—Some applications embed secret codes within large data files, making them difficult to detect. One example is a digital watermark that assigns a serial number to each music or video data file to deter piracy of software. The presence of such secret data within the file is difficult to perceive by merely looking at the file contents.

12.2 CREDIT CARD CODES

At a store or gas station, you have swiped your credit card through a card reader that reads data from the magnetic stripe. The magnetic stripe is a thin layer of magnetic material that stores magnetic patterns on its surface. Data are recorded on the magnetic stripe using a magnetic head that is similar to one used in an audio tape recorder by storing small magnetic spots (domains) having either a north pole $\mathcal{N}$ or south pole $\mathcal{S}$. Magnetic poles are formed by driving current I into a magnetic head to produce magnetic field B. Current flow in one direction forms $\mathcal{N}$ and in the opposite direction forms $\mathcal{S}$, as shown in Figure 12.1.

Figure 12.1

Writing and reading data with magnetic domains as the credit card moves in direction of arrow. Magnetic poles are formed on the magnetic stripe by driving current I into a magnetic head in the appropriate direction through the head. The magnetic head senses changes in the magnetic field to produce a voltage $+V$ when $\mathcal{N} \to \mathcal{S}$, or $-V$ when $\mathcal{S} \to \mathcal{N}$.

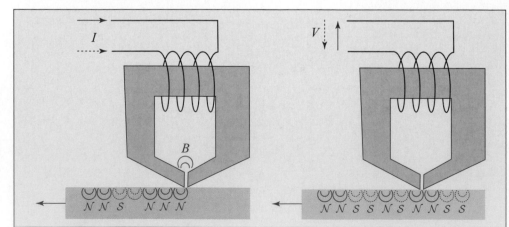

Scanning a Magnetic Stripe

The voltage produced by the magnetic poles passing under the read head is related to the change in the magnetic pole ΔB (equal to either plus or minus $2B$) divided by the time between changes ΔT, as

$$V = \frac{\Delta B}{\Delta T} \tag{12.1}$$

The magnitude of the signal from the magnetic stripe is

$$|V| = \frac{2B}{\Delta T} \tag{12.2}$$

Note that $|V|$ increases as ΔT decreases. To provide a strong signal from a reader, credit-card users are instructed to *remove the card quickly* in order to reduce ΔT, which in turn increases $|V|$. Knowing this, I have always tried to be patient with the careful cashier who does exactly the wrong thing by sliding a credit card ever so slowly through the card reader.

Reading Magnetic Stripe Data

The data signal extracted from the credit-card magnetic stripe provides the information needed to read the recorded data. The challenge is that the signal varies with the card speed, which depends on how quickly the card is pushed or pulled through the reader. In fact, the speed can vary between 3 to 125 inches per second. Furthermore, the speed does not even need to be constant during the time the card is read. To allow the data to be read successfully over such a wide variety of speeds, the data are encoded using a clever *self-clocking* code (also known as the *Manchester code*) that is employed in hard disk drives. The idea is that transitions in the data waveform *always* occur with an almost regular period. These regularly occurring transitions provide the timing necessary to decode the signal into data. Along with these expected periodic transitions, additional transitions can also occur at the mid-period times as well. These mid-period transitions encode the data: the occurrence of an additional transition encodes a 1, while no mid-period transition encodes a 0, as shown Figure 12.2.

For proper operation, the card reader must determine the value of the period and mid-period. To assist in this determination, the beginning of the magnetic stripe produces a waveform that contains only a full-period waveform as shown at the top of Figure 12.2. This waveform allows the card reader to determine the average transition period, which is denoted T_p. The estimate of the next transition time is denoted $\hat{T}_n$ and equals the previous T_p value. The reader uses the value of $\hat{T}_n$ to update the data waveform timing. If a transition (in either direction) does not occur at approximately $\hat{T}_n$, something is wrong, and the message *Please insert the card again* appears on the card reader.

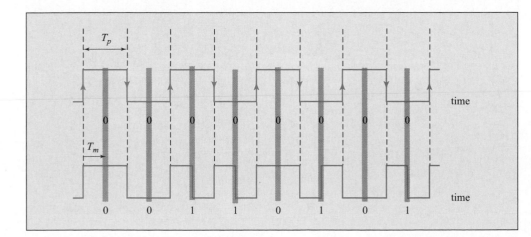

Figure 12.2

Self-clocking code with period T_p used on magnetic stripe. Top waveform corresponding to all 0's. Bottom waveform shows 1's introduce additional transitions at mid-period point T_m (shaded).

Figure 12.3

From the two previous T_p transitions, the mid-period time $\hat{T}_m = T_p/2$ and the time of the next transition $\hat{T}_n = T_p$ are estimated.

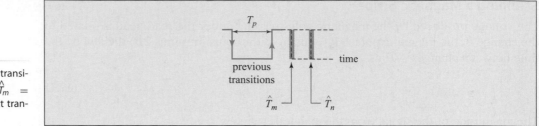

A credit card is typically pulled through the reader at a non-constant speed, which causes T_p to vary slightly. When the card swipe slows down, T_p increases. Because there are 200 transitions per inch (8 transitions per mm) on the magnetic stripe, the card speed does not vary significantly from transition to transition, which causes $\hat{T}_n$ to be close to the previous T_p value. Because $\hat{T}_n$ depends on only the previous T_p value, the $\hat{T}_n$ values track the speed of the card swipe.

Figure 12.3 shows the mid-period time estimate is denoted $\hat{T}_m$ and is estimated from the previous T_p value divided by two. The waveform around $\hat{T}_m$ is monitored for a transition. No transition at time $\hat{T}_m$ indicates a 0 has been encoded and the presence of a transition (in either direction) indicates a 1.

Decoding Magnetic Stripe Data

The data transmitted in credit-card systems use the data packet error-correction method similar to that described in Chapter 10. Most credit-card transactions are accomplished by reading data encoded on a magnetic stripe located on the back of the credit card. The magnetic stripe contains three tracks of information, as shown in Figure 12.4. Track 1 stores up to 79 letters or numbers (alphanumeric characters) and track 2 stores up to 40 numbers. These two tracks specify the owner's name, account number, and expiration date. Track 3 stores up to an additional 105 numbers needed by some systems. Tracks 2 and 3 encode digits using 5-bit code words to encode the digits 0 to 9, a separator, a start sentinel, an end sentinel, and an odd parity for error detection.

Figure 12.4

The magnetic stripe on the back of a credit card contains three data tracks. Five-bit code words are used in tracks 2 and 3.

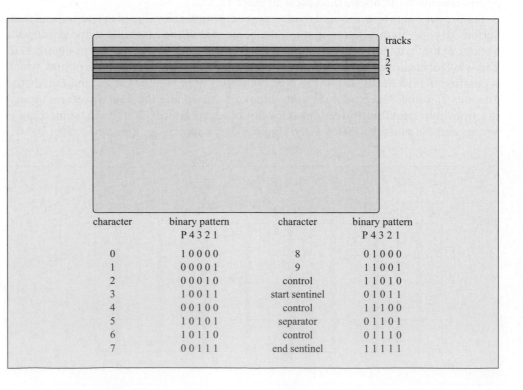

character	binary pattern P 4 3 2 1	character	binary pattern P 4 3 2 1
0	1 0 0 0 0	8	0 1 0 0 0
1	0 0 0 0 1	9	1 1 0 0 1
2	0 0 0 1 0	control	1 1 0 1 0
3	1 0 0 1 1	start sentinel	0 1 0 1 1
4	0 0 1 0 0	control	1 1 1 0 0
5	1 0 1 0 1	separator	0 1 1 0 1
6	1 0 1 1 0	control	0 1 1 1 0
7	0 0 1 1 1	end sentinel	1 1 1 1 1

The data on a magnetic stripe begins with the *start sentinel*, which is a predetermined code word that allows the card reader to verify that it is reading the data correctly. The data are terminated with the *end sentinel*. A *separator* delimits different fields of data (i.e., the account number from the date of expiration). These 5-bit code words are stacked to form a data packet that terminates with a longitudinal redundancy check (LRC) code word.

12.3 BAR CODES

Bar codes were introduced in the early 1970s to implement reliable machine-readable data systems and have quickly spread. They appear on mail, on products as the familiar Universal Product Code (UPC) symbol, and in magazine advertisements and billboards. Bar codes are used for controlling inventory in warehouses, for monitoring books and patrons in libraries, and for identifying patients in hospitals. One major advantage of bar codes over other machine-readable identification devices, such as radio-frequency identification (RFID) tags, is that bar codes can be transmitted electronically and displayed on a smartphone display, such as an airline boarding pass.

To implement a successful and practical bar-code system, the following conflicting requirements must be satisfied.

- The code must have a high density of information.

- The code must be read reliably and quickly.

- The printing process and scanner must be inexpensive.

Trade-offs routinely occur in the design of a bar-code system. For example, some bar codes are designed to have a lower density of information in order to facilitate scanning. Bar codes have evolved through the following four generations.

1. Simple numerical data that are read by machine (U.S. Postal Service codes).

2. Symbols that encode numerical pointers to a data base, or so-called *license plate codes* (UPC symbols).

3. Symbols that contain complete assembly or contextual information, but require specialized readers (PDF codes).

4. Symbols that contain both pointers to web pages and information that can be read with smartphone cameras (QR codes).

Code designers are called *symbologists* and must account for sensor technology, potential users, and applications to implement practical bar-code systems. For example, they must be clever in their designs when they are asked to replace expensive commercial scanners used by trained operators with smartphone cameras operated by the general population.

Bar codes are read by sensing the light reflected from a printed pattern that is black-and-white, while modern bar codes also exploit color coding. Commercial bar-code readers used by package delivery companies are high-speed and read hundreds of codes per second that appear on envelopes. Bar codes allow automation by designing robot delivery systems that automatically locate and read codes on cartons.

We first describe commercial systems that read codes on mail envelopes and products and proceed to smartphone camera bar-code readers.

12.3.1 U.S. Postal Service Bar Code

To speed mail delivery, the United States Postal Service (USPS) implemented the Zoning Improvement Plan (ZIP) code in 1963. A 5-digit ZIP code is sufficient to direct mail to a specific post office anywhere in the United States. The bar code was developed to allow the ZIP code to be read by machine at high speeds—up to ten pieces of mail per second.

Current USPS bar codes appear in several variations.

- The standard 5-number ZIP code, such as 06520, to direct mail to a post office.

- The ZIP+4 code appends a 4-digit number, such as 06520-8284, to sort mail along the delivery route.

- The ZIP+4+2 code, which also includes a 2-digit point-of-delivery location number that is useful when the 9-digit number corresponds to a large apartment complex.

EXAMPLE 12.1 My ZIP code at Yale

The post office that receives mail directed to Yale's School of Engineering and Applied Science has the designation 06520. Mail addressed to me contains the ZIP-plus-4 number 06520-8284. The last four digits allow letters arriving at post office number 06520 to be grouped together with other mail directed to my building to permit efficient delivery.

The USPS currently uses two bar code systems. The first one is the *early USPS* and uses two bars (short and tall) to encode digits. The current one is the *recent USPS* and uses four bars (short, tall, ascending and descending). Figure 12.5 shows both code types that appeared on mail sent to me. The early USPS code is simpler and encodes only the ZIP or Zip+4 code, and we describe its features next. The recent USPS code is more sophisticated and encodes the ZIP code, level of service (First Class, Overnight, Bulk Mail, *ldots*), and other options. The recent code also takes advantage of advances in high-resolution printing and scanning technology, as well as improvements in digital technology.

The early USPS bar code uses the 5-bit code words shown in Figure 12.6. While a 4-bit code would suffice for digits, the 5-bit code words implement a clever error-detection code. Each digit is encoded as two 1's (tall bars) and three 0's (short bars) and any other detected pattern of short and tall bars indicates the occurrence of an error.

Figure 12.7 shows how is the code is arranged in to a bar code for a 5-digit ZIP code. An additional digit is appended to the ZIP code for error correction, as described next.

Figure 12.5

Scanned copies of (a) early USPS and (b) recent USPS bar codes. The early code was produced by a dot-matrix printer, while the recent code used a laser printer. (I have two offices at Yale, which explains the different addresses.)

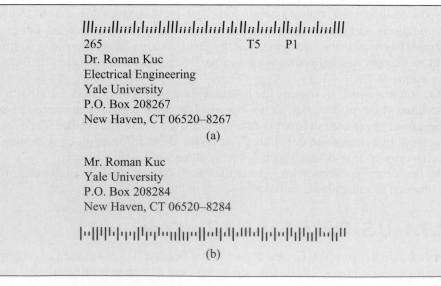

265   T5   P1
Dr. Roman Kuc
Electrical Engineering
Yale University
P.O. Box 208267
New Haven, CT 06520–8267

(a)

Mr. Roman Kuc
Yale University
P.O. Box 208284
New Haven, CT 06520–8284

(b)

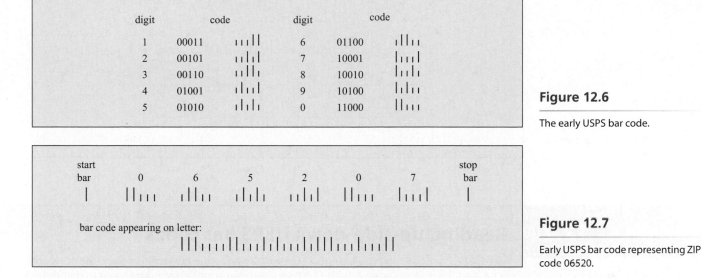

Figure 12.6

The early USPS bar code.

Figure 12.7

Early USPS bar code representing ZIP code 06520.

The set of six 5-bit code words is preceded by a tall *start bar* and terminated with a tall *stop bar* to aid in scanning the bar code.

USPS Bar Code Scanner

The early USPS bar code appears on an envelope as a series of tall and short bars. These are read with an array of optical sensors as the letter moves quickly past the sensor, as shown in Figure 12.8. The emitter shines a small light spot on the code. The black ink forming each bar absorbs the light and produces no reflection, while the white spaces between the bars reflect light back to the detector. As the bar code passes by the array, each sensor registers the presence or absence of reflected light along a narrow horizontal stripe across the bar code. This scanning system automatically aligns the bar code by determining which sensors detect the first tall bar, which is called the *start bar*. Tall *start* and *stop* bars are always present in the bar code to aid in aligning the bar code within the scanner. The sensors above and below the bar code detect reflected light without interruption. These sensors determine the location and the height of the bar code on an envelope.

After alignment, a single pair of sensors differentiates tall from short bars. The *lower* sensor scans the short bars, while the *higher* sensor scans the tall bars. The lower sensor provides timing information, because a short bar is always present in the code. When the lower sensor detects a short bar, the higher sensor is checked to determine if it too detects a black bar. If so, a tall bar is present, otherwise it is a short bar.

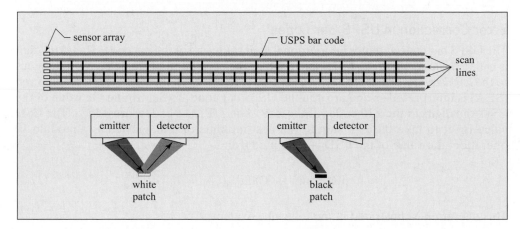

Figure 12.8

An array of optical sensors scans the USPS bar code. The sensor differentiates between black and white patches forming the bar code.

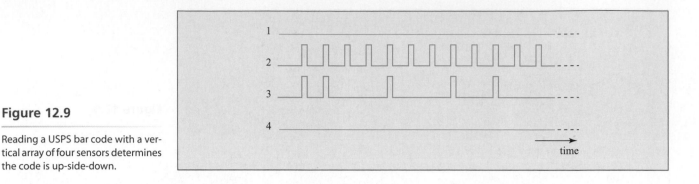

Figure 12.9

Reading a USPS bar code with a vertical array of four sensors determines the code is up-side-down.

EXAMPLE 12.2 — Reading up-side-down USPS bar codes

The USPS bar code is typically printed horizontally on an envelope. But what if the envelope is up-side-down in the scanner? The USPS code contains sufficient information to determine its orientation.

Consider a sensor array that contains four sensors labeled 1 to 4. The vertical array scans the bar code as it passes underneath the scanner. Figure 12.9 shows the waveforms produced by each sensor. The top (labeled 1) and bottom (4) waveforms show no variations, while waveforms 2 and 3 show variations typical of a bar code. This indicates the bar code is contained within the vertical extent of the array.

While short bars always produce pulses in the sensor 2 waveform, tall bars produce a smaller number. The figure shows that these less-frequent tall-bar pulses occur in sensor 3 and occur below the more regular, periodic, short-bar pulses produced by sensor 2. Thus, the tall bars extend below the short bars and indicate to the scanning system that the USPS bar code is up-side-down. Being up-side-down, it is being scanned from right to left and requires a reverse-order correction to determine the correct ZIP code.

Error Detection in USPS Bar Codes

The two-tall and three-short bar-code patten allows single errors to be detected. Small specks of material and other imperfections may convert a short bar into a tall one when read with a scanner. In other cases, imperfections in the paper may prevent the ink from adhering and cause a tall bar to appear as a short bar. Error detection is possible when reading each group of five bars, because only two tall bars should appear. If fewer or more appear, the scanner recognizes that an error has occurred.

Error Correction in USPS Bar Codes

The USPS bar code appends an additional digit at the end of the bar code. This extra digit is called a *checksum digit (CSD)* and enables the correction of single errors analogous to the LRC code word described in Chapter 10. In this case, the single-error assumption (SEA) is interpreted as one error in the USPS bar code. To determine the value of the CSD, the digits in the ZIP code (or the ZIP+4 or ZIP+4+2 codes) are added. The CSD value is set to take the ZIP sum to the next multiple of ten. Viewed as a modulo-10 operation, the value of the CSD is computed from

$$[\text{ZIP sum} + \text{CSD}]_{\text{mod}(10)} = 0 \tag{12.3}$$

This operation is illustrated in the following example.

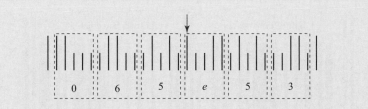

Figure 12.10

Correcting an error in the USPS bar code. The arrow indicates the erroneous bar.

USPS bar code checksum digit EXAMPLE 12.3

To determine the CSD value for the ZIP code 06520, we add the digits in the ZIP code to give

$$0 + 6 + 5 + 2 + 0 = 13$$

The CSD value is set to the number that makes this sum plus the CSD total have a modulo-10 value equal to zero, as

$$[0 + 6 + 5 + 2 + 0 + \text{CSD}]_{\text{mod}(10)} = [13 + \text{CSD}]_{\text{mod}(10)} = 0$$

By inspection (that is, by trial-and-error using different CSD values—there are only 10), CSD = 7. The complete bar code was shown previously in Figure 12.7.

The combination of the error detection and the checksum digit allows single errors in the bar code to be corrected. In this case, SEA means one error in the postal bar code symbol. The error correction procedure is illustrated in the following example.

Error correction in a USPS bar code EXAMPLE 12.4

The USPS bar code shown in Figure 12.10 contains an error. To find the error, decode as many digit code words as possible. *First, remove the start and stop tall bars.* Collect the remaining bars in groups of five. Correctly encoded digits contain two tall bars and three short bars, so any other configuration indicates an error with the erroneous configuration being decoded as an *e*. Applying this procedure to the bar code shown in Figure 12.10 gives the sequence

$$0, \ 6, \ 5, \ e, \ 5, \ \text{and } 3$$

with 3 being the CSD value. The checksum formula gives

$$[0 + 6 + 5 + e + 5 + 3]_{\text{mod}(10)} = [19 + e]_{\text{mod}(10)} = 0$$

Thus, by inspection (that is, by *trial-and-error*) we find $e = 1$. This converts the erroneous bar code into the correct ZIP code 06515. The arrow in Figure 12.10 marks the location of the erroneous tall bar that should have been a short bar.

12.3.2 Universal Product Code

Stores use a clever system of identification tags called the *universal product code (UPC) symbols*, to automatically identify products. A standard UPC symbol is shown in Figure 12.11 and consists of 12 digits that represent a product identification number. This number accesses the price look-up (PLU) data file in the store computer, which transmits the price of the item. These digits are encoded in white and black bars. To allow correct interpretation, the UPC symbol is designed to have the following format.

- A guard pattern forms the 3-unit bar-space-bar (101) sequence that defines the left edge.

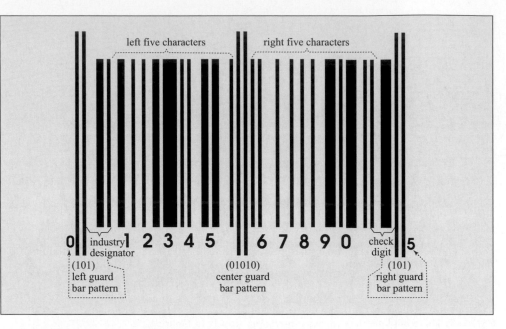

Figure 12.11

Universal Product Code (UPC) symbol.

- Six digits are encoded on the left side of the symbol:

    1. The first digit specifies the industry type, such as 0 for grocery and 3 for pharmaceuticals.

    2. The remaining five digits encode the product manufacturer identity.

- A center guard pattern forms the 5-unit sequence (01010) that separates the two halves of the symbol.

- Six digits encode the information on the right side of the symbol in the following manner:

    1. Five digits encode the item identity.

    2. One digit is used for a checksum value, which is analogous to the CSD in the USPS bar code.

- A guard pattern forms the 3-unit bar-space-bar (101) sequence that defines the right edge.

A UPC symbol encodes each digit using seven bars with each bar being either a void or a black bar having the same thickness. Figure 12.12 shows each code word consists of two pairs of regions with each region being either black (joining black bars) or light (joining void spaces). The code words on the left side have odd-parity and are the bit-wise complements (the logical NOTs) of those on the right. Thus, a code word on the left side of the symbol has an odd number of black bars (odd parity), while the code word for same digit on the right side has an even number (even parity). This coding allows the scanner to determine the pose of the UPC symbol (normal or inverted), as the next example illustrates.

EXAMPLE 12.5   ## UPC symbol pose

Figure 12.13 shows an edge of a UPC symbol. Is the symbol right-side-up or up-side-down?

After removing the 1-0-1 guard band that occurs on both edges, the count of black bars in the first 7-segment unit is four (an even number). Thus, the code word comes from the right side of the UPC symbol and makes the symbol up-side-down.

Figure 12.12

The digit code words in the UPC symbol form complementary pairs on the left and right sides of the symbol. (a) The number of black bars in left-side code words is odd, while on the right side it is even. (b) A typical UPC symbol of a fictitious product.

Figure 12.13

Is this UPC symbol segment right-side-up or up-side-down?

UPC Scanner

Unlike the USPS scanner, which moves the bar code across the sensor array, a UPC scanner contains rotating prisms and mirrors to quickly move a laser light spot across the UPC symbol. Even though the product containing the UPC symbol is typically in motion during the scanning process, its speed is much slower than that of the laser spot, so the symbol can be considered stationary during the scan. The laser in a commercial UPC scanner forms a web pattern, as shown in Figure 12.14. The detector produces

Figure 12.14

A UPC scanner directs a laser to form a web pattern across the UPC symbol for acquiring data regardless of symbol orientation.

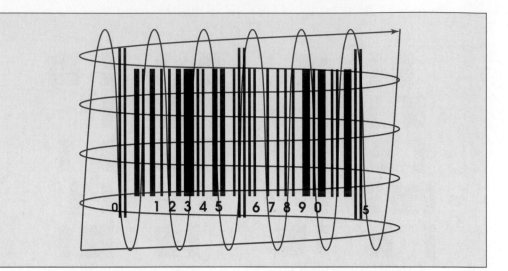

Figure 12.15

Scanning a UPC symbol produces a signal having a variable duration that depends on the scan angle. A black bar produces a 1 and a white space produces a 0.

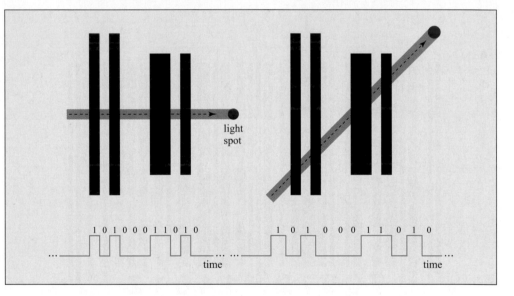

a waveform that contains a specific pattern (a *signature*) that allows the scanner to recognize the symbol and decode it.

One problem in scanning the UPC symbol is the time variation in the signature caused by a variable scan angle shown in Figure 12.15. A light spot moving at right angles to the bars (shown on the left side of the figure) produces the shortest signal. An oblique scan (shown on the right side) elongates the signal duration. To interpret the symbol's signature correctly the amount of time stretching that occurs must be determined. The clever engineering solution to this problem is to insert a bar pattern (1-0-1) that has a known spacing on both sides of the UPC symbol. The width of these bars provides the timing interval that is used to interpret the signals from the remaining bars on the symbol.

EXAMPLE 12.6   Laser spot speed

A UPC symbol has end guard bars that are 1 mm thick and a scanner laser spot has a speed of 100 m per second. When the laser spot passes directly across the UPC symbol, as shown on the left side of Figure 12.15, the duration (denoted by T_x) of the pulse produced by crossing over the first black bar equals the bar thickness divided by the laser speed, as

$$T_x = \frac{1 \text{ mm}}{100 \text{ m/s}} = \frac{10^{-3} \text{ m}}{10^2 \text{ m/s}} = 10^{-5} \text{ s} = 10 \ \mu\text{s}$$

The scanner reads the rest of the symbol by examining the detector waveform every $T_x = 10\ \mu s$ for the presence of a black bar or a space between bars.

When the laser scans the symbol at a 45° angle, as shown on the right side of Figure 12.15, the spot takes longer to cross the black bar, producing a longer pulse. The path length through the black bar when the angle is 45° is

$$1\ \text{mm}/\cos(45°) = 1.414\ \text{mm}$$

making the initial pulse duration equal to

$$T_x = \frac{1.414\ \text{mm}}{100\ \text{m/s}} = \frac{1.414 \times 10^{-3}\ \text{m}}{10^2\ \text{m/s}} = 14.14\ \mu s$$

The scanner reads the rest of the symbol by examining the waveform using a period equal to $T_x = 14.14\ \mu s$.

Error Detection in UPC Symbols

For error correction, each digit is represented as a 7-bar code with each bar being either black or white. The number of black bars used in the 7-bar code words on the left half is odd. Thus, all left-side digits all produce code words having odd parity. For example, the digit 3 appearing on the left side of the symbol consists of five black bars (five 1's) in its 7-bar (7-bit) code word. The number of black bars on the right half of the symbol is an even number. Thus, all right-side digits produce even-parity code words. For example, the digit 3 appearing on the right side of the symbol consists of two black bars (two 1's). Any deviation from these patterns would signal an error.

Error Correction in UPC Symbols

The checksum digit (CSD) is analogous to that in the USPS bar code that can correct single errors. In this case, the single-error assumption (SEA) is interpreted as one error in the UPC symbol. The following formulas compute the value of the checksum digit. If d_1 represents the first digit of the code, d_2 the second, and up to d_{11} represents the eleventh digit, the error-correction method computes the sum

$$c = 210 - 3(d_1 + d_3 + d_5 + d_7 + d_9 + d_{11}) - (d_2 + d_4 + d_6 + d_8 + d_{10}) \qquad \textbf{(12.4)}$$

The CSD equals the modulo-10 value of the sum

$$\text{CSD} = [c]_{\text{mod}(10)} \qquad \textbf{(12.5)}$$

Rather than simply computing the sum of the digits, as was done in the USPS bar code, separating the even- and odd-indexed digits also detects other common problems, such as an accidental interchange of digits by the printer of the label. This is illustrated in the next example.

Check digit calculation   EXAMPLE 12.7

The UPC bar code shown in Figure 12.11 contains the eleven digits

$$d_1 = 0, \quad d_2 = 1, \quad d_3 = 2, \quad d_4 = 3, \quad d_5 = 4, \quad d_6 = 5,$$

$$d_7 = 6, \quad d_8 = 7, \quad d_9 = 8, \quad d_{10} = 9, \quad d_{11} = 0$$

The checksum formula produces the value

$$210 - 3(0 + 2 + 4 + 6 + 8 + 0) - (1 + 3 + 5 + 7 + 9) = 210 - 3(20) - 25 = 125$$

The checksum digit equals

$$\text{CSD} = [125]_{\text{mod}(10)} = 5$$

To demonstrate the significance of the formula for CSD, let two of the digits be mistakenly interchanged by the printer. That is, instead of

$$0\ 12345\ 67890\ 5$$

the mistaken label appears as

$$0\ 12435\ 67890\ 5$$

Clearly, the sum of the digits remains the same, so this error is not detectable by simply computing the sum of all the digits. However, in evaluating the CSD formula, we get

$$210 - 3(0 + 2 + 3 + 6 + 8 + 0) - (1 + 4 + 5 + 7 + 9) = 210 - 3(19) - 26$$
$$= 127$$

and

$$[127]_{\mod(10)} = 7$$

This value does not match the CSD printed on the label (5) and indicates an error occurrence.

EXAMPLE 12.8 **Correcting UPC errors**

A scanner produces the following eleven-digit sequence

$$0\ 7\ 0\ 9\ e\ 2\ 1\ 4\ 0\ 1\ 2\ 7$$

where e corresponds to a error that occurs on the left side of the UPC symbol.

The correct value of e can be found from the checksum formula. Using the observed CSD = 7, the checksum formula gives

$$[210 - 3(0 + 0 + e + 1 + 0 + 2) - (7 + 9 + 2 + 4 + 1)]_{\mod(10)} = 7$$

Doing the arithmetic, we get

$$[210 - 3(e + 3) - 23]_{\mod(10)} = [210 - 3e - 32]_{\mod(10)}$$
$$= [178 - 3e]_{\mod(10)}$$
$$= 7 \quad (\text{CSD on UPC symbol})$$

We need to find the value of e such that $178 - 3e$ ends in a 7. By trial-and-error, we find that $e = 7$ gives

$$[178 - 21]_{\mod(10)} = [157]_{\mod(10)} = 7$$

Thus, the erroneous digit should be a 7.

12.3.3 Two-Dimensional Bar Codes

An extension to the previous one-dimensional bar codes is the two-dimensional (2D) bar codes that contain much more data by exploiting both the horizontal and vertical dimensions. 2D bar codes have become increasingly popular, because smartphones with high-resolution cameras permit scans by merely taking their pictures. This section briefly describes two generations of 2D bar codes that have evolved with the technology.

PDF Symbols

Portable data file (PDF) symbols were designed to contain more data rather than simply identifying the product as with the UPC symbol. The PDF symbol is meant to be read

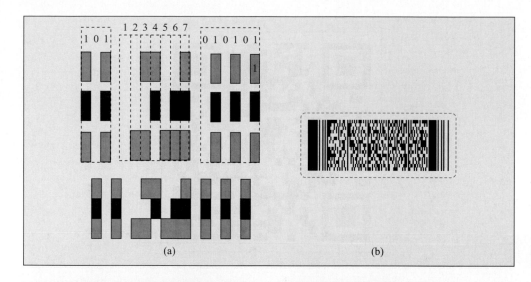

(a)   (b)

Figure 12.16

PDF symbols. (a) Constructing a PDF symbol by stacking UPC symbol rows that contain left-side digits 0 9 8, left-guard band 101, and right-guard band 010101. (b) Commercial PDF symbol.

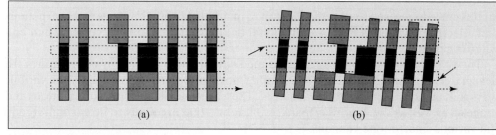

(a)   (b)

Figure 12.17

A PDF symbol is read by a series of horizontal laser scans that are densely spaced in the vertical direction. (a) In the ideal case the symbol is aligned with the scans. (b) Skewed symbol scan (blue arrows indicate scans that cross boundaries and produce non-repeating code words).

using specialized scanners. The code structure can be thought of as a series of short UPC symbols stacked on top of one another. Figure 12.16 (a) shows a simple example where the left-side UPC code previously shown in Figure 12.12 encodes a single digit in each row. The symbol encodes 0 in row 1, 9 in row 2, and 8 in row 3. The UPC coding scheme allows these codes to be scanned from left-to-right or right-to-left and still provide sufficient information in the scanner waveform to decode the digits correctly. A special right-side guard band (010101) identifies the right side of the code in the scanner waveform. Commercial PDF codes are more complex and can contain thousands of alphanumeric characters. An example of a commercial PDF symbol is shown in Figure 14.16b.

This stacked structure requires scanners to perform a sequence of dense horizontal scans across the symbol, which is called a *raster scan*. Figure 12.17 shows a laser raster scan that forms many scan lines across each digit code word in the symbol. Multiple passes occur in each row and the duplicate data can be recognized as such and discarded to yield the codes for each row. The set of row data is combined to form the entire data set represented in the PDF symbol.

Several horizontal scans occur within each row in the symbol to produce the same code words in successive scans. Such repeating code words indicate a valid scan of a row and are reduced to the single code word. Horizontal scans that cross row boundaries produce different code words in adjacent scans that do not repeat the sufficient number of times. Isolated and non-repeating code words are identified as coming from two different rows and are eliminated.

QR Codes

Figure 12.18 shows an example of a *quick response (QR) code*, which is a popular 2D bar code that appears in many advertisements and business cards. The QR code contains information that directs your smartphone browser to a web page where the product details can be viewed. The QR code is finding new applications in manufacturing and

Figure 12.18

Example of QR code that includes the URL of my Web page: http://pantheon.yale.edu/~kuc. (Code generated by Kaywa, see http://qrcode.kaywa.com.)

architecture to replace (bulky and difficult to maintain) paper drawings and instructions with electronic versions that exist in digital databases that can be accessed when and where needed.

The QR code was designed to be scanned with digital cameras in smartphones. To design a reliable smartphone camera scanner, an engineer needs to consider the various ways that an QR image would be acquired. These include the different orientations and ranges—as well as the various camera resolutions—that are likely to be encountered in the various smartphone versions.

To properly decode a camera image of a QR symbol, the symbol contains structural features that are helpful.

- Each informational element in the symbol must occupy at least several pixels in the camera image for robust detection. If an individual QR block occupies less than a pixel in the image, it may fall between two pixels and be missed altogether.

- The symbol must include simple-to-recognize features that indicate the symbol orientation. Camera software can align the image so that it can be scanned horizontally (within the smartphone memory) as in the conventional manner employed by raster scanners.

EXAMPLE 12.9 Reading a QR code with a digital camera

Figure 12.19 shows the considerations with reading a QR symbol with a digital camera. To decode the camera image of a QR symbol, there must be at least one pixel that is situated within a white region and another in an adjacent blue region. Clearly, this minimum resolution requires careful registration of the QR image onto the camera pixel array—both in position and tilt—as shown in the top of the figure. Such a fortuitous arrangement is an unreasonable expectation under normal operation. The high-resolution cameras in current smartphones make such accurate registration unnecessary. If we double the number of pixels to allow four pixels to lie within each region—as shown in the bottom of the figure—even in the worst case there always will be at least one pixel in each region.

Figure 12.20 shows the problems encountered with decoding an image of a QR symbol taken with a digital camera. The pixel information must be sufficient to align and orient the QR code by processing the image. This transformation is simplified by including additional pixels—either by taking the picture at closer range or using a higher-resolution camera. Experience with smartphone cameras indicates that reliable decoding results when a QR symbol occupies at least one quarter of the image.

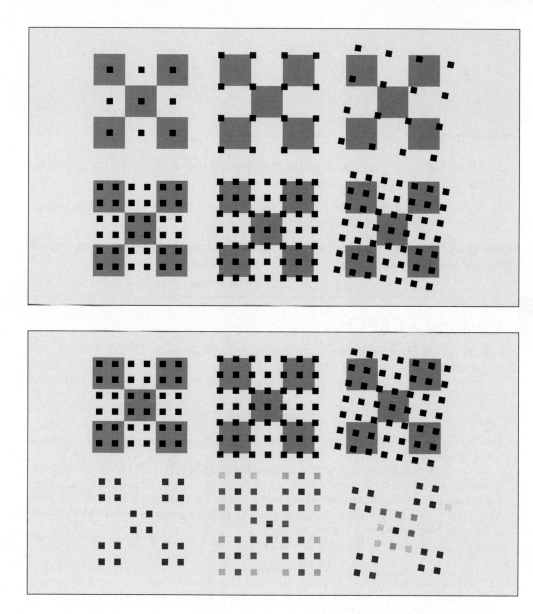

Figure 12.19

Reading a QR code with a digital camera. QR symbol regions are shown in blue and white and camera pixel locations in black. Left-side images show ideal registration of camera to symbol. Other images show problems with registration and camera tilt.

Figure 12.20

Decoding a digital image of a QR symbol.

12.4 MACHINE UNREADABLE CODES

On the other end of the spectrum, how does one make a symbol that *cannot* be read by a computer? One reason is to distinguish a human user from a web robot or *bot*—a program that searches the Web and accesses Web pages much faster than human users can. Malicious bots include those that ticket re-sellers use to buy up good seats at concerts—to be resold later at inflated prices. To prevent such activity, Web sites use text and symbols that humans read easily (using human reasoning) that are called *CAPCHAs*, in order to distinguish human users from computer programs. CAPCHAs are a type of *Turing test* used to distinguish humans from computers that use artificial intelligence (AI).

CAPCHAs are fully automated programs that may contain visual features (or audio features for visually impaired users). The CAPCHA generation process consists of the following steps.

1. A random character generator uses a pseudo-random number generator (PRNG) to form a random string of six keyboard characters.

Figure 12.21

Example of a simple automated CAPCHA.

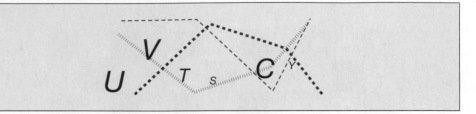

2. A graphics program forms random background that humans readily ignore but that confuse AI pattern-recognition programs.

3. A program randomizes the location and appearance of the characters by using a PRNG to generate random offsets and font sizes.

Figure 12.21 shows an example of a simple CAPCHA using upper-case letters that was generated automatically.

EXAMPLE 12.10   Hacking a CAPCHA

Assume that AI programs are ineffective when trying to recognize the letters in a CAPCHA. The alternative is to use a brute-force technique (that is guessing). A CAPCHA that contains six characters from a 26-character alphabet can generate the number of possible combinations equal to

$$26^6 = 3 \times 10^8$$

On average, a search produces a successful result by guessing half this number of patterns, or 1.5×10^8 queries.

Guessing over an internet connection involves transmitting data packets from your machine to the Web site. Each letter is encoded as a data byte and enclosed within a data packet that has at least a 10-byte overhead (4-byte destination and source addresses, data size byte, and a LRC byte) yielding 16 bytes (128 bits). The time required to transmit each CAPCHA response using a 100 Mbps data rate is approximately

$$\frac{128 \text{ bits}}{10^8 \text{ bits/s}} \approx 1 \ \mu s$$

This transmission time is short, so the limiting factor becomes the response time to transmit an acknowledgment response (ACK) from the Web site. If this response time is 1 ms, the average time to correctly guess a CAPCHA is

$$1.5 \times 10^8 \text{ queries} \times 10^{-3} \text{ seconds/query} = 1.5 \times 10^5 \text{ s} = 41 \text{ hrs}$$

Clearly, this would slow down any ticket-scalper.

▶ 12.5 STEGANOGRAPHY

Encrypted data files draw attention: *This encrypted file contains important (and possibly useful) data.* Encrypted files are usually easy to spot because they occur over special secure connections, such as Web sites that begin with "https" (the "s" indicates *secure*) and between financial institutions.

Another form of transmitting small amounts of data secretly is called *steganography*, which includes (or *embeds*) the data within a large data file (such as image, audio, or video) that does not call attention to itself like an encrypted file does. The idea is that by simply looking at a file's contents, it is difficult to determine if it contains embedded data. This form of transmitting secret data is actually thousands of years old: Ancient rulers were known to shave the head of a loyal servant, tattoo a message on the bare scalp, and let the servant's hair grow back before sending him to the destination. One can only imagine the process at the receiving end (or if the servant was captured by a suspecting enemy).

This section provides two examples of steganography: digital watermarks that indicate ownership of printed material that is distributed illegally, and binary data embedded within audio or image data files.

12.5.1 Digital Watermarks

You have written an original article that you want to distribute to a limited number of clients who pay for your service. You discover that one of your clients has made an unauthorized copy of your article and has distributed it widely. You can determine which client did this through a *digital watermark*.

One digital watermarking technique involves changing the text copy almost imperceptibly by shifting particular lines slightly in the vertical and horizontal directions. This can be done easily with a standard layout program and current laser printers. Each document has a different set of lines that are shifted. The shifted lines are determined by a unique binary code that is assigned to each client. Thus, each client receives a uniquely (and secretly) coded document that can be traced back to the offending client.

Digital watermark **EXAMPLE 12.11**

Figure 12.22 shows example of a digital watermark. The original text is shown in an unshifted version. The title is a landmark used for registering shifted versions. The watermarked copy sent to client 1010 shifts the lines that correspond to the 1's in the binary code assigned to each client. That is, client 1010 is sent a copy in which lines 1 and 3 have been shifted in the vertical and horizontal directions. Viewing the watermarked copy alone makes it difficult to detect the shifts. To illustrate the shifts visually, the original and watermarked copies are superimposed in Figure 12.22. Note that the shifted lines appear darker because the superimposed text prints with more ink. Even electronic photocopies of the watermarked text maintain the shifts—and the *identity* of the dishonest client. (Very clever!)

How large a binary code is necessary? The title is used to register the original and shifted versions. This leaves the remaining lines on the page for the coded shifts. A typical newsletter prints seven lines per inch. A 9-inch column allows for 63 lines for the code or 2^{63} ($= 8 \times 10^{18}$) unique code words—a client base that would satisfy most businesses!

Even though such a digital watermark is subject to attack in an attempt to remove it, the code is still robust. The watermark can be extracted even if the document is photocopied or faxed. My devious colleagues tell me that the way to remove the watermark is to scan the text page, use an optical character recognition program to extract the text from the image, and reprint the document. However, this may involve more expense than simply subscribing to your service in the first place.

As a historical note, mapmakers also had this problem with unauthorized copies. Their solution was to include a fictitious street in the map, establishing their authorship.

12.5.2 Embedding Secret Codes in Data

If the data file size is large, secret data can be embedded with minimal perceptual distortion. For example, consider a data file containing quantized samples of a sinusoidal waveform. If the quantization step size is Δ, changing the least significant bit (LSB)

Original Document

Important Information

This information is very sensitive and for our clients only.

Unauthorized copying of this material is prohibited.

If sent to non-subscribers, your identity is discoverable.

Then our lawyers will be contacting you.

Version sent to client 1010

Important Information

This information is very sensitive and for our clients only.

Unauthorized copying of this material is prohibited.

If sent to non-subscribers, your identity is discoverable.

Then our lawyers will be contacting you.

Superposition of Original and Version sent to client 1010

Important Information

This information is very sensitive and for our clients only.

Unauthorized copying of this material is prohibited.

If sent to non-subscribers, your identity is discoverable.

Then our lawyers will be contacting you.

Figure 12.22

Example of a digital watermark.

affects the waveform amplitude by (at most) a single Δ. If Δ is sufficiently small, the change is imperceptible. Using this method, we can insert secret binary data (say 1010) into the first four sample values by setting their LSB's equal to these to secret values. The receiving party would know to extract the LSBs from the data to discover the secret data. A casual viewer would not notice any difference when viewing the normally decoded sinusoidal waveform.

EXAMPLE 12.12 Adding secret data to the LSBs.

Let the secret code be 1010. We insert this code into the existing binary data in a file with one code bit replacing the existing LSB of each binary value. Let the first four waveform samples be represented by 8-bit sequences

$$0100\ 1101 \quad 0101\ 1001 \quad 0000\ 1100 \quad 1111\ 1111$$

We first set all of the LSBs in the data to 0, as

$$0100\ 1100 \quad 0101\ 1000 \quad 0000\ 1100 \quad 1111\ 1110$$

and add the secret data, as

$$0100\ 1100 + 1 \quad 0101\ 1000 + 0 \quad 0000\ 1100 + 1 \quad 1111\ 1110 + 0$$

to produce the binary values that embed the code. Thus,

$$0100\ 1101 \quad 0101\ 1000 \quad 0000\ 1101 \quad 1111\ 1110$$

Note that adding the secret code leaves some 8-bit data values unchanged, such as in the first byte. Values can change by no more than $\pm$LSB, thus increasing or decreasing the reconstructed amplitude by quantizer step size Δ. If Δ is small, the changes are imperceptible.

To extract the code, we merely extract the LSBs. Thus,

$$0100\ 1101 \rightarrow 1 \quad 0101\ 1000 \rightarrow 0 \quad 0000\ 1101 \rightarrow 1 \quad 1111\ 1110 \rightarrow 0$$

To illustrate the perceptual quality of steganography, we inserted the ASCII code for 'steganography' ('s' = 01110011, 't' = 01110100, 'e' = 01100101, etc.) into samples of a sinusoid. Figure 12.23 shows the results for samples that are encoded with 6 bits and

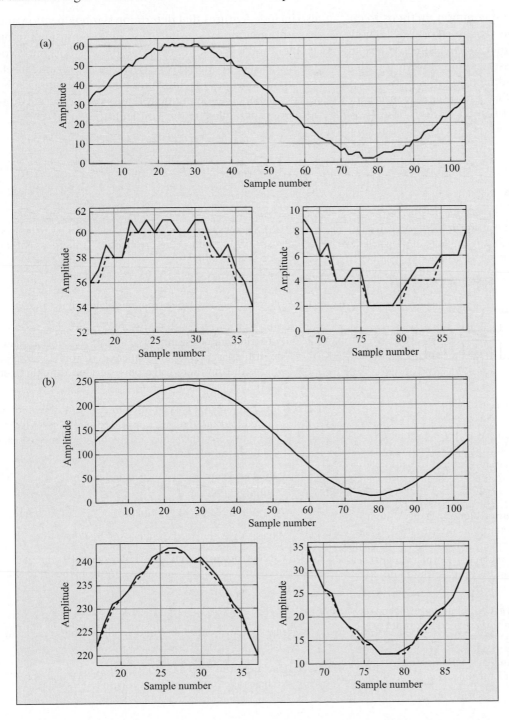

Figure 12.23

Steganography. The ASCII code for "steganography" was inserted into the least-significant bit of the samples of a sinusoidal waveform. (a) Waveform samples encoded with 6 bits. (b) Waveform samples encoded with 8 bits. Bottom curves show detail of differences with dashed curves showing original values.

encoded with 8 bits. The bottom curves show differences in more detail, with dashed curves showing original values. The Δ value for 6-bit quantization is larger than that for 8-bit quantization by a factor of 4, so the changes of $\pm\Delta$ in the waveform are more evident for 6-bit quantization. For a 16-bit quantization (CD-quality audio), Δ is so small that the figure would not display the differences produced by the code, making it imperceptible.

12.6 Summary

This chapter described several common methods to encode digital data in machine-readable forms. Magnetic sensing decodes data on magnetic stripes on credit cards. Techniques that allow robust scanning of various bar codes were described. One-dimensional bar codes typically contain a few digits. UPC symbols are license-plate codes that retrieve data in a computer data base. Two-dimensional bar code symbols contain not only data, but also include features that determine symbol location and orientation. Optical sensing is used in the most common scanners and was described for one-dimensional and two-dimensional bar codes. CAPCHAs were described as efforts to make codes machine unreadable. Steganography was described as a technique for embedding secret codes into larger data files that are difficult to detect.

12.7 Problems

12.1 Credit card self-clocking waveform. A credit card data stripe produces the self-clocking waveform shown in the top half of Figure 12.24 to encode the binary sequence 00000. What is the binary sequence encoded by the other waveform?

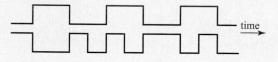

Figure 12.24

Self-clocking waveform in Problem 12.1.

12.2 Generating a self-clocking waveform. Sketch a self-clocking waveform that encodes the binary sequence 00000. Below this sketch, use the above transition times to sketch the waveform that encodes binary sequence 10101.

12.3 Credit Card Code on Tracks 2 and 3. The two-digit sequence 0 9 is very important. Encode these digits in a data block as they may appear on a credit card. Encode each code word on a separate line to form a data packet. Add the sentinel codes and terminate the data packet with a LRC code word.

12.4 Interpreting up-side-down USPS bar codes. Decode as many digits as possible in the ZIP code from the waveforms shown in Figure 12.9.

12.5 USPS bar code checksum. The ZIP code of Mountain View, CA, is 94043. What is the checksum digit?

12.6 USPS error correction. The bar code scanner in the U.S. Post Office produces an erroneous reading from a bar code. The reading is

$$6\ 0\ 6\ 1\ e\ 8$$

where e represents the erroneous digit. What is the value of e?

12.7 Pulse time for 60° scan. In Example 12.6, if the scan angle of the laser spot path is increased from 45° to 60°, what is the duration of the pulse produced by the first bar? Give answer in μs.

12.8 Data rate of a UPC symbol scanner. A UPC symbol is scanned and the data is transmitted in 50 ms. What is the data rate of the scanner? (Ignore the time to scan the UPC symbol.)

12.9 UPC checksum. A product is designated by the UPC symbol number

$$0\ 7\ 2\ 3\ 1\ 0\ 0\ 0\ 0\ 4\ 2$$

What is the value of the checksum digit?

12.10 UPC error correction. The UPC symbol scanner in a store produces an erroneous reading from the UPC symbol. The reading is

$$0\ 7\ 0\ 7\ 3\ e\ 0\ 5\ 3\ 3\ 5\ 9$$

where e represents the erroneous digit. What is the value of e?

12.11 Hacking an advanced CAPCHA. Extend Example 12.10 to the case of upper- and lower-case letters.

12.8 Excel Projects

The following Excel projects illustrate the following features:

- Design a self-clocking code waveform.

- Compute the channel capacity as a function of the signal-to-noise ratio.

Include a Narrative Box, described in Example 13.4, to include observations, conclusions, and answers to questions posed in the projects.

12.1 **Self-clock waveform design.** Following Example 13.56, compose a worksheet that uses $T_p = 16$ to generate and plot the credit card code waveform w_i that encodes the 4-bit sequence 1010 that starts and ends with one additional 0-valued bit for a total of six bits. Let the first waveform value be $w_o = 1$.

CHAPTER

13

14.63
68.54
32.96
14.21
42.54
13.63
55.69
34.12
12.58
13.26
24.28
16.14
89.35
12.15

84.12
12.58
13.26
24.28
16.14
89.35
42.15

12.58
13.26
24.28
16.14
89.35
42.15
84.20
23.00
36.85

EXCEL BEST PRACTICES

LEARNING OBJECTIVES

After completing this chapter, the reader should be able to use Excel at three levels:

- Calculating and graphing—This ability allows us to explore more interesting problems.

- Design—Expressing problems in terms of design parameters allows us to illustrate tradeoffs and achieve optimal designs.

- Programming—The basics of programming are illustrated with
    - *Conditional statements* for highlighting results.
    - *Macros* that store a sequence of steps that are repeated by means of making a simple mouse click.
    - *Visual Basic for Applications (VBA)* programs extend the capabilities of Excel Macros and show how simple programming steps can accomplish elementary tasks. VBA is necessary for implementing a counter and automating Excel procedures.

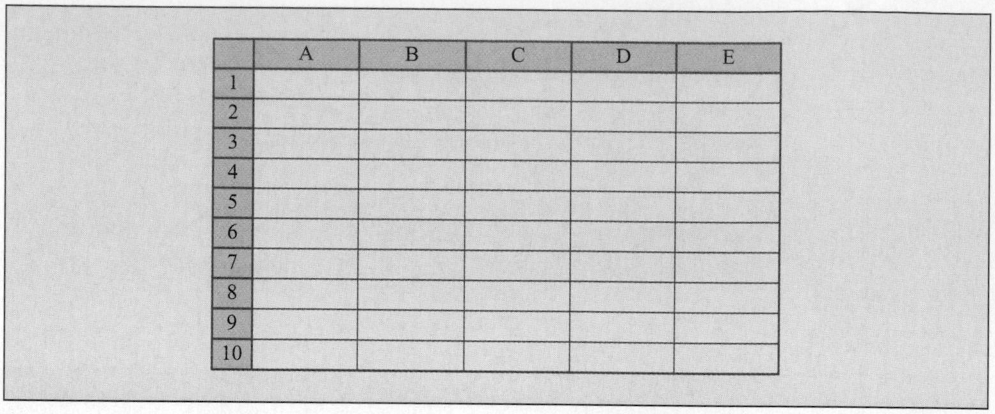

Figure 13.1

Example of a blank Excel worksheet.

	A	B	C	D	E
1					
2					
3					
4					
5					
6					
7					
8					
9					
10					

13.1 INTRODUCTION

Excel is a versatile software for calculating and manipulating data. This chapter illustrates examples of the calculations and designs of electrical engineering systems and refers to illustrating computational methods discussed in referenced chapters in this text.

Microsoft Excel can be considered to be a calculator that has many functions and a flexible display. Excel can evaluate a formula using a set of parameter values and display the results in a meaningful and insightful manner.

Figure 13.1 shows a **worksheet** (also called a **spreadsheet**). Its main feature is a two-dimensional set of cells, each cell specified by a column letter and row number, or its **address**. For example, cell A3 is in the first column (A) and in the third row (3).

13.2 CONTENTS OF CELLS

A cell can contain one of three types of data.

Text—labels that describe and organize content. Excel recognizes that a cell contains text by the first character being other than a number (0–9) or equal sign (=).

Numerical data—a *number* that can be processed and displayed. Excel recognizes that a cell contains a numerical value when the first character is a digit (0–9) or a decimal point (as in .3). Non-digital numerical data are considered later.

Formula—an arithmetic or algebraic expression that can refer to numbers in other cells and displays the result in that cell. Excel recognizes that a cell contains a formula by the first character being the equal sign (=). Formulas vary from a simple calculation of numerical values, either expressed explicitly (= 2 * 114 + 5) or referring to the values of other cells (= A1 + 1).

The numerical value of a cell can be **referenced**, (that is, used by another cell on the worksheet) in one of two ways, with each having its own advantages:

By relative addressing: This is done by using a cell's address (e.g., A3). If this is done by a formula within cell C4, for example =A3, this reference is understood as *value of the cell one row up and two columns to the left* (that is, A3 is one row up and two columns to the left of C4). When the formula in C4 is copied to some other cell, say D5, then the formula refers to the *value of the cell one row up and two columns to the left*, or B4. Addressing by relative location is done for specifying a set of index numbers or for specifying the argument in a function.

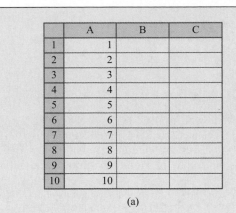

Figure 13.2

Example of an Excel worksheet specifying index values in column A, from A1:A10 (a) worksheet and (b) formulas.

<table>
<tr><td></td><td>A</td><td>B</td><td>C</td></tr>
<tr><td>1</td><td>1</td><td></td><td></td></tr>
<tr><td>2</td><td>2</td><td></td><td></td></tr>
<tr><td>3</td><td>3</td><td></td><td></td></tr>
<tr><td>4</td><td>4</td><td></td><td></td></tr>
<tr><td>5</td><td>5</td><td></td><td></td></tr>
<tr><td>6</td><td>6</td><td></td><td></td></tr>
<tr><td>7</td><td>7</td><td></td><td></td></tr>
<tr><td>8</td><td>8</td><td></td><td></td></tr>
<tr><td>9</td><td>9</td><td></td><td></td></tr>
<tr><td>10</td><td>10</td><td></td><td></td></tr>
</table>

(a)

<table>
<tr><td></td><td>A</td><td>B</td><td>C</td></tr>
<tr><td>1</td><td>1</td><td></td><td></td></tr>
<tr><td>2</td><td>=A1+1</td><td></td><td></td></tr>
<tr><td>3</td><td>=A2+1</td><td></td><td></td></tr>
<tr><td>4</td><td>=A3+1</td><td></td><td></td></tr>
<tr><td>5</td><td>=A4+1</td><td></td><td></td></tr>
<tr><td>6</td><td>=A5+1</td><td></td><td></td></tr>
<tr><td>7</td><td>=A6+1</td><td></td><td></td></tr>
<tr><td>8</td><td>=A7+1</td><td></td><td></td></tr>
<tr><td>9</td><td>=A8+1</td><td></td><td></td></tr>
<tr><td>10</td><td>=A9+1</td><td></td><td></td></tr>
</table>

(b)

Specifying a set of index values   EXAMPLE 13.1

Specifying a set of index values (i.e., 1, 2, 3, ... in column A) with the value in A1 equal to 1 can be done in two ways.

- The hard way: Enter 1 into A1, 2 in A2, 3 into A3, and so on until 10 in A10, with the result shown in the worksheet in Figure 13.2a.

- The better way: Enter 1 into A1, but in A2, enter the formula

$$= A1 + 1$$

which is interpreted by cell A2 as *make the value of this cell equal to the value of the cell one row above plus one*, making A2 equal to 2.

Now the neat part: When A2 is selected, a little square appears in its lower-right corner. Positioning the cursor on this square transforms the cursor into a "+". Holding down the left mouse button while dragging the cursor down to A10 copies (*drag-copies*) the formula into cells A3 to A10, or *A3:A10 in Excel-speak*. This gives exactly the same result as typing the previous tedious method.

The formulas in a worksheet can be displayed by typing **Ctrl '** (the Ctrl key and '– the key located at the upper-left corner of the standard keyboard—simultaneously), producing the formulas in worksheet shown in Figure 13.2b.

By absolute addressing: An absolute address reference is made by including dollar signs in the address (e.g., =$D$3) that are interpreted as *contents of cell D3*—no matter where the formula occurs in the worksheet. One application of the absolute reference is to define a parameter value in a formula in cell D3. Then, changing the value in cell D3 changes the evaluated value of the formula.

Hint: Pressing function key F4 after specifying a cell converts its relative address to an absolute address:

=A5(selected), press *F*4, *to* give =$A$5

The following example illustrates all three data types and two addressing modes.

EXAMPLE 13.2 — Relative and absolute addressing

Figure 13.3 shows a way to generate the linear sequence

$$y_i = mx_i + b \ \text{ for } \ 0 \le i \le 10$$

where the slope m and intercept b are specified constants. Figure 13.3a shows the worksheet, and 13.3b shows the formulas in the cells (viewed by pressing Ctrl-').

Cell A1 contains a text (m=) and B1 contains its value (2) that was entered manually. Cell A2 contains text (b=) and B2 contains its value (4) that was also entered manually. The text cells are filled with distinguishing colors—using the fill icon in the Home ribbon—to differentiate labels from their values.

The x_i array is labeled in A4, and its values (from 0 to 10) are specified in cells A5 to A15 (or in Excel form A5:A15). The first value (0) was entered in A5. The formula given by

=A5+1

was entered into A6 and uses relative addressing to mean the value of this cell equals the value of the cell directly above it plus one. The formula was selected and drag-copied into cells A7:A15. Note that the relative address changes as the formula is copied into the other cells. For example, A11 contains the formula given by

=A10+1

which adds the value of A10 (equal to 5) to 1 to produce the value 6 in A11.

The y_i array is labeled in B4, and its values are computed in cells B5:B15. The formula given by

=$B$1*A5+$B$2

was entered into B5. (Hint: Enter "=", select cell B1, and press F4 to produce "= $B$1".) This formula uses both absolute addressing (using $ symbols) and relative addressing to mean *the value of this cell equals the value of cell B1 (the value of m = 2) multiplied by the value in the cell directly to the left plus the value of cell B2 (the value of the intercept b = 4)*. The formula gives 4 (= 2 × 0 + 4).

The formula in B5 is selected and drag-copied into B6:B15. Note that the relative address changes as the formula is copied into the other cells, but the absolute address remains constant. For example, B11 contains the formula given by

=$B$1*A11+$B$2

Figure 13.3

Linear sequence $y_i = mx_i + b$ using both absolute and relative addressing (a) worksheet and (b) formulas (viewed by pressing Ctrl-').

	A	B
1	m=	2
2	b=	4
3		
4	x_i	y_i
5	0	4
6	1	6
7	2	8
8	3	10
9	4	12
10	5	14
11	6	16
12	7	18
13	8	20
14	9	22
15	10	24

(a)

	A	B
1	m=	2
2	b=	4
3		
4	x_i	y_i
5	0	=$B$1*A5 +$B$2
6	=A5+1	=$B$1*A6 +$B$2
7	=A6+1	=$B$1*A7 +$B$2
8	=A7+1	=$B$1*A8 +$B$2
9	=A8+1	=$B$1*A9 +$B$2
10	=A9+1	=$B$1*A10 +$B$2
11	=A10+1	=$B$1*A11 +$B$2
12	=A11+1	=$B$1*A12 +$B$2
13	=A12+1	=$B$1*A13 +$B$2
14	=A13+1	=$B$1*A14 +$B$2
15	=A14+1	=$B$1*A15 +$B$2

(b)

which multiplies B1 (slope $m = 2$) by the value of A11 (equal to 6) and adds the value in cell B2 (intercept $b = 4$) to give 16 ($= 2 \times 6 + 4$).

The real value of absolute addressing is that the values of m (in B1) and b (in B2) can be changed by entering new values and the y_i values change *automatically*.

13.3 PLOTTING CHARTS WITH EXCEL

Plotting data on a graph (or a chart in Excel-speak) is simple with Excel. Figure 13.4 shows a plot of the worksheet and chart produced by using the following two steps. The locations of the cells that specify the values of m and b have been moved to more convenient locations.

STEP 1. Select the cells that contain the values to be plotted. Place the cursor at one corner of the data to be plotted, in this case cell A2. Hold down the left mouse button and drag the cursor to cell B12.

STEP 2. Select the **Insert** tab and the **Charts/Scatter** icon. The scatter chart uses the first column for the x values and the remaining columns as y values. If more than two columns are selected, multiple y curves are plotted using the same x values.

The chart auto-scales to display the data. The appearance of the chart can be changed and labels added by using *Chart Tools*.

Changing the values of m (in D3) and b (in D7) updates the chart automatically. Figure 13.4 also shows that the parameters m and b can be located anywhere on the worksheet.

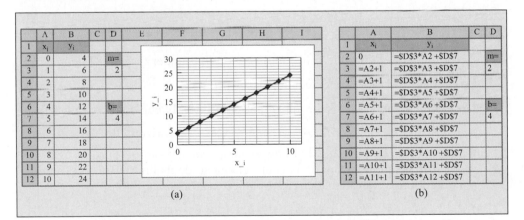

(a)   (b)

Figure 13.4

Plotting the linear sequence, $y_i = mx_i + b$: (a) worksheet and (b) formulas.

13.4 PRINTING AN EXCEL WORKSHEET

The printed version of your work is often the only thing that a viewer (customer or professor) sees, so it is important for it to look good (eventually—by the end of this chapter—*professional*). I have often seen students put in much effort to get the right answer—only to provide a printed worksheet that is hard to read, that contains too many numbers, and that fails to be convincing.

Figure 13.5 compares poor and better printed versions of the same worksheet. IMHO, the poor version is harder to read (no grid lines) and contains extra information (row and column headings). Excel does a *best guess* of what should be printed out, which usually includes all the data and charts on your worksheet. A long list of data will make the data hard to read.

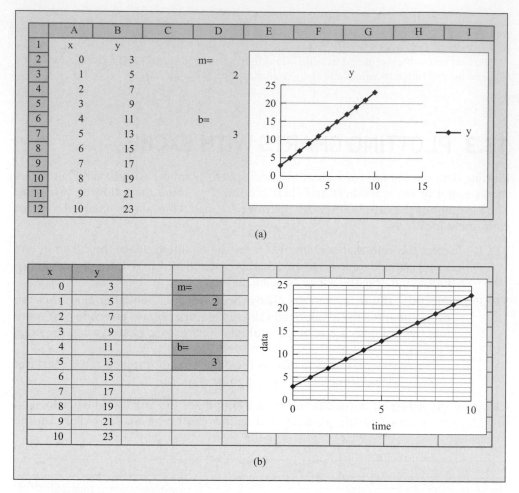

Figure 13.5

Examples of (a) poor and (b) better printed versions of a worksheet. When trouble-shooting a work sheet the row and column labels should be printed to allow easy specification of cells where the problem may occur.

The differences were produced by modifying options under the **_Page Layout_** Tab.

Print area selects the worksheet area to be printed.

Page setup menu modifies the appearance options. My personal favorites are

- **_Page_** (Fit to: 1 page); **_Margins_** (Center on Page—Horizontally).
- **_Sheet_** (Print—Gridlines. Checking _row and column headings_) produces a worksheet that simplifies the process of helping students find mistakes. These last two options are available directly on the **_Page Layout_** ribbon.
- Labels were displayed.
- Important cells were enclosed in borders and shaded.
- **_Chart Tools_** were used to select a more meaningful and labeled chart format.

▶ 13.5 DISPLAYING WORKSHEET FORMULAS

To help you recall how you produced a worksheet, it is helpful to print the formulas you used. To print the formulas you used in your worksheet, use Ctrl' to switch between the data and formula modes. For example, starting with a worksheet that contains formulas, do the following:

- **Ctrl '** displays the formulas on your worksheet.
- Some formulas will not fit into the column cell size. To size the column width to show the entire formula, position the cursor in the _A B C_ display row at the grid line to the right of the column that displays the formulas (when the cursor is

properly positioned, a double-arrow icon appears), and ***Double left click***. Note the column width adjusts to display or enclose the formula.

- Delete charts and non-vital columns to make the formulas readable when printed. (Nothing is so unprofessional as to need a magnifying glass to read a formula.)
- Print the worksheet in the usual manner.
- Pressing Ctrl' again displays the data on your worksheet.

13.6 INSERTING ROWS AND COLUMNS

It is common to need to insert rows (or columns) in your worksheet. Do this by positioning the cursor over a row number (column letter), right click and select ***Insert***.

There are two ways to insert multiple rows.

1. Press F4 (function key *↲ repeat last command*) after your first insert to repeat the insert row command.

2. Select (highlight) the number of row numbers at the row you want the insertion, right click and select ***Insert***.

13.7 CONDITIONAL FORMATTING FOR EMPHASIS

The color of a cell can be set to change depending on its value. This is useful when a particular condition is important, such as having negative funds. (A red cell color or font color may indicate it is time to call the folks!)

Set the cell color with the following five-step sequence.

STEP 1. Select the cell.

STEP 2. On the ***Home*** tab, select ***Style/Conditional Formatting*** to display the pull-down menu.

STEP 3. Select "Highlight cell rules"; then choose one of the options and the cell appearance.

STEP 4. Press "OK."

STEP 5. Enter a formula whose value changes the color. The simplest formula is to reference the contents of another cell (=*A*1 in B1 causes the contents of A1 to control the fill color of B1).

Mistakes in setting rules are easy to make. The best way to fix a mistake is to start again by removing the rules for the cell (or all rules) with the following steps.

STEP 1. Select the cell.

STEP 2. On the ***Home*** tab, select ***Style/Conditional Formatting*** to display the pull-down menu.

STEP 3. Select "Clear rules"; then select "Clear rules from selected cells" (or "Clear rules from entire sheet").

Amplitude indication EXAMPLE 13.3

Excel's pseudo-random number generator, ***RAND()***, produces a random number between 0 and 1. Conditional formatting will give an indication of the size of this number. Figure 13.6 shows a bar pattern whose height increases with random number size.

Figure 13.6

Sheet that uses conditional formatting to fill cell with a distinguishing color to visually indicate the size of a random number with bars: (a) worksheet and (b) formulas.

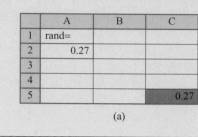

	A	B	C
1	rand=		
2	0.27		
3			
4			
5			0.27

(a)

	A	B	C
1	rand=		
2	=RAND()		=$A$2
3			=$A$2
4			=$A$2
5			=$A$2

(b)

Cell C5 was selected, and the ***Home/Conditional Formatting*** triangle was selected to display the pull-down menu. Selecting ***Highlight Cells Rules*** displays a list of options, from which **Greater Than ...** was chosen. The window that appears asks for the GREATER THAN value. For C5 0.2 was entered. The Format option that was chosen was *Light red fill*. Clicking on **OK** completed the operation.

A similar process was performed for C4 with GREATER THAN value 0.4, for C3 with GREATER THAN value 0.6, and for C2 with GREATER THAN value 0.8.

The formulas entered into C2:C5 were the same and referenced the value in A2.

The display shows a rand value 0.27 that lights only C5. The numbers in C2:C5 can be made less apparent by changing the font color.

13.8 INCLUDING A NARRATIVE IN EXCEL

A typical Excel worksheet can contain a narrative text that notes important observations and provides a conclusion based on the data processing results. The next example illustrates the process.

EXAMPLE 13.4 Inserting a Narrative Box

Figure 13.7 shows a narrative text box that was formed using the following four steps.

STEP 1. Select cells to define a region that is large enough to enclose your narrative. The example shows B2:E5.

STEP 2. In the *Alignment* Box on the *Home* tab, do the following.

- Merge the region by clicking on the *Merge & Center* icon,
- Click on the *Wrap Text* icon.
- Select *upper justify* and *left justify* cell contents.

Figure 13.7

Worksheet that contains a narrative text box.

	A	B	C	D	E	F
1						
2		This is an example of including a narrative in your Excel worksheet. Include observations and a conclusion based on the data processing results.				
3						
4						
5						
6						

STEP 3. Enclose the new box within an outline using the cell outline icon on the *Font* box on the *Home* tab.

STEP 4. Select the box and begin typing your narrative.

If the box is too small, define a larger box in your worksheet below your existing box, and copy your narrative from the formula bar and insert it into the new box. Delete your old box and move your new box to its location. In a pinch, you have the option of making the font size smaller.

13.9 EXCEL TIPS

Everyone occasionally hits the wrong key or makes a typing mistake. The following tips are useful for fixing mistakes.

TIP 1. Periodically (no, ***often***!) save your work by entering "Ctrl + s" ("Control s" in geek-speak—Control key depressed while typing s), the shortcut for "file save" (file pull-down menu, selecting "save").

TIP 2. Enter "Ctrl + z" to undo the previous entry or command.

TIP 3. Enter "Esc" ("escape") when something strange is happening in Excel, such as a Macro that is running or a cell that is flickering. ***Esc*** terminates the last command Excel thought it received.

TIP 4. A cell value can be changed in two ways:

1. Hard way for long entries: left-click on the cell and retype the whole thing to make correction.
2. Easy alternative: Make the change in the formula box labeled "fx" above the worksheet. The cursor can be positioned between letters or numbers to add or delete single letters.

TIP 5. The worksheet appearance can be improved by moving rows. The easiest way to move a row is given here. A column is moved in a similar manner.

1. To insert an empty row at a desired location, position the cursor at the desired row label to the left of column A, right-click to display the drop-down menu, and select "Insert".
2. To cut an existing row, right-click on the row number and select "cut." The row boundary will flash.
3. To move the row, place the cursor on the newly inserted row number, right-click, select "Paste Options," and choose the first option.

TIP 6. Copying worksheets. Worksheets need to be copied *with care*: Charts of data, although appearing to be copied to a new worksheet, actually display data from the original worksheet. This can be quite confusing!

There are two ways of copying a worksheet.

1. Through copy and paste:
   - On the original worksheet, right-click the box between Column A and Row 1. This selects the entire worksheet.
   - Select "Copy" to copy the entire worksheet.
   - Select a blank worksheet, right-click the box between Column A and Row 1 and select the first paste option to produce a copy.
   - *IMMEDIATELY* delete all charts on the new worksheet, because they display the data on the original worksheet and not the data on the new worksheet.
   - Insert plots of the desired data on the new worksheet.

2. Through copying worksheet tabs:

- Right-click on the worksheet tab below the worksheet, and select "Move or Copy"
- On the menu that appears, check the "Make a copy" box. This produces a new worksheet named worksheet_name(2). The charts on this worksheet display the data on the copied worksheet.
- Rename the worksheet by right-clicking on the new tab.

TIP 7. Coping with unusual or unexpected numerical values.

When #VALUE! appears This is the Excel notation indicating the result is not a valid number. This can happen if a result is the square root of a negative number. You need to find the computational mistake that led to this error.

When ##### appears This is the Excel notation indicating that the numerical result is too large to fit into the cell dimension. You can fix the problem by making the column wider. Place your cursor in the "A B C" row to the line separating columns, causing the symbol $\leftarrow | \rightarrow$ to appear. Double left-clicking widens the column.

Many of these tips were found by typing the description of the encountered problem into a search engine. The search engine returns many links that contain suggestions for possible solutions.

▶ 13.10 RECORDING EXCEL MACROS

A *Macro* is a sequence of Excel operations, such as data entry, mouse-clicking, etc., that is stored as a file and can be repeated with a single key stroke or mouse click. The Macro is created by clicking on a *start record* button, performing the operations that are to be repeated, and terminating with a *stop record* button click. The stored file contains Visual Basic statements that encode the operations you performed.

Developer Tab on PC and Mac

You will need the Developer Tab on the Excel ribbon to work with Macros and VBA. The *out-of-the-box* installation does not include the Developer tab. This section gives instructions on how to make the Developer tab visible.

Installing Developer on PC

To install Developer on your PC, Excel Help suggests the following steps.

STEP 1. Click on the *File* tab.

STEP 2. Under *Help*, click *Options*.

STEP 3. Click *Customize Ribbon*.

STEP 4. Under *Customize the Ribbon*, select the *Developer* check box.

STEP 5. Click *OK*, and the Developer tab will appear.

Installing Developer on a Mac

Unfortunately, Microsoft decided not to include Developer in Office 2008 for the Mac. It does exist on Office 2004 and Office 2011, so you need either one of these versions.

To install Developer on your Mac, Excel Help suggests the following steps.

STEP 1. Start Excel 2011. When it starts, the *Excel* menu item shows on the menu bar that appears along the top of the screen.

STEP 2. Click on the *Excel* menu and select *Preferences*.

STEP 3. In the *Excel Preferences* box, under *Sharing and Privacy*, select *Ribbon*.

STEP 4. In the *Ribbon* box, scroll to the bottom of the list and select the *Developer* check box.

STEP 5. Click *OK*, and the Developer tab will appear.

Recording a Macro

Once the Developer tab is visible, recording a Macro is accomplished using the following steps:

STEP 1. Select toolbar tab item *Developer*. On the left side there is a *Record Macro* button. Clicking on it will open a *Record Macro* box.

STEP 2. In the *Macro name* text box enter a descriptive name, such as "Save."

STEP 3. In the "Shortcut key" text box after Ctrl+, enter "b" and the Macro will repeat every time Ctrl-b is pressed. We are ready to record our Macro after we click on the OK button. Every key stroke thereafter is stored in memory.

STEP 4. Perform the desired operations, including key strokes, data entry, cell selections, copy, paste special, ...

STEP 5. When recording, a *stop recording* button replaces the *Record Macro* button. When finished with the desired keystroke sequence, click on the *stop recording* button.

A Macro executes when an ***Event*** happens. While there are many events that cause a Macro to execute, the two most common are:

Shortcut key—A Macro is assigned to a shortcut key sequence, and when pressing that sequence initiates the Macro execution.

Clicking on a shape—Assigning a Macro (or VBA subroutine, described next) to a shape on the worksheet, causes the Macro to execute when the shape is left-clicked.

> *Factoid:* Be careful when using Macros that use relative address references. For example, if a column or row is inserted after the Macro is defined, as when putting your name on the top row of the worksheet, the Macro will not shift its operation to the new locations, but refer to the original cells, which may now contain different data. In other words, the Macro will not do what was intended when it was recorded. The best thing to do in this case is to delete the old Macro and record a new one, after the desired rows have been inserted and the worksheet format has been finalized.

▶ 13.11 INSERTING SHAPES

Shapes are inserted into a worksheet using the **Insert/Shapes** tab and selecting your favorite shape. The shape can be labeled and the text can be centered in the **Home/Alignment** tab and resized with the $A^\triangle$ and $A^\triangledown$ icons in the **Home/Font** tab.

The following points are useful for forming nice-looking shapes.

■ Shape sizes can be adjusted or made uniform using **Drawing Tools/Size**. Choose a convenient size, such as 0.8" for both height and width.

■ Shapes can be positioned using **Drawing Tools/Arrange/Align** and snapping to grid or to shape. The latter is useful for connecting lines to shapes.

- Text within the shape can be scaled using the $A^\triangle$ or $A^\triangledown$ icons. Text can also be centered both horizontally and vertically.

- Shape locations can be changed by selecting object and using the arrow keys. Small movements are accomplished by simultaneous pressing the Ctrl key with the arrows.

- Separate shapes can be *grouped* into a single shape for copying and moving. Click on the first shape, then hold down Ctrl key while selecting additional shapes. In the *Drawing Tools/Arrange* select *Group* to combine the shapes into one group. *Ungroup* separates a combined shape into its individual components.

Assigning a Macro to a Shape

Right-clicking on a shape opens a menu that contains **Assign Macro**. Selecting this opens the **Assign Macro** text box. Click **Record** to display the **Record Macro** text box. Simply click **OK** to begin recording the Macro named (Shape)1_Click where (Shape1) is the name of your shape (Oval, Rectangle, ...). All keyboard strokes and mouse clicks are stored until **Developer/Stop Recording** is selected. The following steps form a simple Macro.

STEP 1. Select the large display cell.
STEP 2. Type the button cell text (i.e., A1).
STEP 3. Enter key.
STEP 4. Select **Developer/Stop Recording**.

If done correctly, left-clicking on the A1 button shows "A1" in the display cell.

If you make a mistake: Click on **Developer/Stop Recording**, then click **Macros** to display the list of Macro names. Find and delete the Macro with the mistake and record the Macro again.

13.12 EXTENDING EXCEL MACROS WITH VBA

Excel does not permit certain operations that may be useful. For such cases we use *Visual Basic for Applications* (VBA), which is a programming language within Excel. This book is not meant to be a comprehensive VBA programming guide, but only to illustrate how VBA can be used to implement simple tasks. VBA in Excel tutorials can be found on the Web, (http://www.excel-vba.com/excel-vba-contents.htm). Figure 13.8 shows the shortcut keys for accessing the VBA Macro editor.

Figure 13.8

Shortcut keys for accessing VBA Macro editor.

Command	PC	Mac
Open VBA Window	Alt+F11	Alt+Fn+F11
View Macros	Alt+F8	Alt+Fn+F11

EXAMPLE 13.5

Implementing a counter in VBA

Let us implement the counter shown in Figure 13.9 that records the number of times the *Count* shape is clicked. The number is stored in B1. Clicking on *Reset* resets the value of B1 to zero. This is a useful device that Excel cannot implement with its generic commands and Macros, so VBA is required.

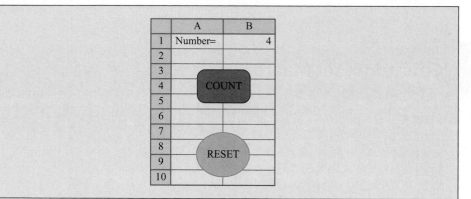

Figure 13.9

Sheet showing Counter and shapes that are assigned to Macros. Clicking on the *Count* shape increments the number in B1. Clicking on *Reset* resets the value of B1 to zero.

When you record a Macro, Excel stores the actions as VBA code, called a **subroutine**, containing "statements" that perform particular actions, one statement per line. The code starts with "sub Macro name," which identifies the code with a Macro, and ends with "End Sub." We will examine this code and modify it to implement a counter.

We start by creating a new Macro named *Counter* by using the following steps.

STEP 1. Select the **Developer** tab and click on the **Macros** icon. This will display the names of the current Macros in your workbook.

STEP 2. Type "Counter" in the **Macro name:** box. If a Macro with that name already exists, type another name, such as "NewCounter."

STEP 3. Left-click the **Create** button. This opens the *Microsoft Visual Basic for Applications* Editor window. Alternatively, Alt+F11 opens the Visual Basic Editor. Pressing Alt-F11 again will return to the Excel worksheet. Hence, Alt+F11 *toggles* the VBA Editor.

STEP 4. In the large text window, the following text should appear:

```
Sub Counter()


End Sub
```

Sub denotes the beginning of the *subroutine* named *Counter* (the name of the Macro we created). In fact, all Macros that you record in the usual way are stored as such subroutines. The () is a standard way of denoting a subroutine title line.

We implement a counter in B1 by entering VBA instructions so the result appears as

```
Sub Counter()
' Counter Macro 11/1/2013
 Range("B1").Value = Range("B1").Value + 1
End Sub
```

The VBA Macro contain the following elements:

- The apostrophe (') indicates a comment that contains useful information. VBA ignores the text on lines beginning with an apostrophe and text that occurs after an apostrophe.

- *Range* indicates the cell that is affected in the instruction. The cell address is specified within quotation marks. In this case, cell B1 is selected.

- *Value* is the cell *attribute* that specifies the numerical value of the cell.

- The equals sign (=) *assigns* the left side to be equal to the right side, which contains the expression `Range("B1").Value + 1`. In words, take the value of cell B1 and add 1 to it. The = assigns this value to cell B1.

Each time this Macro is executed, the contents of B1 increases by one. Once defined in this manner, the Counter Macro can be included in any other Macro in your workbook by merely typing "Counter" within the Macro.

That was fun. Let us create a new Macro named *Reset*.

EXAMPLE 13.6 Counter reset Macro

The following steps implement a Macro that resets the counter in the previous example.

STEP 1. Select the *Developer* tab and click on the *Macros* icon.

STEP 2. Type "Reset" in the *Macro name:* box.

STEP 3. Left-click the *Create* button.

STEP 4. In the large text window, the following text should appear:

```
Sub Reset()

End Sub
```

The Reset Macro is implemented with the code:

```
Sub Reset()
Range("B1").Value = 0
End Sub
```

The meaning of the statement **Range("B1").Value = 0** is to select B1, specify the value attribute of the cell, and set the value of B1 to zero.

We now insert shapes into our worksheet and assign a Macro to each shape. A rectangle is inserted and named COUNT. Right-clicking on the shape after it is inserted produces a menu, from which we select *Assign Macro*, and specify *Counter*. An oval is inserted and named RESET. Right-clicking on the shape after it is formed produces a menu, from which we select *Assign Macro* and specify *Reset*.

The RESET button should be pressed before the COUNT button for proper operation. If B1 contains text, rather than a number, an error will occur. Try typing an "a" into B1 and clicking COUNT. You should get an error message indicating a "Type mismatch" (a number was expected, but text was found).

This error causes Excel to suspend operation and display a box that will help you fix the error. You have to select one of the following two options:

END—This ignores the error condition. You should click on RESET to insert a zero into B1. A numerical value in B1 allows COUNT to increment it each time it is clicked.

DEBUG—This displays the VBA Editor showing the statement that produced the error. In this case, it was count = Range("B1").Value because "count" was specified to be an integer. This is the *DEBUGGER* program that is showing the error. If you made a typing mistake, you can fix it now. After making changes, save the Macro. Click on the "X" to exit the Debugger.

Useful Macro Tips

The following is a list of useful tips regarding VBA Macros.

TIP 1. Each Macro is a subroutine that starts with Sub name() and ends with End Sub.

TIP 2. Comments in VBA are indicated with an apostrophe

TIP 3. The value of cell A1 is obtained with the code Range("A1").Value.

TIP 4. Do While loop repeats instructions within the loop as long as the While condition is true.

TIP 5. A Macro calls other Macros by merely writing their name with the Macro.

TIP 6. A delay is introduced using the Application.Wait instruction (see Example 13.19). Without this delay, Excel may not chart the result of each loop, but only the final result. On a Mac use a one second delay to observe the behavior on a chart.

TIP 7. Assign a VBA Macro to a shape on the worksheet for running the Macro.

TIP 8. A workbook that uses Macros must be saved as a ".xlsm" file (Macro enabled Excel file).

Dealing with VBA problems

You will make errors in entering VBA code. Everyone does. Here's how to fix some common problems.

Typo: A *Compile error:* box will appear and the line in your code that contains a typing error will be shown in red, along with a little yellow arrow. Click *OK* and fix the error.

Run-time error: A *Run-time error:* box will appear. Click *Debug* and the line expecting the mistyped variable will appear. Fix the error in a previous line and then click on the *blue square* in the Editor ribbon to stop the Debugger and resume normal operation.

Program running but nothing is happening: Hit *Esc* and a *Code execution has been interrupted* box appears, Click *Debug* to make the current VBA instruction appear in yellow. A This is typically due to an error in your logic in writing the code, and requires some thinking about your code and what is missing. After modifying the code, click the *blue square* in the Editor ribbon to stop the Debugger and resume normal operation.

13.13 EXCEL BUILT-IN FUNCTIONS

Excel has many useful functions for programming tasks. This section describes several useful ones.

RAND() returns a random number in the range [0,1). The formula is written as

$$=\texttt{RAND()}$$

NORM.S.INV(RAND()) returns a standard Gaussian random number with zero mean and unit variance (and SD). The formula is written as

$$=\texttt{NORM.S.INV(RAND())}$$

MOD(number, divisor) returns the remainder after number is divided by divisor. For example, the formula written as

$$=\texttt{MOD(A1,2)}$$

gives 1 if value of A1 is an odd integer, or 0 if even. If A1 contains a number that is positive but not an integer, this function gives A1/2. If A1 contains a negative number, the function gives 0.

CEILING(number, significance) returns the integer that is the number rounded upward to the nearest multiple of the significance value. Examples include the following:

$$=\texttt{CEILING(2.5,1)} \quad (=3)$$

$$=\texttt{CEILING(41,3)} \quad (=42)$$

The last value equals the nearest multiple of 3 that is greater than the number.

FLOOR(number, significance) returns the integer that is the number rounded downward to the nearest multiple of the significance value. Examples include the following:

$$=\texttt{FLOOR(2.5,1)} \quad (=2)$$

$$=\texttt{FLOOR(41,3)} \quad (=39)$$

The last value equals the nearest multiple of 3 that is smaller than the number.

ROUND(number, num_digits) returns the integer that is the number rounded to the nearest multiple of the significance value. Examples include the following:

$$=\texttt{ROUND(2.5,0)} \quad (= 2)$$

$$=\texttt{ROUND(2.55,1)} \quad (= 2.6)$$

13.14 PC AND MAC EXCEL COMMANDS

The PC and Mac use different keyboard shortcuts. General information and a list can be found by clicking on the Excel question button (?) and searching for "Excel keyboard shortcuts." Figure 13.10 shows a list of common shortcut keys for the PC and Mac.

Figure 13.10

Common shortcut keys for the PC and Mac.

Command	PC	Mac
Re-calculate	F9	command =
Display Formulas	Ctrl+~ or (')	control '
Set Absolute Address	F4	command t
Open VBA Window	Alt+F11	Alt+Fn+F11
View Macros	Alt+F8	Alt+Fn+F8

13.15 SAMPLE EXCEL WORKSHEETS

This section provides examples that illustrate approaches to complete the Excel projects at the end of each chapter.

13.15.1 Introduction

EXAMPLE 13.7

Generating data with a linear trend

Consider the slope-intercept formula

$$y_i = m x_i + b$$

where m is the slope of the linear trend and b is the intercept.

We evaluate this formula by specifying a set of points x_i and computing the corresponding values of y_i, employing cells using relative addressing. We use specified parameter values of m and b by employing formulas using absolute addressing. Figure 13.11 shows the resulting worksheet.

Figure 13.11

Example of an Excel worksheet specifying a linear function, $y_i = m x_i + b$ where the value of m is in D3 and b in D7: (a) worksheet and (b) formulas.

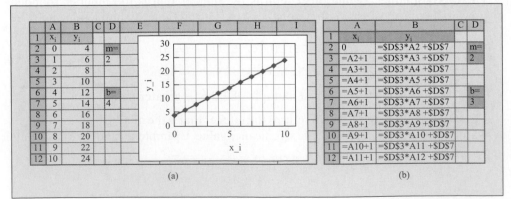

Set up the worksheet with the following steps.

STEP 1. In a new Excel worksheet, select A1 and enter "x_i". For a *professional look*, enter xi, highlight "i", and select *Font* in the Home tab, and check the *Subscript* box. Similarly, enter "y_i" into B1, "m=" into D2, and "b=" into D6.

STEP 2. Enter the slope value "2" into D3, and intercept value "3" into D7.

STEP 3. Enter "0" into A2, and

$$=A2+1$$

into A3. Copy A3 into cells A4:A12 by selecting and dragging the A3 cell.

STEP 4. Enter

$$= \$D\$3 * A2 + \$D\$7$$

into B2 and copy by dragging B2 into B3:B12.

Let's examine the copied formulas more closely. The right worksheet in Figure 13.11 shows the formulas by entering **Ctrl '**. The formula in B5 is

$$= \$D\$3 * A5 + \$D\$7$$

This formula differs slightly from that in B6, which is = $\$D\$3 * A6 + \$D\$7$. Note that while the absolute referenced addresses $\$D\$3$ (slope value) and $\$D\$7$ (intercept value) remain the same, the relative referenced location A5, ("cell one column to the left") in B5, changed to A6 in B6.

Changing the Parameter Values   EXAMPLE 13.8

Change the Parameter Values with the following steps.

STEP 1. Select D3, the slope value, and enter a "3" ($y_i = 3x_i + 3$) and hit Enter. Note all y_i values change to $3x_i + 3$.

STEP 2. Select D3 and enter a "10" ($y_i = 10x_i + 3$) and hit Enter. Note all y_i values change accordingly.

STEP 3. Select D7, the intercept value, and enter a "10" ($y_i = 10x_i + 10$) and hit Enter. Note all y_i values change to $10x_i + 10$.

Moore's Law   EXAMPLE 13.9

Consider Moore's law

$$N(t_i) = N_0 \, 2^{\frac{t_i - t_0}{1.5}}$$

where N_0 is the number of transistors at time t_0 measured in years. We evaluate this formula by specifying a set of year values t_i and computing the corresponding values of $N(t_i)$. Figure 13.12 shows the resulting worksheet. A scatter plot chart was used initially. When the large values of $N(t_i)$ made it difficult to read, the plot was modified by left-clicking on the y-axis scale and selecting a *logarithmic scale* with Base $= 10$.

(a) Worksheet

	A	B	C	D
1	N_0=	2500	t_0=	1971
2				
3	t_i(year)	N(t_i)		
4	1971	2500		
5	1973	6300		
6	1975	15874		
7	1977	40000		
8	1979	100794		
9	1981	253984		
10	1983	640000		
11	1985	1612699		
12	1987	4063747		
13	1989	10240000		
14	1991	25803183		
15	1993	65019947		
16	1995	163840000		
17	1997	412850930		
18	1999	1040319153		
19	2001	2621440000		
20	2003	6605614874		
21	2005	16645106455		
22	2007	41943040000		
23	2009	105689837985		
24	2011	266321703275		
25	2013	671088640000		
26	2015	1691037407763		

Chart (overlaid columns C–J): y-axis "Number of transistors" with gridlines 1.E+03, 1.E+04, 1.E+05, 1.E+06, 1.E+07, 1.E+08, 1.E+09, 1.E+10, 1.E+11, 1.E+12, 1.E+13; x-axis "Year" with values 1970, 1980, 1990, 2000, 2010, 2020.

(b) Formulas

	A	B	C	D
1	N_0=	2500	t_0=	1971
2				
3	t_i(year)	N(t_i)		
4	1971	=$B$1*2^((A4-$D$1)/1.5)		
5	=A4+2	=$B$1*2^((A5-$A$4)/1.5)		
6	=A5+2	=$B$1*2^((A6-$A$4)/1.5)		
7	=A6+2	=$B$1*2^((A7-$A$4)/1.5)		
8	=A7+2	=$B$1*2^((A8-$A$4)/1.5)		
9	=A8+2	=$B$1*2^((A9-$A$4)/1.5)		
10	=A9+2	=$B$1*2^((A10-$A$4)/1.5)		
11	=A10+2	=$B$1*2^((A11-$A$4)/1.5)		
12	=A11+2	=$B$1*2^((A12-$A$4)/1.5)		
13	=A12+2	=$B$1*2^((A13-$A$4)/1.5)		
14	=A13+2	=$B$1*2^((A14-$A$4)/1.5)		
15	=A14+2	=$B$1*2^((A15-$A$4)/1.5)		
16	=A15+2	=$B$1*2^((A16-$A$4)/1.5)		
17	=A16+2	=$B$1*2^((A17-$A$4)/1.5)		
18	=A17+2	=$B$1*2^((A18-$A$4)/1.5)		
19	=A18+2	=$B$1*2^((A19-$A$4)/1.5)		
20	=A19+2	=$B$1*2^((A20-$A$4)/1.5)		
21	=A20+2	=$B$1*2^((A21-$A$4)/1.5)		
22	=A21+2	=$B$1*2^((A22-$A$4)/1.5)		
23	=A22+2	=$B$1*2^((A23-$A$4)/1.5)		
24	=A23+2	=$B$1*2^((A24-$A$4)/1.5)		
25	=A24+2	=$B$1*2^((A25-$A$4)/1.5)		
26	=A25+2	=$B$1*2^((A26-$A$4)/1.5)		

Figure 13.12

Excel worksheet that plots Moore's law: (a) worksheet and (b) formulas.

13.15.2 Sensors and Actuators

EXAMPLE 13.10   Click Counter

This example shows how to compose a VBA Macro to implement a counter that indicates the number of times a shape is clicked.

Starting with a blank worksheet, perform the following steps.

STEP 1. Open the Macro command window with Alt+F8 (PC) or Alt+Fn+F8 (Mac).

STEP 2. Enter Macro name *counter* in the text box and select *Create*.

STEP 3. Type text shown as *Counter Macro* in Figure 13.13.

STEP 4. Save workbook as type *Excel Macro-Enabled Workbook (*.xlsm)* with name *Counter.xlsm*.

STEP 5. Insert a shape and assign it to the Macro using the following steps

  (a) On the *Insert* tab of the Excel Window, select *Shapes* and place a rectangle in region bordered by B1 and C3. Use the right side of the the *Drawing Tools* tab to adjust the *Height* and *Width* to *a nice size* using the up/down arrows. Try a 0.4 in. by 0.4 in. shape.

(b) Type *Count* within the shape. Center text using *Alignment* options in the *Home* tab.

(c) Double right-click on shape and select *Assign Macro* with left click.

(d) In *Assign Macro* window, left-click on *counter*, and OK.

(e) Test by left-clicking on shape and observing the count in A1.

STEP 6. To insert the second shape, open the Macro command window with Alt+F8 (PC) or Alt+Fn+F8 (Mac).

STEP 7. Enter Macro name *reset* in the text box and select *Create*.

STEP 8. Type text shown as *Reset Macro* in Figure 13.13.

STEP 9. Insert a shape and assign it to the reset Macro using the steps above.

Counter Macro

```
Sub counter()
' counts the number of times shape is clicked
Range("A1").Value = Range("A1").Value + 1    ' increments A1
End Sub
```

Reset Macro

```
Sub reset()
' resets the counter to 0
Range("A1").Value = 0       ' set A1=0
End Sub
```

	A	B	C
1	0		
2		Count	
3			
4		Reset	
5			

Figure 13.13

VBA Macro that implements a counter and reset operation. The VBA Macro codes are included.

Switch Array  **EXAMPLE 13.11**

Figure 13.14 implements a switch array and displays the binary row and column address of the clicked switch. Shapes were inserted and assigned to VBA Macros as described in Example 13.10. Each Macro assigns a shape-specific binary code (displayed in E2 and E3) to each shape.

```
Sub square()
Range("E2").Value = 0
Range("E3").Value = 0
End Sub
-------
Sub star()
Range("E2").Value = 0
Range("E3").Value = 1
End Sub
-------
Sub smiley()
Range("E2").Value = 1
Range("E3").Value = 0
End Sub
-------
Sub heart()
Range("E2").Value = 1
Range("E3").Value = 1
End Sub
```

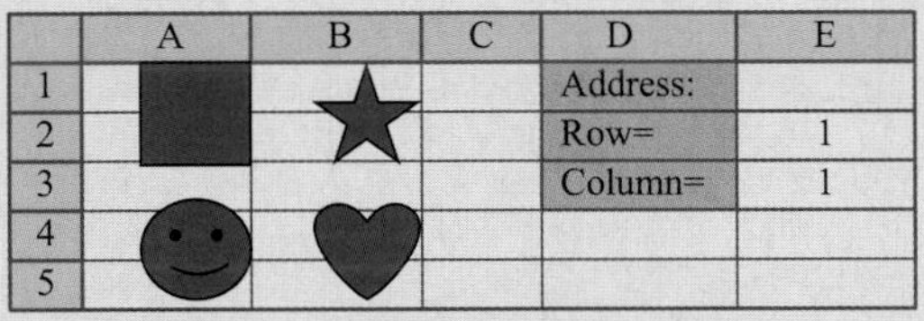

Figure 13.14

Worksheet that implements a 2 × 2 switch array and provides the binary row and column address of the shape that is clicked. The VBA Macro codes are included.

EXAMPLE 13.12

Setting your favorite RGB color

Figure 13.15 adjusts the color of a cell according to the specified RGB values. The VBA Macro codes for the red component are shown. Shapes were inserted and assigned to VBA Macros as described in Example 13.10. This example shows the following new features.

- The range of cells is selected using the ":", For example, Range("A1:B3") selects the six cells defined by corners A1 and B3.
- The cell color is adjusted by using small shapes that increase (∧) and decrease (∨) each RGB component by ten. Identical shapes are obtained by copying and pasting the first shape that is sized to fit within one cell. Multiple shape alignment is performed by selecting a set of shape and using the *Align* tool in the *Arrange* box under the *Page Layout* tab.

- The range of valid color is from 0 to 255. The VBA If statement assures that value remains valid.
- Macro *CellColor* is called from another Macro by merely typing its name as a separate instruction (the instruction following *End If*).

```
Sub CellColor()
Range("A1:B3").Interior.Color =
     RGB(Range("D1").Value, Range("D2").Value, Range("D3").Value)
End Sub
-------
Sub More_R()
If Range("D1").Value < 245 Then
     Range("D1").Value = Range("D1").Value + 10
End If
CellColor
End Sub
-------
Sub Less_R()
If Range("D1").Value > 10 Then
     Range("D1").Value = Range("D1").Value - 10
End If
CellColor
End Sub
```

	A	B	C	D	E	F
1			R–	157	∧	∨
2			G=	117	∧	∨
3			B=	97	∧	∨

Figure 13.15

Worksheet that adjusts the cell color according to its RGB values. The VBA Macro codes for increasing and decreasing the red component are shown.

13.15.3 Combinatorial Logic Circuits

AND Gate   **EXAMPLE 13.13**

Figure 13.16 illustrates the use of logic equation and symbols to draw a 2-input AND gate. Logic values can be entered into G3 and G5 and the AND operation result appears in K4.

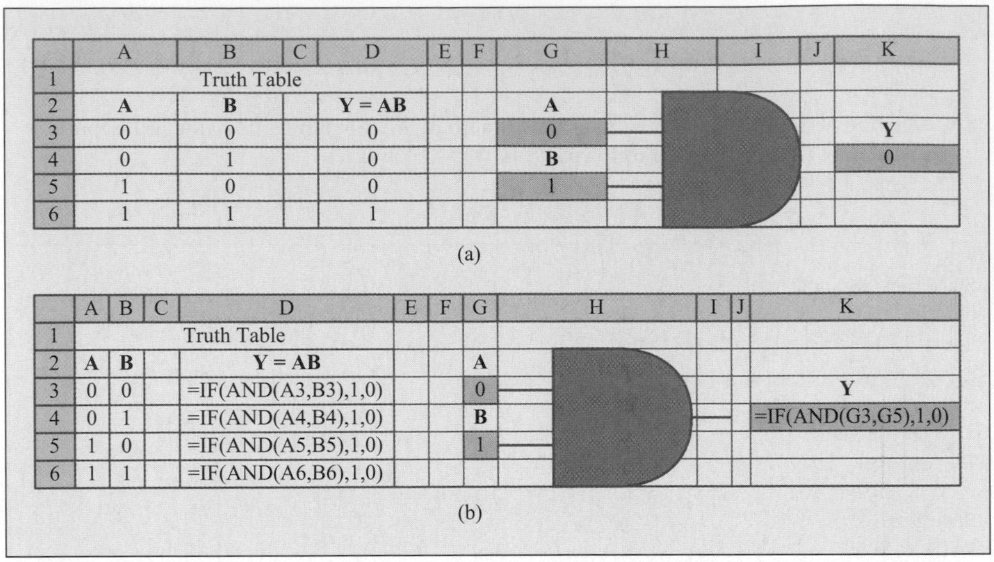

Figure 13.16

Worksheet that implements a 2-input AND logic gate: (a) worksheet and (b) formulas.

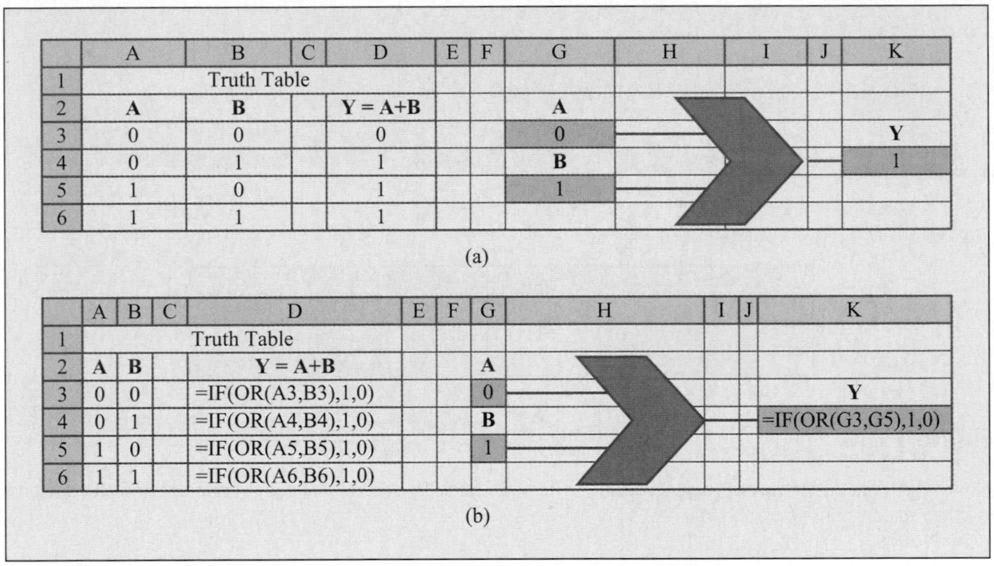

Figure 13.17

Worksheet that implements a 2-input OR logic gate: (a) worksheet and (b) formulas.

EXAMPLE 13.14

OR Gate

Figure 13.17 illustrates the use of logic equations and symbols to draw a 2-input OR gate. Logic values can be entered into G3 and G5 and the OR operation result appears in K4.

EXAMPLE 13.15

NOT Gate

Figure 13.18 illustrates the use of logic equations and symbols to draw a NOT gate. A logic value can be entered into F3 and the NOT operation result appears in J3.

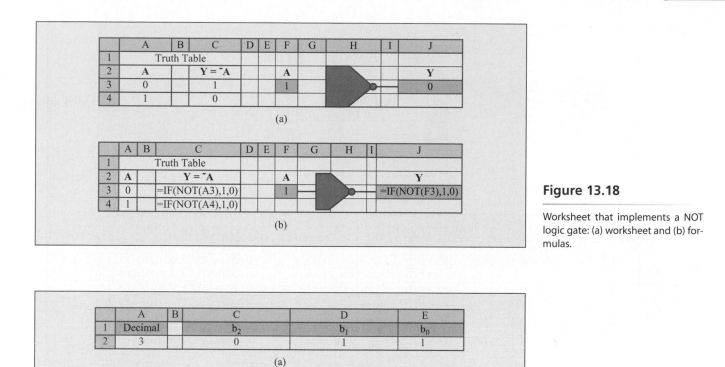

Figure 13.18

Worksheet that implements a NOT logic gate: (a) worksheet and (b) formulas.

Figure 13.19

Worksheet that converts a decimal number to a 3-bit binary value: (a) worksheet and (b) formulas.

Decimal to binary conversion   EXAMPLE 13.16

Figure 13.19 illustrates how to convert a decimal number into its binary representation, with each bit in its separate column. The decimal number in A2 is divided by decreasing powers of two, with the power corresponding to the bit position. The INT function forms the integer value of each division and the MOD function forms the binary value.

Verifying a logic equation by implementing the truth table   EXAMPLE 13.17

Figure 13.20 illustrates the evaluation of the logic equation

$$Y = A \cdot \overline{B} + C$$

by implementing the truth table. The truth table input section in columns C, D, and E is formed from the binary equivalent of the row number in column A. The truth table output in column G is computed from the Excel formula that implements the logic equation.

	A	B	C	D	E	F	G
1	row		A	B	C		Y
2	0		0	0	0		0
3	1		0	0	1		1
4	2		0	1	0		0
5	3		0	1	1		1
6	4		1	0	0		1
7	5		1	0	1		1
8	6		1	1	0		0
9	7		1	1	1		1

(a)

	A	B	C	D	E	F	G
1	row		A	B	C		Y
2	0		=MOD(INT(A2/4),2)	=MOD(INT(A2/2),2)	=MOD(A2,2)		=IF(OR(AND(C2,NOT(D2)),E2),1,0)
3	=1+A2		=MOD(INT(A3/4),2)	=MOD(INT(A3/2),2)	=MOD(A3,2)		=IF(OR(AND(C3,NOT(D3)),E3),1,0)
4	=1+A3		=MOD(INT(A4/4),2)	=MOD(INT(A4/2),2)	=MOD(A4,2)		=IF(OR(AND(C4,NOT(D4)),E4),1,0)
5	=1+A4		=MOD(INT(A5/4),2)	=MOD(INT(A5/2),2)	=MOD(A5,2)		=IF(OR(AND(C5,NOT(D5)),E5),1,0)
6	=1+A5		=MOD(INT(A6/4),2)	=MOD(INT(A6/2),2)	=MOD(A6,2)		=IF(OR(AND(C6,NOT(D6)),E6),1,0)
7	=1+A6		=MOD(INT(A7/4),2)	=MOD(INT(A7/2),2)	=MOD(A7,2)		=IF(OR(AND(C7,NOT(D7)),E7),1,0)
8	=1+A7		=MOD(INT(A8/4),2)	=MOD(INT(A8/2),2)	=MOD(A8,2)		=IF(OR(AND(C8,NOT(D8)),E8),1,0)
9	=1+A8		=MOD(INT(A9/4),2)	=MOD(INT(A9/2),2)	=MOD(A9,2)		=IF(OR(AND(C9,NOT(D9)),E9),1,0)

(b)

Figure 13.20

Worksheet that evaluates the logic equation $Y = A \cdot \overline{B} + C$ by implementing the truth table: (a) worksheet and (b) formulas.

EXAMPLE 13.18 Designing a 7-segment display

Figure 13.21 shows the design of a 7-segment display. The logic equation for each segment letter in column C uses the decimal digit entered in A2. The font size of the digit was increased for emphasis using $A^\wedge$ in the Font ribbon in the Home Tab. The segments in columns D, E, and F refer to their respective values in column C. Conditional formatting sets the segment color by its binary value. The heights of rows 2 and 4 were increased to form the segments.

13.15.4 Sequential Logic Circuits

EXAMPLE 13.19 VBA Macro to generate a 1-second pulse

This example shows how to compose a VBA Macro to generate a 1-second long pulse ($0 \rightarrow 1 \rightarrow 0$) in A1 and assign it to a shape. The delay is implemented with the Application.Wait instruction. The shape was inserted and assigned to the VBA Macro as described in Example 13.10 and is shown in Figure 13.22.

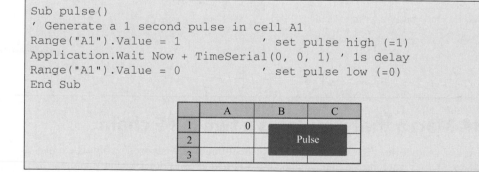

	A	B	C	D	E	F	G
1	TYPE digit				0		
2	4			1		1	
3					1		
4				0		1	
5					0		
6		a=	0				
7		b=	1				
8		c=	1				
9		d=	0				
10		e=	0				
11		f=	1				
12		g=	1				
13							

(a)

	A	B	C	D	E	F
1	TYPE digit				=C6	
2	4			=C11		=C7
3					=C12	
4				=C10		=C8
5					=C9	
6		a=	=IF(OR(A2=0,A2=2,A2=3,A2=5,A2=7,A2=8,A2=9),1,0)			
7		b=	=IF(OR(A2=0,A2=1,A2=2,A2=3,A2=4,A2=7,A2=8,A2=9),1,0)			
8		c=	=IF(OR(A2=0,A2=1,A2=3,A2=4,A2=5,A2=6,A2=7,A2=8,A2=9),1,0)			
9		d=	=IF(OR(A2=2,A2=3,A2=5,A2=6,A2=8,A2=0),1,0)			
10		e=	=IF(OR(A2=0,A2=2,A2=6,A2=8),1,0)			
11		f=	=IF(OR(A2=4,A2=5,A2=6,A2=8,A2=9,A2=0),1,0)			
12		g=	=IF(OR(A2=2,A2=3,A2=4,A2=5,A2=6,A2=8,A2=9),1,0)			
13						

(b)

Figure 13.21

Worksheet that implements a 7-segment display: (a) worksheet and (b) formulas.

```
Sub pulse()
' Generate a 1 second pulse in cell A1
Range("A1").Value = 1          ' set pulse high (=1)
Application.Wait Now + TimeSerial(0, 0, 1) ' 1s delay
Range("A1").Value = 0          ' set pulse low (=0)
End Sub
```

	A	B	C
1	0		
2		Pulse	
3			

Figure 13.22

VBA Macro assigned to shape that when clicked generates a one-second pulse ($0 \rightarrow 1 \rightarrow 0$) in A1.

VBA Macro that simulates SR-FF   EXAMPLE 13.20

Example 13.19 explains how to write a Macro to produce a 1-second pulse and assign it to a shape. This example composes one Macro to generate a pulse that sets the SR-FF and a second Macro that resets it. Figure 13.23 shows the VBA code and the worksheet that was composed using the following steps starting with a blank worksheet. Shapes were inserted and assigned to VBA Macros as described in Example 13.10.

Set Macro

```
Sub SET_SR_FF()
' Sets the Set-Reset FF with pulse on Set input
' B1 is set, B2 is reset, E1 is Q, E2 is ~Q
Range("B1").Value = 1 ' Set 0 -> 1
Range("E1").Value = 1 ' Q -> 1
Range("E2").Value = 0 ' ~Q -> 0
Application.Wait Now + TimeSerial(0, 0, 1) ' 1s delay
Range("B1").Value = 0 ' Set 1 -> 0
End Sub
```

Reset Macro

```
Sub Reset_SR_FF()
' Resets the Set-Reset FF with pulse on Reset input
' B1 is set, B2 is reset, E1 is Q, E2 is ~Q
Range("B2").Value = 1    ' Reset 0 -> 1
Range("E1").Value = 0    ' Q -> 0
Range("E2").Value = 1    ' ~Q -> 1
Application.Wait Now + TimeSerial(0, 0, 1) ' 1s delay
Range("B2").Value = 0    ' Reset 1 -> 0
End Sub
```

	A	B	C	D	E
1	S=	0		Q=	1
2	R=	0		~Q=	0
3					
4			Set	Reset	
5					

Figure 13.23

VBA Macros that simulate a set-reset flip-flop (SR-FF).

EXAMPLE 13.21 **VBA Macro that simulates a two-T-FF chain**

Figure 13.24 shows the VBA code and the worksheet that was composed to simulate a two-T-FF chain. Shapes were inserted and assigned to VBA Macros as described in Example 13.10. This example introduces the Boolean variable type in VBA that takes on the logical values of True and False. The program was composed using the following spreadsheet cells and tested accordingly.

1. Enter labels A input in A1, $Q0$ output in C1, $Q1$ output in C1, and Clear C in D4.

2. Test the two T-FF simulation by clicking on A and *Clear* in any order and observing how the $Q0$ and $Q1$ values change.

T-FF Macro

```
Sub Pulse_2TFF_Chain()
' A2 is Input A & FF0-T, C2 is FF0-Q & FF1-T, D2 is FF1-Q & FF2-T(in any)
Dim DW0 As Boolean ' downward transition on Q0-T
Dim DW1 As Boolean ' downward transition on Q1-T
Application.Wait Now + TimeSerial(0, 0, 1) ' 1s delay
Range("A2").Value = 1    ' FF0-T = 1
Application.Wait Now + TimeSerial(0, 0, 1) ' 1s delay
Range("A2").Value = 0    ' FF0-T 1 -> 0
DW0 = True               ' DW transition on FF0-T
If DW0 = True Then  ' Toggle FF0 (First FF always True but makes code simpler)
    If Range("C2").Value = 0 Then
        Range("C2").Value = 1   ' 0 -> 1
        DW1 = False             ' upward transition
    Else
        Range("C2").Value = 0   ' 1 -> 0 DW transition
        DW1 = True              ' downward transition on FF1-T
    End If
End If

If DW1 = True Then  ' Toggle FF1
    If Range("D2").Value = 0 Then
        Range("D2").Value = 1   ' 0 -> 1
        DW2 = False             ' upward transition
    Else
        Range("D2").Value = 0   ' 1 -> 0 DW transition
        DW2 = True              ' downward transition on FF2-T (if any)
    End If
End If
End Sub
```

Clear 2 T-FF Macro

```
Sub Clear_2TFF()
' Clear signal in D5
Range("D5").Value = 1   ' C -> 1
Range("C2").Value = 0   ' Q0 -> 0
Range("D2").Value = 0   ' Q1 -> 0
Application.Wait Now + TimeSerial(0, 0, 1) ' 1s delay
Range("D5").Value = 0   ' C -> 0
End Sub
```

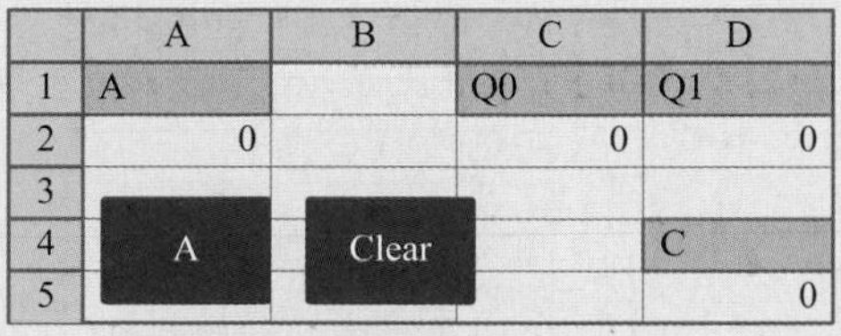

	A	B	C	D
1	A		Q0	Q1
2	0		0	0
3				
4	A	Clear		C
5				0

Figure 13.24

VBA Macros simulate a two-toggle flip-flop chain.

EXAMPLE 13.22 | ## Modulo-5 counter

Figure 13.25 shows the VBA code and the worksheet that was composed to implement a modulo-5 counter. This example requires that Example 13.21 be extended to form a three T-FF chain by modifying Macros *Sub Pulse_2TFF_Chain()* to *Sub Pulse_3TFF_Chain()* and *Sub Clear_2TFF* to *Sub Clear_3TFF*. Shapes were inserted and assigned to VBA Macros as described in Example 13.10.

```
Sub Mod5Counter()
Dim Q0 As Boolean
Dim Q1 As Boolean
Dim Q2 As Boolean
Dim C As Boolean
Clear_3TFF
Do While Range("A2").Value < 2  ' infinite loop
    Pulse_3TFF_Chain
    ' check state
    Q0 = True           ' default value
    If Range("C2").Value = 0 Then ' change if needed
     Q0 = False
    End If
    Q1 = True           ' default value
    If Range("D2").Value = 0 Then ' change if needed
     Q1 = False
    End If
    Q2 = True           ' default value
    If Range("E2").Value = 0 Then ' change if needed
     Q2 = False
    End If
    ' compute C
    C = Q0 And Q2       ' logic equation for clearing on 101
    If C = True Then
        Range("D5").Value = 1
    Else
        Range("D5").Value = 0
    End If
    If Range("D5").Value = 1 Then   ' check if Clear = 0
        Clear_3TFF                  ' reset T-FF chain
    End If
Loop
End Sub
```

	A	B	C	D	E
1	A		Q0	Q1	Q2
2	0		0	0	0
3					
4	Demo Mod-5 Counter			C	
5				0	

Figure 13.25

VBA Macros simulate a modulo-5 counter.

To demonstrate the operation of the modulo-5 counter, the VBA Macro Sub Mod5Counter() implements an infinite loop contained within the two instructions, as

```
Do While Range("A2").Value < 2   ' infinite loop
...
Loop
End Sub
```

This loop repeatedly calls *Pulse_3TFF_Chain* to produce an input pulse to the T-FF chain. The Macro defines Boolean variables $Q0$, $Q1$, $Q2$, and C to form the valid logic equation

```
C = Q0 And Q2        ' logic equation for clearing on 101
```

Note that C=Q0 And NOT(Q1) And Q2, while being also correct, is not necessary (why?).

13.15.5 Converting Between Analog and Digital Signals

Sampling sinusoids

EXAMPLE 13.23

When generating a time series of sample values, it is often convenient to let the sampling period $T_s = 1$, with arbitrary units. Then, the frequency f_o is defined in terms of the reciprocal of the number of points per period. For example, $f_o = 1/16$ means that 16 samples occur in one period. It is also conventional to have the index i start at $i = 0$. Figure 13.26 illustrates the following points.

	A	B
1	$f_o=$	0.0625
2		
3	i	$\sin(2\pi f_o i)$
4	0	0.000
5	1	0.383
6	2	0.707
7	3	0.924
8	4	1.000
9	5	0.924
10	6	0.707
11	7	0.383
12	8	0.000
13	9	−0.383
14	10	−0.707
15	11	−0.924
16	12	−1.000
17	13	−0.924
18	14	−0.707
19	15	−0.383
20	16	0.000

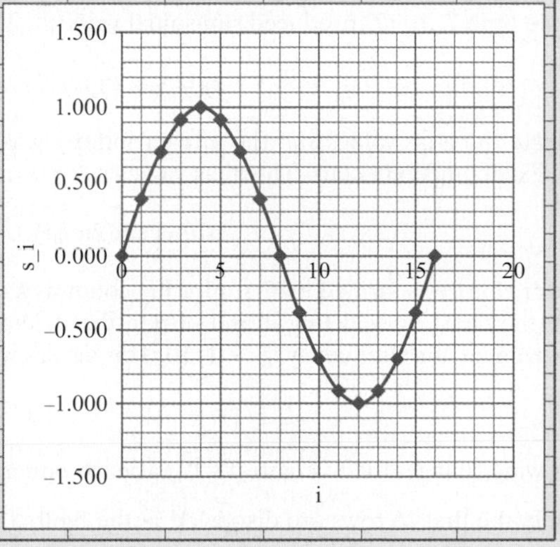

(a)

	A	B
1	$f_o=$	=1/16
2		
3	i	$\sin(2\pi f_o i)$
4	0	=SIN(2*PI()*$B$1*A4)
5	=1+A4	=SIN(2*PI()*$B$1*A5)
6	=1+A5	=SIN(2*PI()*$B$1*A6)
7	=1+A6	=SIN(2*PI()*$B$1*A7)
8	=1+A7	=SIN(2*PI()*$B$1*A8)
9	=1+A8	=SIN(2*PI()*$B$1*A9)
10	=1+A9	=SIN(2*PI()*$B$1*A10)
11	=1+A10	=SIN(2*PI()*$B$1*A11)
12	=1+A11	=SIN(2*PI()*$B$1*A12)
13	=1+A12	=SIN(2*PI()*$B$1*A13)
14	=1+A13	=SIN(2*PI()*$B$1*A14)
15	=1+A14	=SIN(2*PI()*$B$1*A15)
16	=1+A15	=SIN(2*PI()*$B$1*A16)
17	=1+A16	=SIN(2*PI()*$B$1*A17)
18	=1+A17	=SIN(2*PI()*$B$1*A18)
19	=1+A18	=SIN(2*PI()*$B$1*A19)
20	=1+A19	=SIN(2*PI()*$B$1*A20)

(b)

Figure 13.26

Worksheet that generates samples of a sinusoidal waveform: (a) worksheet and (b) formulas.

- The frequency $f_o = 1/16$ means 16 points are sampled per period. Note that the sinusoid is evaluated for 17 points, or $0 \le i \le 16$. The 17th point, having index $i = 16$, is actually the first sample of the next period, but is convenient because it brings the sequence back to 0.

- The index i is conveniently generated by setting the first value $i = 0$, and writing the relative address formula =A4+1 in A5, which is drag-copied from A6 to A20.

- The formula that evaluates the sine values uses the Excel value for π, written as PI(). The values are conveniently generated by using absolute addressing for the frequency (=$B$1). The simple way to include the $ signs is to type B1 followed by the function key $F4$. Relative addressing is used in B4 for the index i given in A4.

- The sine values are shown expressed with three places to the right of the decimal point, accomplished with the precision icon in the *Number* box in the *Home* tab.

- In the scatter plot, I selected my favorite scale that uses a grid, by choosing that option in the *Chart Layouts* box in the *Chart Tools* tab. Legends and labels can be deleted when they are obvious to leave more room for the plot.

EXAMPLE 13.24 — Demonstration of Aliasing

Consider sampling period $T_s = 1$ considered in Example 13.23, making $f_s = 1/T_s = 1$. The Nyquist criterion states that aliasing will not occur for $f_o < f_s/2 = 1/2$. Example 13.23 illustrated that $f_o = 1/16$ produces samples that clearly follow a sinusoidal trend. To illustrate aliasing we make $f_o = 1.1$. We show eleven periods at this f_o by computing eight samples per period using the finer time increment T_o specified in E1. Figure 13.27 shows this calculation for the first 13 points in columns A and B, although a total of 88 samples were computed, where the sample time $T_j = jT_o$ produced sinusoidal value s_j. Thus,

$$s_j = \sin(2\pi f_o j T_o)$$

To illustrate the alias values sa_i, the integer index i was formed by downward rounding T_j with the Excel function FLOOR. The alias values were formed by

$$sa_i = \sin(2\pi f_o i)$$

The scatter chart was formed by first plotting columns A through D, and then deleting the "i" curve on the chart by selecting it and hitting the delete key. The result plot clearly shows samples from an alias frequency $f_a = 1/10$. This agrees with the calculation

$$f_a = |f_s - f_o| = |1 - 1.1| = 0.1$$

The following changes made Figure 13.27 easier to understand.

- Only the first 16 rows are displayed in the figure. Rows 17 to 100 were merely copied versions of the formulas in row 16.

- The formats of the horizontal and vertical scales were changed from the default values assigned by Excel.

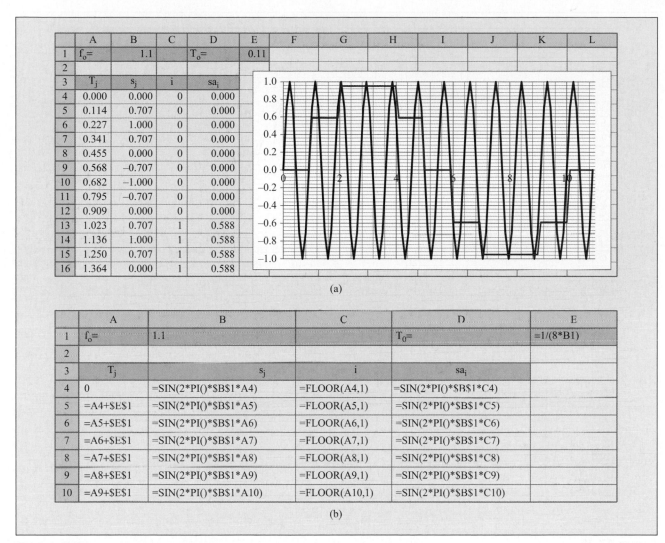

	A	B	C	D	E	F	G	H	I	J	K	L
1	$f_o=$	1.1		$T_o=$	0.11							
2												
3	T_j	s_j	i	sa_i								
4	0.000	0.000	0	0.000								
5	0.114	0.707	0	0.000								
6	0.227	1.000	0	0.000								
7	0.341	0.707	0	0.000								
8	0.455	0.000	0	0.000								
9	0.568	−0.707	0	0.000								
10	0.682	−1.000	0	0.000								
11	0.795	−0.707	0	0.000								
12	0.909	0.000	0	0.000								
13	1.023	0.707	1	0.588								
14	1.136	1.000	1	0.588								
15	1.250	0.707	1	0.588								
16	1.364	0.000	1	0.588								

(a)

	A	B	C	D	E
1	$f_o=$	1.1		$T_o=$	=1/(8*B1)
2					
3	T_j	s_j	i	sa_i	
4	0	=SIN(2*PI()*$B$1*A4)	=FLOOR(A4,1)	=SIN(2*PI()*$B$1*C4)	
5	=A4+$E$1	=SIN(2*PI()*$B$1*A5)	=FLOOR(A5,1)	=SIN(2*PI()*$B$1*C5)	
6	=A5+$E$1	=SIN(2*PI()*$B$1*A6)	=FLOOR(A6,1)	=SIN(2*PI()*$B$1*C6)	
7	=A6+$E$1	=SIN(2*PI()*$B$1*A7)	=FLOOR(A7,1)	=SIN(2*PI()*$B$1*C7)	
8	=A7+$E$1	=SIN(2*PI()*$B$1*A8)	=FLOOR(A8,1)	=SIN(2*PI()*$B$1*C8)	
9	=A8+$E$1	=SIN(2*PI()*$B$1*A9)	=FLOOR(A9,1)	=SIN(2*PI()*$B$1*C9)	
10	=A9+$E$1	=SIN(2*PI()*$B$1*A10)	=FLOOR(A10,1)	=SIN(2*PI()*$B$1*C10)	

(b)

Figure 13.27

Worksheet that illustrates aliasing: (a) worksheet and (b) formulas.

Quantization EXAMPLE 13.25

Figure 13.28 illustrates quantizing a sinusoidal waveform having frequency $f_o(= 1/32)$ to a specified number of bits $n_b = 3$. The step size $\Delta = 1$, making the quantizer range extend from 0 to $V_{max} = 2^{n_b} - 1$. The sine waveform was converted to signal s_i to have positive values by adding one (producing values between 0 and 2) and multiplying by $V_{max}/2$. The quantized signal sq_i is formed by rounding s_i to the nearest integer with Excel function ROUND. The corresponding n_b-bit binary code is obtained with Excel function DEC2BIN.

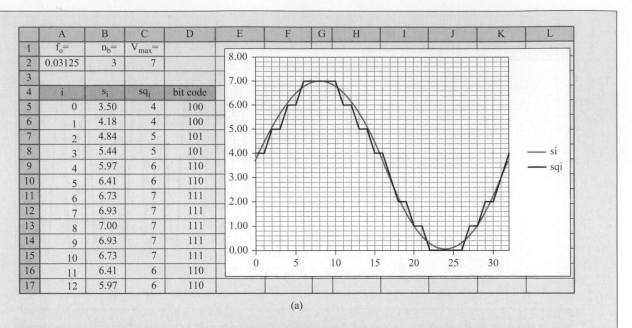

	A	B	C	D	E	F	G	H	I	J	K	L
1	f_o=	n_b=	V_{max}=									
2	0.03125	3	7									
3												
4	i	s_i	sq_i	bit code								
5	0	3.50	4	100								
6	1	4.18	4	100								
7	2	4.84	5	101								
8	3	5.44	5	101								
9	4	5.97	6	110								
10	5	6.41	6	110								
11	6	6.73	7	111								
12	7	6.93	7	111								
13	8	7.00	7	111								
14	9	6.93	7	111								
15	10	6.73	7	111								
16	11	6.41	6	110								
17	12	5.97	6	110								

(a)

	A	B	C	D
1	f_o=	n_b=	V_{max}=	
2	=1/32	3	=2^B2–1	
3				
4	i	s_i	sq_i	bit code
5	0	=($C$2/2)*(1+SIN(2*PI()*$A$2*A5))	=ROUND(B5,0)	=DEC2BIN(C5,$B$2)
6	=1+A5	=($C$2/2)*(1+SIN(2*PI()*$A$2*A6))	=ROUND(B6,0)	=DEC2BIN(C6,$B$2)
7	=1+A6	=($C$2/2)*(1+SIN(2*PI()*$A$2*A7))	=ROUND(B7,0)	=DEC2BIN(C7,$B$2)
8	=1+A7	=($C$2/2)*(1+SIN(2*PI()*$A$2*A8))	=ROUND(B8,0)	=DEC2BIN(C8,$B$2)
9	=1+A8	=($C$2/2)*(1+SIN(2*PI()*$A$2*A9))	=ROUND(B9,0)	=DEC2BIN(C9,$B$2)
10	=1+A9	=($C$2/2)*(1+SIN(2*PI()*$A$2*A10))	=ROUND(B10,0)	=DEC2BIN(C10,$B$2)

(b)

Figure 13.28

Worksheet that quantizes a sinusoidal waveform to a specified number of bits: (a) worksheet and (b) formulas.

EXAMPLE 13.26 Boxcar DAC

This example uses the binary codewords produced by Example 13.25 to reconstruct the quantized waveform sq_i. Figure 13.29 shows the results. The bit code in column B was copied and pasted (values) from the worksheet produced in Example 13.25. The sample value was reconstructed in column C by multiplying the step size specified in B2 by the decimal equivalent of column B using Excel function BIN2DEC.

EXAMPLE 13.27 PWM DAC

Figure 13.30 illustrates how a PWM DAC forms the quantized waveform value sq_i from the average of the PWM waveform over one period.

The desired sq_i value is entered in A2, the number of quantization bits n_b in B2. From these, the values of $V_{max} = 2^{n_b} - 1$ and the PWM period (also $= 2^{n_b} - 1$) are computed.

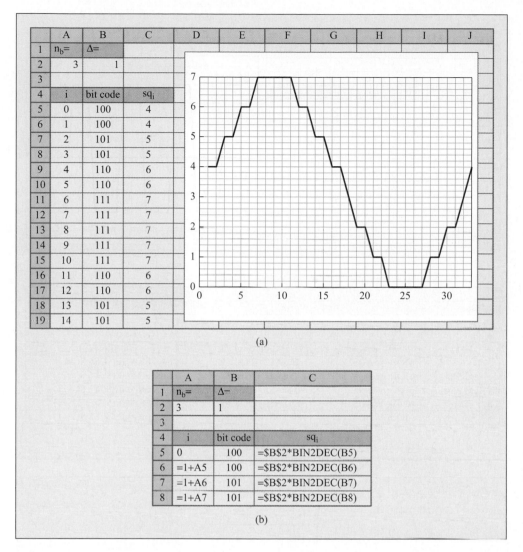

Figure 13.29

Worksheet that illustrates a boxcar DAC that reconstructs the quantized waveform from binary codewords containing $n_b = 3$ bits: (a) worksheet and (b) formulas.

The index i is specified to produce 28 values, corresponding to four PWM periods, which are plotted in the scatter chart, although the figure shows only the first eleven values.

The PWM interval is the resolution cell number, 0 through *PWM Period*−1, computed using Excel function MOD to obtain the modulo-(PWM Period) value of i. If the $sq_i > PWM$ *interval*, the interval value $= V_{max}$, otherwise it equals 0. Column D computes the reconstructed sq_i value from the average over the first PWM interval using Excel function AVERAGE.

13.15.6 Modeling Random Data and Noise

Excel has two useful pseudo-random number generators (*PRNGs*) described below.

> **Uniformly distributed random numbers RAND()**—gives random numbers denoted by Y that are equally distributed over the interval between 0 and 1 as shown in Figure 13.31. These random numbers have theoretical mean $\mu_Y = 0.5$ and variance $\sigma_Y^2 = 1/12$. A worksheet produces realizations denoted Y_i for $0 \le i \le n_y - 1$. The index starts at $i = 0$ to conform with the signal sequence notation used in this book. Statisticians typically start their index value at $i = 1$ to indicate the *first* sample.

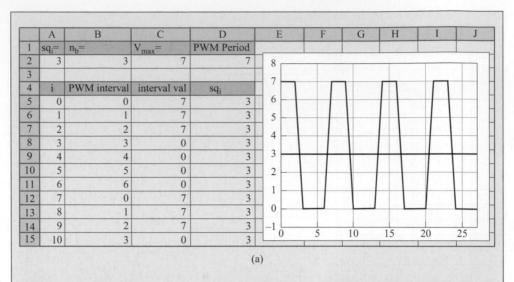

	A	B	C	D	E	F	G	H	I	J
1	sq$_i$=	n$_b$=	V$_{max}$=	PWM Period						
2	3	3	7	7						
3										
4	i	PWM interval	interval val	sq$_i$						
5	0	0	7	3						
6	1	1	7	3						
7	2	2	7	3						
8	3	3	0	3						
9	4	4	0	3						
10	5	5	0	3						
11	6	6	0	3						
12	7	0	7	3						
13	8	1	7	3						
14	9	2	7	3						
15	10	3	0	3						

(a)

	A	B	C	D
1	sq$_i$=	n$_b$=	V$_{max}$=	PWM Period
2	3	3	=2^B2 – 1	=C2
3				
4	i	PWM interval	interval val	sqi
5	0	=MOD(A5,$D$2)	=IF($A$2>B5,$C$2,0)	=AVERAGE($C$5:$C$11)
6	=1+A5	=MOD(A6,$D$2)	=IF($A$2>B6,$C$2,0)	=AVERAGE($C$5:$C$11)
7	=1+A6	=MOD(A7,$D$2)	=IF($A$2>B7,$C$2,0)	=AVERAGE($C$5:$C$11)
8	=1+A7	=MOD(A8,$D$2)	=IF($A$2>B8,$C$2,0)	=AVERAGE($C$5:$C$11)
9	=1+A8	=MOD(A9,$D$2)	=IF($A$2>B9,$C$2,0)	=AVERAGE($C$5:$C$11)

(b)

Figure 13.30

Worksheet that illustrates how a PWM DAC forms the quantized waveform value sq_i from the average value of the PWM waveform over one period: (a) worksheet and (b) formulas.

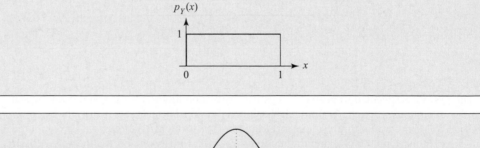

Figure 13.31

Random variable Y is uniformly distribution between 0 and 1, having mean $\mu_Y = 0.5$ and variance $\sigma_Y^2 = 1/12$.

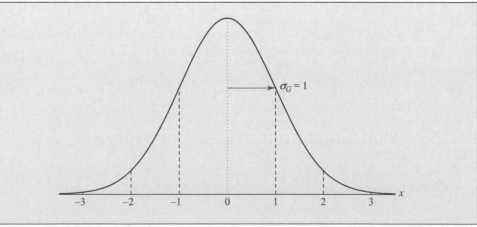

Figure 13.32

Standard Gaussian distribution (the bell-shaped curve) having zero mean and standard deviation σ_x equal to one.

Gaussian random numbers *NORM.S.INV(RAND())* —gives standard Gaussian random numbers, denoted by G, that have zero mean and unit standard deviation $\sigma_G = 1$. Figure 13.32 shows the standard Gaussian distribution.

Factoid:

- On a PC, Excel computes a new set of random numbers when the function key `F9` is pressed.
- On a Mac, this happens by pressing the "open-apple" (*meta-apple*) and equal sign (=).

Random numbers using RAND() — EXAMPLE 13.28

Figure 13.33 shows a worksheet that uses `RAND()` to generate $n_Y = 10$ random numbers Y_i for $0 \le i \le 9$, that are uniformly distributed over [0,1). The worksheet computes the sample average, sample variance, and sample SD.

Sample average. The worksheet estimates the mean of Y_i for $1 \le i \le 9$ by computing the sample average in `D4` using

```
= AVERAGE(B2:B11)
```

Even though the random numbers were generated using a uniform distribution with $\mu_Y = 0.5$, the particular set of random number actually observed will typically have an average value that close to 0.5.

Sample SD. The sample SD of the observe Y_i sequence is computed in `D7` using

```
=STDEV.S(B2:B11)
```

The `.S` indicates it is the standard deviation value computed from the observed sample of random numbers. For this small sample size, it is approximately equal to the theoretical value $\sigma_Y = \sqrt{1/12} \approx 0.3$.

Sample variance. The sample variance is computed in D10 using

```
=VAR(B2:B11)
```

For this small sample size, it is approximately equal to the theoretical value $\sigma_Y^2 = 1/12 \approx 0.08$.

Factoid: The number of digits displayed in a cell varies with the application. Typically, it is hard to read and appreciate values that have four or more digits in a cell. So, my Excel worksheets typically limit the precision of the displayed values to three digits. Internally, Excel performs computations with a precision of 15 digits, regardless of the displayed precision.

	A	B	C	D
1	i	Y_i		
2	0	0.602		
3	1	0.401		Sample Ave=
4	2	0.496		0.642
5	3	0.736		
6	4	0.940		Sample SD=
7	5	0.644		0.212
8	6	0.291		
9	7	0.588		Sample Var=
10	8	0.824		0.045
11	9	0.901		

(a)

	A	B	C	D
1	i	Y_i		
2	0	=RAND()		
3	=A2+1	=RAND()		Sample Ave=
4	=A3+1	=RAND()		=AVERAGE(B2:B11)
5	=A4+1	=RAND()		
6	=A5+1	=RAND()		Sample SD=
7	=A6+1	=RAND()		=STDEV.S(B2:B11)
8	=A7+1	=RAND()		
9	=A8+1	=RAND()		Sample Var=
10	=A9+1	=RAND()		=VAR(B2:B11)
11	=A10+1	=RAND()		

(b)

Figure 13.33

Sheet that generates random numbers Y_i uniformly distributed between 0 and 1, and computes the sample average, sample variance, and sample SD: (a) worksheet and (b) formulas.

EXAMPLE 13.29

Simulating random binary data

The *half-open* interval notation $[0,1)$ for the uniform PRNG indicates that 0 is a valid number produced by this PRNG, but 1 is not, although it can be approached (i.e., 0.9999...).

Figure 13.34 shows how RAND() simulates random binary data by dividing this $[0,1)$ interval into two equal sub-intervals, with $[0,0.5)$ producing data $D_i = 0$, and $[0.5,1)$ $D_i = 1$. Excel makes the decision using the conditional function IF(logical_test, [value_if_true], [value_if_false]).

EXAMPLE 13.30

Simulating random die toss

Figure 13.35 shows how RAND() simulates a random die toss. The die produces random integers in the range from 1 to 6 that correspond to the number of dots that appear on the top face. Note the use of the Excel function CEILING that rounds a number up to the next integer value.

Figure 13.34

Worksheet uses RAND() to produce Y_i that is used to simulate random binary data B_i: (a) worksheet and (b) formulas.

	A	B	C
1	i	Y_i	B_i
2	0	0.601	1
3	1	0.216	0
4	2	0.251	0
5	3	0.493	0
6	4	0.985	1
7	5	0.249	0
8	6	0.676	1
9	7	0.380	0
10	8	0.021	0
11	9	0.369	0

(a)

	A	B	C
1	i	Y_i	B_i
2	0	=RAND()	=IF(B2<0.5,0,1)
3	=A2+1	=RAND()	=IF(B3<0.5,0,1)
4	=A3+1	=RAND()	=IF(B4<0.5,0,1)
5	=A4+1	=RAND()	=IF(B5<0.5,0,1)
6	=A5+1	=RAND()	=IF(B6<0.5,0,1)
7	=A6+1	=RAND()	=IF(B7<0.5,0,1)
8	=A7+1	=RAND()	=IF(B8<0.5,0,1)
9	=A8+1	=RAND()	=IF(B9<0.5,0,1)
10	=A9+1	=RAND()	=IF(B10<0.5,0,1)
11	=A10+1	=RAND()	=IF(B11<0.5,0,1)

(b)

Figure 13.35

Worksheet uses RAND() to produce Y_i that is used to simulate a random die toss D_i: (a) worksheet and (b) formulas.

	A	B	C
1	i	Y_i	D_i
2	0	0.626	4
3	1	0.611	4
4	2	0.521	4
5	3	0.538	4
6	4	0.522	4
7	5	0.268	2
8	6	0.063	1
9	7	0.924	6
10	8	0.691	5
11	9	0.518	4

(a)

	A	B	C
1	i	Y_i	D_i
2	0	=RAND()	=CEILING(6*B2,1)
3	=A2+1	=RAND()	=CEILING(6*B3,1)
4	=A3+1	=RAND()	=CEILING(6*B4,1)
5	=A4+1	=RAND()	=CEILING(6*B5,1)
6	=A5+1	=RAND()	=CEILING(6*B6,1)
7	=A6+1	=RAND()	=CEILING(6*B7,1)
8	=A7+1	=RAND()	=CEILING(6*B8,1)
9	=A8+1	=RAND()	=CEILING(6*B9,1)
10	=A9+1	=RAND()	=CEILING(6*B10,1)
11	=A10+1	=RAND()	=CEILING(6*B11,1)

(b)

	A	B	C	D
1	i	G_i		
2	0	1.552		
3	1	−0.714		Sample Ave=
4	2	1.001		−0.055
5	3	1.781		
6	4	−0.828		Sample SD=
7	5	−1.291		1.178
8	6	−0.894		
9	7	0.620		Sample Var=
10	8	−0.487		1.389
11	9	−1.290		

(a)

	A	B	C	D
1	i	G_i		
2	0	=NORM.S.INV(RAND())		
3	=A2+1	=NORM.S.INV(RAND())		Sample Ave=
4	=A3+1	=NORM.S.INV(RAND())		=AVERAGE(B2:B11)
5	=A4+1	=NORM.S.INV(RAND())		
6	=A5+1	=NORM.S.INV(RAND())		Sample SD=
7	=A6+1	=NORM.S.INV(RAND())		=STDEV.S(B2:B11)
8	=A7+1	=NORM.S.INV(RAND())		
9	=A8+1	=NORM.S.INV(RAND())		Sample Var=
10	=A9+1	=NORM.S.INV(RAND())		=VAR(B2:B11)
11	=A10+1	=NORM.S.INV(RAND())		

(b)

Figure 13.36

Sheet that generates standard Gaussian random numbers and computes the sample average, sample variance, and sample SD: (a) worksheet and (b) formulas.

Standard Gaussian random numbers **EXAMPLE 13.31**

Figure 13.36 shows a worksheet that uses NORM.S.INV(RAND()) to generate $n_G = 10$ standard Gaussian random numbers G_i for $0 \leq i \leq 9$. The worksheet computes the sample average, sample variance, and sample SD.

Histogram **EXAMPLE 13.32**

Figure 13.37 shows a worksheet that generates a simple histogram from the Gaussian PRNG. Excel can generate a histogram of values displayed in a worksheet, but this method is not practical when the number of values is large. This example shows how to compute the histogram of 10,000 values with a VBA Macro. The bell-shaped curve is evident in the number of digits in column D.

The random number X_i is generated in A2. In this example, the PRNG that generates standard Gaussian random numbers is used. The index value of X_i displayed in A8 is reset and incremented in the Macro *hist*. The number of values generated n_x is specified in A8.

The bin number is computed in A11 with

=ROUND(2*(A2+4),0)+2

using the following steps:

STEP 1. A2 contains a standard Gaussian random number, which falls in the range $[−4,4]$ almost always.

STEP 2. Adding 4 to A2 gives a number typically in the range $[0,8]$.

STEP 3. Multiplying by 2 gives a number typically in the range $[0,16]$. This number sets the histogram resolution.

STEP 4. Adding 2 gives a number typically in the range $[2,18]$, which is a valid bin in terms of the worksheet row number.

Macro *hist* verifies the bin number is valid (>1), and increments the count of that bin. The Macro defines integer Val to simplify the code using DIM Val As Integer.

```
Sub hist()
Dim Val As Integer          ' define an integer variable
Range("D2:D18").Value = 0    ' reset bin counts
Range("A5").Value = 0        ' reset counter i
Do While Range("A5").Value < Range("A8").Value  ' loop for nX times
    Val = Range("A11").Value                     ' get bin value
    If Val > 1 Then                              '  if valid bin value
        Range("D" & Val).Value = Range("D" & Val).Value + 1  ' incr bin count
    End If
    Range("A5").Value = Range("A5").Value + 1 ' increment counter i
Loop
End Sub
```

	A	B	C	D
1	X$_i$=		bin	count
2	−2.0176914		−4	0
3			−3.5	5
4	i=		−3	23
5	10000		−2.5	87
6			−2	284
7	n$_x$=		−1.5	690
8	10000		−1	1227
9			−0.5	1778
10	bin#=		0	1920
11	6		0.5	1744
12			1	1186
13			1.5	638
14			2	284
15	Hist		2.5	97
16			3	27
17			3.5	7
18			4	3

(a)

	A
1	X$_i$=
2	=NORM.S.INV(RAND())
3	
4	i=
5	10000
6	
7	n$_x$=
8	10000
9	
10	bin#=
11	=ROUND(2*(A2+4),0)+2
12	

(b)

Figure 13.37

Worksheet that generates a simple histogram: (a) worksheet and (b) formulas.

EXAMPLE 13.33   Gaussian noise having specified SD

Figure 13.38 shows a worksheet that generates $n_N = 10$ Gaussian random noise values N_i for $0 \le i \le 9$ having zero mean and specified SD value, in this case equal to $\sigma_N = 4$.

Figure 13.38 (a)

	A	B	C	D	E
1	$\sigma_N=$	4			
2					
3	i	G_i	N_i		
4	0	1.002	4.007		
5	1	−0.120	−0.480		Sample Ave=
6	2	0.042	0.169		0.534
7	3	−0.392	−1.567		
8	4	−0.076	−0.305		Sample SD=
9	5	0.611	2.445		4.051
10	6	1.272	5.086		
11	7	−2.160	−8.639		Sample Var=
12	8	−0.095	−0.380		16.413
13	9	1.252	5.009		

Figure 13.38 (b)

	A	B	C	D	E
1	$\sigma_N=$	4			
2					
3					
4	0	=NORM.S.INV(RAND())	=B4*$B$1		
5	=A4+1	=NORM.S.INV(RAND())	=B5*$B$1		Sample Ave=
6	=A5+1	=NORM.S.INV(RAND())	=B6*$B$1		=AVERAGE(C4:C13)
7	=A6+1	=NORM.S.INV(RAND())	=B7*$B$1		
8	=A7+1	=NORM.S.INV(RAND())	=B8*$B$1		Sample SD=
9	=A8+1	=NORM.S.INV(RAND())	=B9*$B$1		=STDEV.S(C4:C13)
10	=A9+1	=NORM.S.INV(RAND())	=B10*$B$1		
11	=A10+1	=NORM.S.INV(RAND())	=B11*$B$1		Sample Var=
12	=A11+1	=NORM.S.INV(RAND())	=B12*$B$1		=VAR(C4:C13)
13	=A12+1	=NORM.S.INV(RAND())	=B13*$B$1		

Figure 13.38

Sheet that generates Gaussian random noise values with specified $\sigma_N = 4$ and computes the sample average, sample variance, and sample SD: (a) worksheet and (b) formulas.

Random binary data with added Gaussian random noise

EXAMPLE 13.34

Figure 13.39 shows a worksheet that generates $n_B = 10$ random binary values with the uniform PRNG and $n_N = 10$ Gaussian random noise values having $\sigma_N = 0.1$ with the Gaussian PRNG. The data and noise components are added to produce random signals

$$S_i = B_i + N_i \text{ for } 0 \leq i \leq 9$$

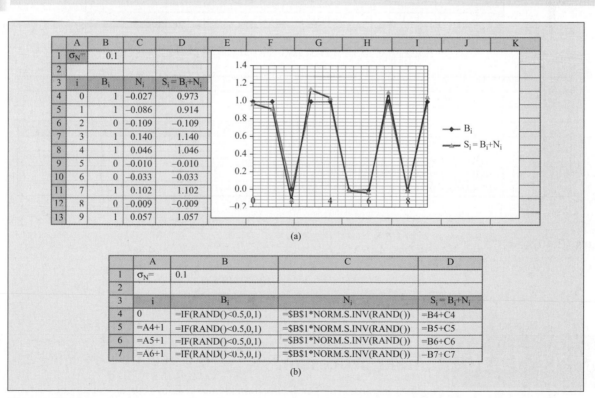

Figure 13.39 (a)

	A	B	C	D
1	$\sigma_N=$	0.1		
2				
3	i	B_i	N_i	$S_i=B_i+N_i$
4	0	1	−0.027	0.973
5	1	1	−0.086	0.914
6	2	0	−0.109	−0.109
7	3	1	0.140	1.140
8	4	1	0.046	1.046
9	5	0	−0.010	−0.010
10	6	0	−0.033	−0.033
11	7	1	0.102	1.102
12	8	0	−0.009	−0.009
13	9	1	0.057	1.057

Figure 13.39 (b)

	A	B	C	D
1	$\sigma_N=$	0.1		
2				
3	i	B_i	N_i	$S_i=B_i+N_i$
4	0	=IF(RAND()<0.5,0,1)	=$B$1*NORM.S.INV(RAND())	=B4+C4
5	=A4+1	=IF(RAND()<0.5,0,1)	=$B$1*NORM.S.INV(RAND())	=B5+C5
6	=A5+1	=IF(RAND()<0.5,0,1)	=$B$1*NORM.S.INV(RAND())	=B6+C6
7	=A6+1	=IF(RAND()<0.5,0,1)	=$B$1*NORM.S.INV(RAND())	=B7+C7

Figure 13.39

Sheet that generates random data and adds Gaussian random noise: (a) worksheet and (b) formulas.

EXAMPLE 13.35 Threshold detection of noisy data

Figure 13.40 shows a worksheet that attempts to retrieve the data from signals that contain noise by applying a threshold equal to $\tau = 0.5$ (half way between a data 0 and data 1). The binary data are estimated with $\tilde{B}_i$, where

$$\tilde{B}_i = 0 \text{ if } S_i < \tau$$

$$\tilde{B}_i = 1 \text{ if } S_i \geq \tau$$

Column E indicates an error

$$\text{Error} = 1 \text{ if } \tilde{B}_i \neq B_i$$

$$= 0 \text{ if } \tilde{B}_i = B_i$$

To make errors more easily observable, *conditional formatting* was applied to the error column causing cells that contain a 1 to be highlighted.

	A	B	C	D	E	F
1	$\sigma_N=$	0.3				
2						
3	i	B_i	N_i	$S_i = B_i + N_i$	$\sim B_i$	Error
4	0	0	−0.095	−0.095	0	0
5	1	0	0.119	0.119	0	0
6	2	1	0.449	1.449	1	0
7	3	1	0.144	1.144	1	0
8	4	1	0.000	1.000	1	0
9	5	1	−0.135	0.865	1	0
10	6	1	0.402	1.402	1	0
11	7	0	0.272	0.272	0	0
12	8	1	−0.515	0.485	0	1
13	9	1	−0.175	0.825	1	0

(a)

	A	B	C	D	E	F
1	$\sigma_N=$	0.3				
2						
3	i	B_i	N_i	$S_i = B_i + N_i$	$\sim B_i$	Error
4	0	=IF(RAND()<0.5,0,1)	=$B$1*NORM.S.INV(RAND())	=B4+C4	=IF(D4<0.5,0,1)	=IF(B4<>E4,1,0)
5	=A4+1	=IF(RAND()<0.5,0,1)	=$B$1*NORM.S.INV(RAND())	=B5+C5	=IF(D5<0.5,0,1)	=IF(B5<>E5,1,0)
6	=A5+1	=IF(RAND()<0.5,0,1)	=$B$1*NORM.S.INV(RAND())	=B6+C6	=IF(D6<0.5,0,1)	=IF(B6<>E6,1,0)
7	=A6+1	=IF(RAND()<0.5,0,1)	=$B$1*NORM.S.INV(RAND())	=B7+C7	=IF(D7<0.5,0,1)	=IF(B7<>E7,1,0)

(b)

Figure 13.40

Worksheet that generates random data and adds Gaussian random noise: (a) worksheet and (b) formulas.

13.15.7 Detecting Data Signals in Noise

EXAMPLE 13.36 Linear processor

Figure 13.41 shows a worksheet that implements a linear processor having coefficient set

$$c_i = i + 1 \text{ for } i = 0, 1, 2, \ldots, n_x - 1 \quad (n_x = 5)$$

	A	B	C
1	c_i	X_i	$c_i X_i$
2	1	1.01	1.01
3	2	0.14	0.27
4	3	−1.74	−5.23
5	4	0.12	0.47
6	5	0.64	3.19
7			
8		V=	−0.29

(a)

	A	B	C
1	c_i	X_i	
2	1	=NORM.S.INV(RAND())	=A2*B2
3	2	=NORM.S.INV(RAND())	=A3*B3
4	3	=NORM.S.INV(RAND())	=A4*B4
5	4	=NORM.S.INV(RAND())	=A5*B5
6	5	=NORM.S.INV(RAND())	=A6*B6
7			
8		V=	=SUM(C2:C6)

(b)

Figure 13.41

Worksheet that forms a linear processor: (a) worksheet and (b) formulas.

A Gaussian random input sequence X_i for $0 \leq i \leq n_x - 1$ is applied to this processor. The term-by-term products are computed in column C, and the sum of products is computed in C8 to implement

$$V = \sum_{i=0}^{4} c_i X_i$$

Matched processor   EXAMPLE 13.37

Figure 13.42 shows a worksheet that implements a matched processor having coefficient set that is matched to the input signal. Column B defines an arbitrary signal

$$s_i = 2i + 1 \quad \text{for } i = 0, 1, 2, \ldots, n_x - 1 \quad (n_x = 10)$$

The filter coefficients in column D are set equal to the signal

$$c_i = s_i \quad \text{for } i = 0, 1, 2, \ldots, n_x - 1$$

To demonstrate the output of the matched filter when the input in column E equals the signal

$$X_i = s_i \quad \text{for } i = 0, 1, 2, \ldots, n_x - 1$$

The term-by-term products are computed in column F, and summed in F13 to form the matched filter output, which equals the signal energy $\mathcal{E}_S$.

	A	B	C	D	E	F
1	i	s_i		c_i	X_i	$c_i X_i$
2	0	1		1	1	1
3	1	3		3	3	9
4	2	5		5	5	25
5	3	7		7	7	49
6	4	9		9	9	81
7	5	11		11	11	121
8	6	13		13	13	169
9	7	15		15	15	255
10	8	17		17	17	289
11	9	19		19	19	361
12						
13					V=	1330

(a)

	A	B	C	D	E	F
1	i	S_i		c_i	X_i	$c_i X_i$
2	0	1		=B2	=B2	=D2*E2
3	=1+A2	=B2+2		=B3	=B3	=D3*E3
4	=1+A3	=B3+2		=B4	=B4	=D4*E4
5	=1+A4	=B4+2		=B5	=B5	=D5*E5
6	=1+A5	=B5+2		=B6	=B6	=D6*E6
7	=1+A6	=B6+2		=B7	=B7	=D7*E7
8	=1+A7	=B7+2		=B8	=B8	=D8*E8
9	=1+A8	=B8+2		=B9	=B9	=D9*E9
10	=1+A9	=B9+2		=B10	=B10	=D10*E10
11	=1+A10	=B10+2		=B11	=B11	=D11*E11
12						
13					V=	=SUM(F2:F11)

(b)

Figure 13.42

Worksheet that implements a matched processor: (a) worksheet and (b) formulas.

EXAMPLE 13.38 · Matched processor applied to complementary signals

Figure 13.43 shows a worksheet that implements a matched processor having coefficient set that is matched to the input signal and processes complementary signals. Column B defines an arbitrary signal

$$s_i = 2i + 1 \text{ for } i = 0, 1, 2, \dots, n_x - 1 \quad (n_x = 10)$$

The filter coefficients in columns D and H are set equal to the signal

$$c_i = s_i \text{ for } i = 0, 1, 2, \dots, n_x - 1$$

To demonstrate the output of the matched filter when the input X_i in column E equals the signal that encodes a 1, as

$$X_i = s1_i = s_i \text{ for } i = 0, 1, 2, \dots, n_x - 1$$

The term-by-term products are computed in column F, and summed in F14 to form the matched filter output, which equals the signal energy ($V|_{X=s1} = \mathcal{E}_s$).

To demonstrate the output of the matched filter when the input X_i in column I equals the signal that encodes a 0

$$X_i = s0_i = -s_i \text{ for } i = 0, 1, 2, \dots, n_x - 1$$

The term-by-term products are computed in column J, and summed in J14 to form the matched filter output, which equals the negative of the signal energy ($V|_{X=s0} = -\mathcal{E}_s$).

(a)

	A	B	C	D	E	F	G	H	I	J
1					$X_i = s1_i = s_i$				$X_i = s0_i = -s_i$	
2	i	s_i		c_i	X_i	c_iX_i		c_i	$X_i = s0_i$	c_iX_i
3	0	1		1	1	1		1	−1	−1
4	1	3		3	3	9		3	−3	−9
5	2	5		5	5	25		5	−5	−25
6	3	7		7	7	49		7	−7	−49
7	4	9		9	9	81		9	−9	−81
8	5	11		11	11	121		11	−11	−121
9	6	13		13	13	169		13	−13	−169
10	7	15		15	15	225		15	−15	−225
11	8	17		17	17	289		17	−17	−289
12	9	19		19	19	361		19	−19	−361
13										
14					$V|_{X=s1}=$	1330			$V|_{X=s0}=$	−1330

(b)

	A	B	C	D	E	F	G	H	I	J
1					$X_i = s1_i = s_i$				$X_i = s0_i = -s_i$	
2	i	s_i		c_i	X_i	c_iX_i		c_i	$X_i = s0_i$	c_iX_i
3	0	1		=B3	=B3	=D3*E3		=B3	=−B3	=H3*I3
4	=1+A3	=B3+2		=B4	=B4	=D4*E4		=B4	=−B4	=H4*I4
5	=1+A4	=B4+2		=B5	=B5	=D5*E5		=B5	=−B5	=H5*I5
6	=1+A5	=B5+2		=B6	=B6	=D6*E6		=B6	=−B6	=H6*I6
7	=1+A6	=B6+2		=B7	=B7	=D7*E7		=B7	=−B7	=H7*I7
8	=1+A7	=B7+2		=B8	=B8	=D8*E8		=B8	=−B8	=H8*I8
9	=1+A8	=B8+2		=B9	=B9	=D9*E9		=B9	=−B9	=H9*I9
10	=1+A9	=B9+2		=B10	=B10	=D10*E10		=B10	=−B10	=H10*I10
11	=1+A10	=B10+2		=B11	=B11	=D11*E11		=B11	=−B11	=H11*I11
12	=1+A11	=B11+2		=B12	=B12	=D12*E12		=B12	=−B12	=H12*I12
13										
14					$V|_{X=s1}=$	=SUM(F3:F12)			$V|_{X=s0}=$	=SUM(J3:J12)

Figure 13.43

Worksheet that shows matched processor processing complementary signals: (a) worksheet and (b) formulas.

EXAMPLE 13.39 · Matched processor applied to complementary signals in noise

Figure 13.44 shows a worksheet that implements a matched processor having coefficient set that is matched to the input signal and processes the complementary signals that contain Gaussian noise having a specified SD value. Column B defines an arbitrary signal

$$s_i = 2i + 1 \text{ for } i = 0, 1, 2, \dots, n_x - 1 \quad (n_x = 10)$$

	A	B	C	D	E	F	G	H	I	J	K
1	$\sigma_N=$	20				$X_i = s1_i + N_i = s_i + N_i$				$X_i = s0_i + N_i = -s_i + N_i$	
2	i	s_i	N_i		c_i	X_i	$c_i X_i$		c_i	$X_i = s0_i$	$c_i X_i$
3	0	1	−9.70		1	−8.70	−8.70		1	−10.70	−10.70
4	1	3	−15.69		3	−12.69	−38.06		3	−18.69	−56.06
5	2	5	11.80		5	16.80	84.02		5	6.80	34.02
6	3	7	−17.52		7	−10.52	−73.65		7	−24.52	−171.65
7	4	9	−8.19		9	0.81	7.29		9	−17.19	−154.71
8	5	11	−27.88		11	−16.88	−185.71		11	−38.88	−427.71
9	6	13	−18.63		13	−5.63	−73.17		13	−31.63	−411.17
10	7	15	16.47		15	31.47	472.00		15	1.47	22.00
11	8	17	17.65		17	34.65	589.07		17	0.65	11.07
12	9	19	24.40		19	43.40	824.67		19	5.40	102.67
13											
14						$V\|_{X=s1+N}=$	1597.76			$V\|_{X=s0+N}=$	−1062.24
15											
16						$\tilde{B}=$	1			$\tilde{B}=$	0

(a)

	A	B	C	D	E	F	G	H	I	J	K
1	$\sigma_N=$	20				$X_i = s1_i + N_i = s_i + N_i$				$X_i = s0_i + N_i = -s_i + N_i$	
2	i	s_i	N_i		c_i	X_i	$c_i X_i$		c_i	$X_i = s0_i$	$c_i X_i$
3	0	1	=−$B$1*NORM.S.INV(RAND())		=B3	=B3+C3	=E3*F3		=B3	=−B3+C3	=I3*J3
4	=1+A3	=B3+2	=−$B$1*NORM.S.INV(RAND())		=B4	=−B4+C4	=E4*F4		=B4	=B4+C4	=I4*J4
5	=1+A4	=B4+2	=$B$1*NORM.S.INV(RAND())		=B5	=B5+C5	=E5*F5		=B5	=B5+C5	=I5*J5
6	=1+A5	=B5+2	=$B$1*NORM.S.INV(RAND())		=B6	=B6+C6	=E6*F6		=B6	=B6+C6	=I6*J6
7	=1+A6	=B6+2	=$B$1*NORM.S.INV(RAND())		=B7	=B7+C7	=E7*F7		=B7	=B7+C7	=I7*J7
8	=1+A7	=B7+2	=$B$1*NORM.S.INV(RAND())		=B8	=B8+C8	=E8*F8		=B8	=B8+C8	=I8*J8
9	=1+A8	=B8+2	=$B$1*NORM.S.INV(RAND())		=B9	=B9+C9	=E9*F9		=B9	=B9+C9	=I9*J9
10	=1+A9	=B9+2	=$B$1*NORM.S.INV(RAND())		=B10	=B10+C10	=E10*F10		=B10	=B10+C10	=I10*J10
11	=1+A10	=B10+2	=$B$1*NORM.S.INV(RAND())		=B11	=B11+C11	=E11*F11		=B11	=B11+C11	=I11*J11
12	=1+A11	=B11+2	=$B$1*NORM.S.INV(RAND())		=B12	=B12+C12	=E12*F12		=B12	=B12+C12	=I12*J12
13											
14						$V\|_{X=s1+N}=$	=SUM(G3:G12)			$V\|_{X=s0+N}=$	=SUM(K3:K12)
15											
16						$\tilde{B}=$	=IF(G14<0,0,1)			$\tilde{B}=$	=IF(K14<0,0,1)

(b)

Figure 13.44

Worksheet that shows matched processor processing complementary signals corrupted with noise: (a) worksheet and (b) formulas.

The filter coefficients in columns E and I are set equal to the signal

$$c_i = s_i \quad \text{for } i = 0, 1, 2, \ldots, n_x - 1$$

A noise sequence is generated in column C with SD specified in B1. To demonstrate the output of the matched filter when the input X_i in column F equals the signal that encodes a signal (corresponding to $D_t = 1$) plus the noise

$$X_i = s1_i + N_i = s_i + N_i \quad \text{for } i = 0, 1, 2, \ldots, n_x - 1$$

The term-by-term products are computed in column G, and summed in G14 to form the matched filter output $V|_{X=s1+N}$. The output is applied to a threshold detector having threshold value equal to zero in G16 to estimate what binary value was transmitted:

$$\tilde{B} = 0 \quad \text{if } V|_{X=s1+N} < 0$$

$$\tilde{B} = 1 \quad \text{if } V|_{X=s1+N} \geq 0$$

To demonstrate the output of the matched filter when the input X_i in column J equals the signal that encodes a signal (corresponding to $D_t = 0$) plus the noise

$$X_i = s0_i + N_i = -s_i + N_i \quad \text{for } i = 0, 1, 2, \ldots, n_x - 1$$

The term-by-term products are computed in column K, and summed in K14 to form the matched filter output, $V|_{X=s0+N}$. The output is applied to a threshold detector having threshold value equal to zero in K16 to estimate what binary value was transmitted:

$$\tilde{B} = 0 \quad \text{if} \quad V|_{X=s0+N} < 0$$

$$\tilde{B} = 1 \quad \text{if} \quad V|_{X=s0+N} \geq 0$$

Conditional formatting is used to make errors evident. When G16 (containing the result of threshold detection of $V|_{X=s1+n}$) equals 0, an error has occurred. Conditional formatting causes the result G16=0 to highlight the cell. Similarly, the result K16=1 uses conditional formatting to highlight the error.

EXAMPLE 13.40   Viewing complementary signals in noise

Figure 13.45 shows a worksheet that displays the signal s_i and its noise-corrupted version

$$X_i = s_i + N_i \quad \text{for} \quad i = 0, 1, 2, \ldots, n_x - 1$$

for a specified noise SD.

It is instructive to see the waveform and the degradation caused by additive Gaussian random noise. *If you cannot see the signal component in X_i, the processor will not produce good performance. It is not magic!*

	A	B	C	D
1	$\sigma_N=$	3		
2	i	s_i	N_i	X_i
3	0	0.00	2.98	2.98
4	1	6.43	−0.34	6.08
5	2	9.85	3.03	12.88
6	3	8.66	−1.93	6.73
7	4	3.42	−0.84	2.58
8	5	−3.42	−1.05	−4.47
9	6	−8.66	−1.84	−10.50
10	7	−9.85	−5.79	−15.64
11	8	−6.43	1.29	−5.14
12	9	0.00	−3.39	−3.39

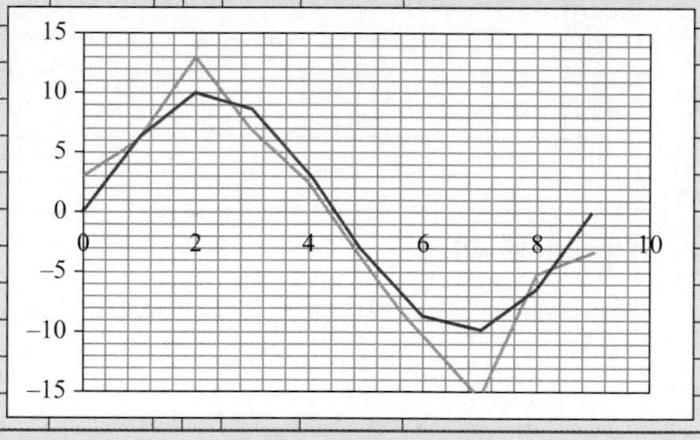

(a)

	A	B	C	D
1	$\sigma_N=$	3		
2	i	s_i	N_i	X_i
3	0	=10*SIN(2*PI()*A3/9)	=$B$1*NORM.S.INV(RAND())	=B3+C3
4	=1+A3	=10*SIN(2*PI()*A4/9)	=$B$1*NORM.S.INV(RAND())	=B4+C4
5	=1+A4	=10*SIN(2*PI()*A5/9)	=$B$1*NORM.S.INV(RAND())	=B5+C5
6	=1+A5	=10*SIN(2*PI()*A6/9)	=$B$1*NORM.S.INV(RAND())	=B6+C6
7	=1+A6	=10*SIN(2*PI()*A7/9)	=$B$1*NORM.S.INV(RAND())	=B7+C7
8	=1+A7	=10*SIN(2*PI()*A8/9)	=$B$1*NORM.S.INV(RAND())	=B8+C8
9	=1+A8	=10*SIN(2*PI()*A9/9)	=$B$1*NORM.S.INV(RAND())	=B9+C9
10	=1+A9	=10*SIN(2*PI()*A10/9)	=$B$1*NORM.S.INV(RAND())	=B10+C10
11	=1+A10	=10*SIN(2*PI()*A11/9)	=$B$1*NORM.S.INV(RAND())	=B11+C11
12	=1+A11	=10*SIN(2*PI()*A12/9)	=$B$1*NORM.S.INV(RAND())	=B12+C12

(b)

Figure 13.45

Worksheet that displays a signal corrupted with noise: (a) worksheet and (b) formulas.

Estimating probability of error   EXAMPLE 13.41

This example employs VBA Macros to implement counters to tally the number of transmissions and errors to estimate the probability of error. The following steps produce the results shown in Figure 13.46.

STEP 1. A signal s_i is defined in B6:B15.

STEP 2. The signal energy $\mathcal{E}_s$ is computed in B1 as the sum of the squares of B6:B15 using Excel function SUMSQ.

STEP 3. The noise σ_N is specified in B2.

STEP 4. The signal-to-noise ratio is computed in B3 as $\mathcal{E}_s/\sigma_N^2$ (note the square of σ_N compares noise variance to signal energy).

STEP 5. A random bit D is generated in E1.

STEP 6. The transmitted signal sT_i in C6:C15 is determined from the D value

$$sT_i = s1_i = s_i \ \ \text{if} \ \ D = 1$$
$$sT_i = s0_i = -s_i \ \ \text{if} \ \ D = 0$$

STEP 7. Gaussian noise N_i with σ_N is computed in D6:D15.

STEP 8. Detected signal in G6:G15 equals

$$X_i = sT_i + N_i$$

is multiplied with matched processor coefficients $c_i = s_i$ in F6:F15 to produce products in H6:H15 that are summed in H16 to produce the matched processor output V.

STEP 9. The bit value $\tilde{D}$ detected with threshold detection ($\tau = 0$) occurs in E2.

STEP 10. VBA Macro *start*, assigned to shape *Start*, resets the count of transmitted signals in H1 and errors in H2, and suspends the automatic recalculation typically performed by Excel with VBA statement

```
Application.Calculation = xlCalculationManual
```

STEP 11. VBA Macro *xmit*, assigned to shape *Xmit*, increments the transmit count, transmits a single data bit, determines if an error occurred and, if so, increments the error count. The Macro forces a single recalculation of the worksheet with the instruction

```
Calculate
```

Excel appears to calculate the PRNGs first (D and N_i), followed by the cells that contain these values (X_i), followed by the functions that evaluate theses values ($c_i X_i$), followed by the functions that use those values ($V|_{X=ST+N}$), and, finally, by evaluating the determination of the transmitted bit ($\tilde{D}$). The Macro then compares D and $\tilde{D}$. If these two differ, the error count is incremented.

STEP 12. The probability of error is estimated in H3 with

$$P[\text{err}] = \frac{n_{\text{errors}}}{n_{\text{transmits}}}$$

only when $n_{\text{transmits}} > 0$ (to prevent a division by 0 indication).

STEP 13. VBA Macro *restore*, assigned to shape *Restore*, restores the automatic calculation mode with

```
Application.Calculation = xlCalculationAutomatic
```

```
Sub start()      ' resets counters to zero
Application.Calculation = xlCalculationManual ' Stop Auto Calculations
Range("H1").Value = 0 ' Reset # transmits
Range("H2").Value = 0 ' Reset # errors
End Sub
-------
Sub xmit()        ' transmits another data signal
Range("H1").Value = Range("H1").Value + 1         ' Increment # transmits
Calculate        ' Force one recalculation, New D
If Range("E1").Value <> Range("E2").Value Then     ' ~D not equal D -> error
    Range("H2") = Range("H2") + 1                   ' Increment # errors
End If
End Sub
-------
Sub restore()      ' restore Auto Calculations
Application.Calculation = xlCalculationAutomatic ' Restore Auto Calculations
End Sub
```

	A	B	C	D	E	F	G	H
1	E_s=	1330		D=	0		#Transmits=	29
2	σ_N=	30		~D=	0		#Errors=	3
3	SNR=	1.48					P[err]=	0.10
4								
5	i	s_i	sT_i	N_i		c_i	X_i	c_iX_i
6	0	1	−1	−15.38		1	−16.38	−16.38
7	1	3	−3	−29.36		3	−32.36	−97.09
8	2	5	−5	4.71		5	−0.29	−1.46
9	3	7	−7	31.76		7	24.76	173.34
10	4	9	−9	−35.79		9	−44.79	−403.13
11	5	11	−11	25.94		11	14.94	164.38
12	6	13	−13	16.63		13	3.63	47.21
13	7	15	−15	44.19		15	29.19	437.91
14	8	17	−17	42.00		17	25.00	425.03
15	9	19	−19	−42.22		19	−61.22	−1163.16
16	Start	Xmit		Restore			$V|_{X=sT+N}$=	−433.34

(a)

	A	B	C	D	E	F	G	H
1	E_s=	=SUMSQ(B6:B15)		D=	=IF(RAND()<0.5,0,1)		#Transmits=	29
2	σ_N=	30		~D=	=IF(H16<0,0,1)		#Errors=	3
3	SNR=	=B1/B2^2					P[err]=	=IF(H1>0,H2/H1,0)
4								
5	i	s_i	sT_i	N_i		c_i	X_i	c_iX_i
6	0	1	=IF($E$1=1,B6,−B6)	=$B$2*NORM.S.INV(RAND())		=B6	=C6+D6	=F6*G6
7	=1+A6	=B6+2	=IF($E$1=1,B7,−B7)	=$B$2*NORM.S.INV(RAND())		=B7	=C7+D7	=F7*G7
8	=1+A7	=B7+2	=IF($E$1=1,B8,−B8)	=$B$2*NORM.S.INV(RAND())		=B8	=C8+D8	=F8*G8
9	=1+A8	=B8+2	=IF($E$1=1,B9,−B9)	=$B$2*NORM.S.INV(RAND())		=B9	=C9+D9	=F9*G9
10	=1+A9	=B9+2	=IF($E$1=1,B10,−B10)	=$B$2*NORM.S.INV(RAND())		=B10	=C10+D10	=F10*G10
11	=1+A10	=B10+2	=IF($E$1=1,B11,−B11)	=$B$2*NORM.S.INV(RAND())		=B11	=C11+D11	=F11*G11
12	=1+A11	=B11+2	=IF($E$1=1,B12,−B12)	=$B$2*NORM.S.INV(RAND())		=B12	=C12+D12	=F12*G12
13	=1+A12	=B12+2	=IF($E$1=1,B13,−B13)	=$B$2*NORM.S.INV(RAND())		=B13	=C13+D13	=F13*G13
14	=1+A13	=B13+2	=IF($E$1=1,B14,−B14)	=$B$2*NORM.S.INV(RAND())		=B14	=C14+D14	=F14*G14
15	=1+A14	=B14+2	=IF($E$1=1,B15,−B15)	=$B$2*NORM.S.INV(RAND())		=B15	=C15+D15	=F15*G15
16	Start		Xmit		Restore		$V|_{X=sT+N}$=	=SUM(H6:H15)

(b)

Figure 13.46

Worksheet that employs VBA Macro to estimate the probability of error when processing complementary signals: (a) worksheet and (b) formulas.

Automating the probability of error

EXAMPLE 13.42

Example 13.41 transmits a new data signal each time the *Xmit* shape is clicked. This example automates this process with the simple VBA Macro. Figure 13.47 shows Macro *auto* assigned to the *Auto* shape that produces 1,000 transmissions to compute an accurate probability of error value. The Macro introduces the *Do While ... Loop* statement to monitor the number of transmissions in H1 and performs the statements within the loop. Macro *xmit* increments H1 at each transmission, and Macro *auto* monitors H1 and ends the loop when the number in H1 reaches 1,000.

```
Sub auto()
start
Do While Range("H1").Value < 1000    ' 1000 transmissions
    xmit
Loop
restore
End Sub
```

	A	B	C	D	E	F	G	H
1	Es=	1330		D=	1		#Transmits=	1000
2	σ_N=	30		~D=	1		#Errors=	0
3	SNR=	1.48					P[err]=	0.00
4								
5	i	s_i	sT_i	N_i		c_i	X_i	c_iX_i
6	0	1	1	15.33		1	16.33	16.33
7	1	3	3	46.34		3	49.34	148.02
8	2	5	5	21.26		5	26.26	131.30
9	3	7	7	-5.97		7	1.03	7.19
10	4	9	9	-46.01		9	-37.01	-333.08
11	5	11	11	-2.35		11	8.65	95.12
12	6	13	13	31.13		13	44.13	573.67
13	7	15	15	9.61		15	24.61	369.09
14	8	17	17	2.15		17	19.15	325.53
15	9	19	19	-12.91		19	6.09	115.65
16	Start		Xmit	Restore		Auto	$V\|_{x=sT+N}$=	1448.80

Figure 13.47

Worksheet that employs VBA Macro *auto* to automate the probability of error computation for transmitting 1,000 bit values.

Probability of error as a function of SNR

EXAMPLE 13.43

This example illustrates how to compute the probability of error as the signal-to-noise ratio $\mathcal{E}_s/\sigma_N^2$ varies by composing an additional VBA Macro *SNR* to extend to operation described in Example 13.42. This simplifies the task of comparing the performance of different specified signal sequences, as shown in Figure 13.48.

The value of $\mathcal{E}_s$ is computed in B1 by summing the squared values of s_i in B6:B15 using the Excel function *SUMSQ*. The *SNR* in B3 equals the ratio of signal energy to noise variance

$$\text{SNR} = \frac{\mathcal{E}_s}{\sigma_N^2}$$

```
Sub SNR()
Dim RowNum As Integer                    ' specify an integer variable
RowNum = 2                               ' starting row of data compilation
Range("B2").Value = 10                   ' initial noise SD value
Do While Range("B2").Value < 1000        ' end noise SD value
    auto
Range("J" & RowNum).Value = Range("B3").Value    ' enter SNR
    Range("K" & RowNum).Value = Range("H3").Value    ' enter Prob[err]
    RowNum = RowNum + 1                          ' increment row
    Range("B2").Value = Range("B2").Value * Sqr(2)  ' increase noise SD
Loop
End Sub
```

	A	B	C	D	E	F	G	H	I	J	K
1	E_s=	28.284		D=	0		#Transmits=	1000		SNR	P[err]
2	σ_N=	1280		$\tilde{}$D=	1		#Errors=	495		0.2828	0.0000
3	SNR=	2E-05					P[err]=	0.495		0.1414	0.0040
4										0.0707	0.0400
5	i	s_i	sT_i	N_i		c_i	X_i	c_iX_i		0.0354	0.0951
6	0	1	−1	−1586.30		1	−1587.30	−1587.30		0.0177	0.1632
7	1	3	−3	1585.74		3	1582.74	4748.21		0.0088	0.2683
8	2	5	−5	1172.95		5	1167.95	5839.75		0.0044	0.3123
9	3	7	−7	39.33		7	32.33	226.33		0.0022	0.3744
10	4	9	−9	255.92		9	246.92	2222.28		0.0011	0.3964
11	5	11	−11	−355.25		11	−366.25	−4028.71		0.0006	0.4164
12	6	13	−13	−376.36		13	−389.36	−5061.74		0.0003	0.4525
13	7	15	−15	1129.25		15	1114.25	16713.79		0.0001	0.4605
14	8	17	−17	572.00		17	555.00	9434.92		0.0001	0.4665
15	9	19	−19	−184.73		19	−203.73	−3870.91		0.0000	0.4955
16	Start	Xmit	Restore		Auto		$V\big|_{x=sT+N}$=	24636.62			
17											
18		P[err]as SNR									

(a)

	A	B	C	D	E	F	G	H
1	E_s=	28.2842712474619		D=	=IF(RAND()<0.5,0,1)		#Transmits=	1000
2	σ_N=	1280		$\tilde{}$D=	=IF(H16<0,0,1)		#Errors=	495
3	SNR=	=B1/B2^2					P[err]=	=IF(H1>0,H2/H1,0)
4								
5	i	s_i	sT_i	N_i		c_i	X_i	c_iX_i
6	0	1	=IF($E$1=1,B6,-B6)	=$B$2*NORM.S.INV(RAND())		=B6	=C6+D6	=F6*G6
7	=1+A6	=B6+2	=IF($E$1=1,B7,-B7)	=$B$2*NORM.S.INV(RAND())		=B7	=C7+D7	=F7*G7
8	=1+A7	=B7+2	=IF($E$1=1,B8,-B8)	=$B$2*NORM.S.INV(RAND())		=B8	=C8+D8	=F8*G8
9	=1+A8	=B8+2	=IF($E$1=1,B9,-B9)	=$B$2*NORM.S.INV(RAND())		=B9	=C9+D9	=F9*G9
10	=1+A9	=B9+2	=IF($E$1=1,B10,-B10)	=$B$2*NORM.S.INV(RAND())		=B10	=C10+D10	=F10*G10
11	=1+A10	=B10+2	=IF($E$1=1,B11,-B11)	=$B$2*NORM.S.INV(RAND())		=B11	=C11+D11	=F11*G11
12	=1+A11	=B11+2	=IF($E$1=1,B12,-B12)	=$B$2*NORM.S.INV(RAND())		=B12	=C12+D12	=F12*G12
13	=1+A12	=B12+2	=IF($E$1=1,B13,-B13)	=$B$2*NORM.S.INV(RAND())		=B13	=C13+D13	=F13*G13
14	=1+A13	=B13+2	=IF($E$1=1,B14,-B14)	=$B$2*NORM.S.INV(RAND())		=B14	=C14+D14	=F14*G14
15	=1+A14	=B14+2	=IF($E$1=1,B15,-B15)	=$B$2*NORM.S.INV(RAND())		=B15	=C15+D15	=F15*G15
16	Start	Xmit		Restore		Auto	$V\big|_{x=sT+N}$=	=SUM(H6:H15)
17								
18			P[err]as SNR					

(b)

Figure 13.48

Worksheet that employs VBA Macro to compute and list the probability of error as a function of SNR: (a) worksheet and (b) formulas.

VBA Macro *SNR* introduces the following new features.

- *Specifying a new integer variable*—Integer *RowNum* specifies the row number in the worksheet to enter results.
- *Specifying a variable row number*—The VBA instruction Range("B" & RowNum) allows the row number to depend on the value of RowNum. For example, if RowNum = 2, Range("B" & RowNum) is equivalent to Range("B2"). The statement

$$RowNum = RowNum + 1$$

increments the row number to form a list of empirical values.

- *Sqr(.)*—The VBA instruction that computes the square root is Sqr.

The list of *SNR* values and the *P[err]* produced by this simulation is given in columns J and K. Notice the following for the signal chosen.

- *For large SNR*, such as SNR > 1, the probability of error is very small, as should be expected.
- *For small SNR*, such as SNR < 0.002, the probability of error=0.5, that is, the noise is so much greater than the signal that the processor performs as well as randomly guessing.

13.15.8 Designing Signals for Multiple-Access Systems

Designing orthogonal signals EXAMPLE 13.44

Figure 13.49 shows a worksheet that generates pairs of signals that are orthogonal in time, frequency, and code. All signals have the same amplitude, in this case equal to ten. An orthogonal signal pair is copied from this worksheet to worksheets that compare their performance.

Probability of error for orthogonal signals EXAMPLE 13.45

The simplest method of computing the probability of error using orthogonal signals is use one of the Examples 13.41, 13.42, or 13.43 and replace s_i in those sheets with the orthogonal signal assigned to a single user. Figure 13.50 shows this using a TDMA signal sequence that serves two users. Clearly, with $n_x/2$ of the sample values equal to zero, its $\mathcal{E}_s$ is half that of a signal that is constant over the entire n_x interval.

13.15.9 Source Coding

Forming random source symbols EXAMPLE 13.46

Figure 13.51 shows a worksheet that forms n_T random symbols from a vocabulary $m = 6$. Column C forms a text string, such as "X1" by concatenating a text string with a number to text conversion. Excel function TEXT converts a number into a text string, and CONCATENATE combines two text strings into one.

	A	B	C	D	E	F	G	H	I	J	K
1	A_{max}=	n_x=			FA=	FB=					
2	10	16			0.0625	0.125					
3					ωA=	ωB=					
4					0.393	0.785					
5											
6		TDMA			FDMA			CDMA			
7	i	sA_i	sB_i		sA_i	sB_i		sA_i	sB_i		sA_isB_i
8	0	10	0		0.00	0.00		−10	10		−100
9	1	10	0		3.83	7.07		−10	10		−100
10	2	10	0		7.07	10.00		10	10		100
11	3	10	0		9.24	7.07		10	−10		−100
12	4	10	0		10.00	0.00		−10	10		−100
13	5	10	0		9.24	−7.07		10	−10		−100
14	6	10	0		7.07	−10.00		−10	−10		100
15	7	10	0		3.83	−7.07		−10	10		−100
16	8	0	−10		0.00	0.00		−10	−10		100
17	9	0	−10		−3.83	7.07		−10	10		−100
18	10	0	−10		−7.07	10.00		10	10		100
19	11	0	−10		−9.24	7.07		−10	10		−100
20	12	0	−10		−10.00	0.00		10	10		100
21	13	0	−10		−9.24	−7.07		10	−10		100
22	14	0	−10		−7.07	−10.00		−10	10		−100
23	15	0	−10		−3.83	−7.07		−10	10		−100
24										Sum=	−400

(a)

	A	B	C	D	E	F	G	H	I	J	K
1	A_{max}=	n_x=			FA=	FB=					
2	10	16			=1/B2	=2/B2					
3					ωA=	ωB=					
4					=(2*PI()*E2)	=(2*PI()*F2)					
5											
6		TDMA			FDMA			CDMA			
7	i	sA_i	sB_i		sA_i	sB_i		sA_i	sB_i		sA_isB_i
8	0	=$A$2	0		=$A$2*SIN($E$4*A8)	=$A$2*SIN($F$4*A8)		=IF(RAND()<0.5,-$A$2,$A$2)	=IF(RAND()<0.5,-$A$2,$A$2)		=H8*I8
9	=1+A8	=$A$2	0		=$A$2*SIN($E$4*A9)	=$A$2*SIN($F$4*A9)		=IF(RAND()<0.5,-$A$2,$A$2)	=IF(RAND()<0.5,-$A$2,$A$2)		=H9*I9
10	=1+A9	=$A$2	0		=$A$2*SIN($E$4*A10)	=$A$2*SIN($F$4*A10)		=IF(RAND()<0.5,-$A$2,$A$2)	=IF(RAND()<0.5,-$A$2,$A$2)		=H10*I10
11	=1+A10	=$A$2	0		=$A$2*SIN($E$4*A11)	=$A$2*SIN($F$4*A11)		=IF(RAND()<0.5,-$A$2,$A$2)	=IF(RAND()<0.5,-$A$2,$A$2)		=H11*I11
12	=1+A11	=$A$2	0		=$A$2*SIN($E$4*A12)	=$A$2*SIN($F$4*A12)		=IF(RAND()<0.5,-$A$2,$A$2)	=IF(RAND()<0.5,-$A$2,$A$2)		=H12*I12
13	=1+A12	=$A$2	0		=$A$2*SIN($E$4*A13)	=$A$2*SIN($F$4*A13)		=IF(RAND()<0.5,-$A$2,$A$2)	=IF(RAND()<0.5,-$A$2,$A$2)		=H13*I13
14	=1+A13	=$A$2	0		=$A$2*SIN($E$4*A14)	=$A$2*SIN($F$4*A14)		=IF(RAND()<0.5,-$A$2,$A$2)	=IF(RAND()<0.5,-$A$2,$A$2)		=H14*I14
15	=1+A14	=$A$2	0		=$A$2*SIN($E$4*A15)	=$A$2*SIN($F$4*A15)		=IF(RAND()<0.5,-$A$2,$A$2)	=IF(RAND()<0.5,-$A$2,$A$2)		=H15*I15
16	=1+A15	0	=-$A$2		=$A$2*SIN($E$4*A16)	=$A$2*SIN($F$4*A16)		=IF(RAND()<0.5,-$A$2,$A$2)	=IF(RAND()<0.5,-$A$2,$A$2)		=H16*I16
17	=1+A16	0	=-$A$2		=$A$2*SIN($E$4*A17)	=$A$2*SIN($F$4*A17)		=IF(RAND()<0.5,-$A$2,$A$2)	=IF(RAND()<0.5,-$A$2,$A$2)		=H17*I17
18	=1+A17	0	=-$A$2		=$A$2*SIN($E$4*A18)	=$A$2*SIN($F$4*A18)		=IF(RAND()<0.5,-$A$2,$A$2)	=IF(RAND()<0.5,-$A$2,$A$2)		=H18*I18
19	=1+A18	0	=-$A$2		=$A$2*SIN($E$4*A19)	=$A$2*SIN($F$4*A19)		=IF(RAND()<0.5,-$A$2,$A$2)	=IF(RAND()<0.5,-$A$2,$A$2)		=H19*I19
20	=1+A19	0	=-$A$2		=$A$2*SIN($E$4*A20)	=$A$2*SIN($F$4*A20)		=IF(RAND()<0.5,-$A$2,$A$2)	=IF(RAND()<0.5,-$A$2,$A$2)		=H20*I20
21	=1+A20	0	=-$A$2		=$A$2*SIN($E$4*A21)	=$A$2*SIN($F$4*A21)		=IF(RAND()<0.5,-$A$2,$A$2)	=IF(RAND()<0.5,-$A$2,$A$2)		=H21*I21
22	=1+A21	0	=-$A$2		=$A$2*SIN($E$4*A22)	=$A$2*SIN($F$4*A22)		=IF(RAND()<0.5,-$A$2,$A$2)	=IF(RAND()<0.5,-$A$2,$A$2)		=H22*I22
23	=1+A22	0	=-$A$2		=$A$2*SIN($E$4*A23)	=$A$2*SIN($F$4*A23)		=IF(RAND()<0.5,-$A$2,$A$2)	=IF(RAND()<0.5,-$A$2,$A$2)		=H23*I23
24										Sum=	=SUM(K8:K23)

(b)

Figure 13.49

Worksheet that designs orthogonal signals in time, frequency, and code: (a) worksheet and (b) formulas.

	A	B	C	D	E	F	G	H
1	E_s=	500		D=	1		#Transmits=	1000
2	σ_N=	30		~D=	1		#Errors=	220
3	SNR=	0.56					P[err]=	0.22
4								
5	i	s_i	sT_i	N_i		c_i	X_i	c_iX_i
6	0	10	10	27.76		10	37.76	377.61
7	1	10	10	15.34		10	25.34	253.35
8	2	10	10	−23.17		10	−13.17	−131.67
9	3	10	10	−2.91		10	7.09	70.94
10	4	10	10	21.07		10	31.07	310.68
11	5	0	0	20.11		0	20.11	0.00
12	6	0	0	−1.76		0	−1.76	0.00
13	7	0	0	−9.96		0	−9.96	0.00
14	8	0	0	30.09		0	30.09	0.00
15	9	0	0	18.59		0	18.59	0.00
16	Start	Xmit	Restore		Auto		$V|_{X=sT+N}$=	880.92

Figure 13.50

Worksheet that estimates the probability of error for TDMA orthogonal signals.

(a)

	A	B	C
1	m=		n_T=
2	6		10
3			
4	i	Rand i	Symbol
5	1	5	X5
6	2	2	X2
7	3	1	X1
8	4	1	X1
9	5	5	X5
10	6	5	X5
11	7	1	X1
12	8	3	X3
13	9	5	X5
14	10	6	X6

(b)

	A	B	C
1	m=		n_T=
2	6		10
3			
4	i	Rand i	Symbol
5	1	=FLOOR($A$2*RAND()+1,1)	=CONCATENATE("X",TEXT(B5,0))
6	=1+A5	=FLOOR($A$2*RAND()+1,1)	=CONCATENATE("X",TEXT(B6,0))
7	=1+A6	=FLOOR($A$2*RAND()+1,1)	=CONCATENATE("X",TEXT(B7,0))
8	=1+A7	=FLOOR($A$2*RAND()+1,1)	=CONCATENATE("X",TEXT(B8,0))
9	=1+A8	=FLOOR($A$2*RAND()+1,1)	=CONCATENATE("X",TEXT(B9,0))
10	=1+A9	=FLOOR($A$2*RAND()+1,1)	=CONCATENATE("X",TEXT(B10,0))
11	=1+A10	=FLOOR($A$2*RAND()+1,1)	=CONCATENATE("X",TEXT(B11,0))
12	=1+A11	=FLOOR($A$2*RAND()+1,1)	=CONCATENATE("X",TEXT(B12,0))
13	=1+A12	=FLOOR($A$2*RAND()+1,1)	=CONCATENATE("X",TEXT(B13,0))
14	=1+A13	=FLOOR($A$2*RAND()+1,1)	=CONCATENATE("X",TEXT(B14,0))

Figure 13.51

Worksheet that forms $n_T = 10$ random symbols from a vocabulary $m = 6$: (a) worksheet and (b) formulas.

Computing effective entropy of a data file EXAMPLE 13.47

Figure 13.52 shows a worksheet that computes the effective entropy of a data file. The data file contains the symbols that were formed in Example 13.46, copies and paste/special/values into column A. Pasting *values* forms a data file of n_T symbols that does not change with worksheet calculations. To compute the effective probability, a column is formed for each of the m symbols and an "IF" statement determines the presence of that symbol. Row 13 computes the count of occurrences for each symbol. The sum total of the counts equals n_T. The effective probability is computed in column I from the Row 13 values. The log_2 values are computed in column J, and the effective entropy of the data file is computed in K2.

	A	B	C	D	E	F	G	H	I	J	K
1	Symbol	X1?	X2?	X3?	X4?	X5?		X	$P_e[X]$	$P_e \log_2 P_e =$	$H_e =$
2	X1	1	0	0	0	0		X1	0.2	−0.464	22.464
3	X4	0	0	0	1	0		X2	0.1	−0.332	
4	X5	0	0	0	0	1		X3	0.2	−0.464	
5	X5	0	0	0	0	1		X4	0.2	−0.464	
6	X2	0	1	0	0	0		X5	0.3	−0.521	
7	X4	0	0	0	1	0					
8	X3	0	0	1	0	0					
9	X1	1	0	0	0	0					
10	X3	0	0	1	0	0					
11	X5	0	0	0	0	1					
12											
13	$n_{Xi} =$	2	1	2	2	3					

(a)

	A	B	C	D	E	F	G	H	I	J	K
1	Symbol	X1?	X2?	X3?	X4?	X5?		X	$P_e[X]$	$P_e \log_2 P_e =$	$H_e =$
2	X1	=IF(A2="X1",1,0)	=IF(A2="X2",1,0)	=IF(A2="X3",1,0)	=IF(A2="X4",1,0)	=IF(A2="X5",1,0)		X1	=B13/SUM($B$13:$F$13)	=I2*LOG(I2,2)	=−SUM(J2:J6)*SUM(B13:F13)
3	X4	=IF(A3="X1",1,0)	=IF(A3="X2",1,0)	=IF(A3="X3",1,0)	=IF(A3="X4",1,0)	=IF(A3="X5",1,0)		X2	=C13/SUM($B$13:$F$13)	=I3*LOG(I3,2)	
4	X5	=IF(A4="X1",1,0)	=IF(A4="X2",1,0)	=IF(A4="X3",1,0)	=IF(A4="X4",1,0)	=IF(A4="X5",1,0)		X3	=D13/SUM($B$13:$F$13)	=I4*LOG(I4,2)	
5	X5	=IF(A5="X1",1,0)	=IF(A5="X2",1,0)	=IF(A5="X3",1,0)	=IF(A5="X4",1,0)	=IF(A5="X5",1,0)		X4	=E13/SUM($B$13:$F$13)	=I5*LOG(I5,2)	
6	X2	=IF(A6="X1",1,0)	=IF(A6="X2",1,0)	=IF(A6="X3",1,0)	=IF(A6="X4",1,0)	=IF(A6="X5",1,0)		X5	=F13/SUM($B$13:$F$13)	=I6*LOG(I6,2)	
7	X4	=IF(A7="X1",1,0)	=IF(A7="X2",1,0)	=IF(A7="X3",1,0)	=IF(A7="X4",1,0)	=IF(A7="X5",1,0)					
8	X3	=IF(A8="X1",1,0)	=IF(A8="X2",1,0)	=IF(A8="X3",1,0)	=IF(A8="X4",1,0)	=IF(A8="X5",1,0)					
9	X1	=IF(A9="X1",1,0)	=IF(A9="X2",1,0)	=IF(A9="X3",1,0)	=IF(A9="X4",1,0)	=IF(A9="X5",1,0)					
10	X3	=IF(A10="X1",1,0)	=IF(A10="X2",1,0)	=IF(A10="X3",1,0)	=IF(A10="X4",1,0)	=IF(A10="X5",1,0)					
11	X5	=IF(A11="X1",1,0)	=IF(A11="X2",1,0)	=IF(A11="X3",1,0)	=IF(A11="X4",1,0)	=IF(A11="X5",1,0)					
12											
13	$n_{Xi} =$	=SUM(B2:B11)	=SUM(C2:C11)	=SUM(D2:D11)	=SUM(E2:E11)	=SUM(F2:F11)					

(b)

Figure 13.52

Worksheet that computes the effective entropy of a data file: (a) worksheet and (b) formulas.

EXAMPLE 13.48 Huffman Code

Figure 13.53 shows a worksheet that illustrates a manual Huffman coding procedure. Column A lists the $m = 5$ symbols that represent the source vocabulary. Column B gives their respective probabilities, which can be either the theoretical values or the effective probabilities computed from the symbol relative frequencies. Columns D and E show the first sort by decreasing probability values, and column F shows the first (least significant) code bit assignment. The second through fourth sorts are shown, along with the code bit assignments. The resulting code table shown in columns P and Q was determined manually by forming the Huffman code tree from the bit assignments.

EXAMPLE 13.49 Pseudo-random number generator

Figure 13.54 shows a worksheet that generates pseudo-random integers and bits used for encryption. The PRNG parameter and seed values are labeled in row 1 and specified in row 2. Column B starts with the seed value ($i = 0$) in row 5 to simplify the PRNG equation that produces X_i that starts in B6. The random bit column C equals 0 if X_i falls in the first half of the interval, $[0, h/2)$, and equals 1 otherwise.

	A	B	C	D	E	F	G	H	I	J	K	L	M	N	O	P	Q	R
1	Original			First Sort				Second Sort				Third Sort				Fourth Sort		
2	X	P_e(X)		X	P_e(X)	code		X	P_e(X)	code		X	P_e(X)	code		X	P_e(X)	code
3	X1	0.2		X5	0.3			X5	0.3			X1-X3	0.4			X2-X4-X5	0.6	1
4	X2	0.1		X1	0.2			X2-X4	0.3			X5	0.3	1		X1-X3	0.4	0
5	X3	0.2		X3	0.2			X1	0.2	1		X2-X4	0.3	0				
6	X4	0.2		X4	0.2	1		X3	0.2	0						Code Table		
7	X5	0.3		X2	0.1	0										X1	01	
8																X2	100	
9																X3	00	
10																X4	101	
11																X5	11	

Figure 13.53

Worksheet that shows the Huffman coding procedure.

	A	B	C	D
1	A=	B=	h=	seed=
2	778456	45232	10000	13
3				
4	i	X_i	Rand Bit	
5	0	13		
6	1	5160	1	
7	2	8192	1	
8	3	6784	1	
9	4	736	0	

(a)

	A	B	C	D
1	A=	B=	h=	seed=
2	778456	45232	10000	13
3				
4	i	X_i		Rand Bit
5	0	=D2		
6	=1+A5	=MOD($A$2*B5+$B$2,$C$2)	=IF(B6<$C$2/2,0,1)	
7	=1+A6	=MOD($A$2*B6+$B$2,$C$2)	=IF(B7<$C$2/2,0,1)	
8	=1+A7	=MOD($A$2*B7+$B$2,$C$2)	=IF(B8<$C$2/2,0,1)	
9	=1+A8	=MOD($A$2*B8+$B$2,$C$2)	=IF(B9<$C$2/2,0,1)	

(b)

Figure 13.54

Worksheet that generates pseudo-random integers and bits: (a) worksheet and (b) formulas.

Encryption   EXAMPLE 13.50

This example illustrates the encryption of a data waveform. Figure 13.55 shows 14 of the 33 samples of a sinusoidal waveform s_i in column B. Column C and the chart show the quantizer levels after an offset and scaling factor were applied to s_i to form values between 0 and 15. The 4-bit data in columns E through H is the binary representation of the corresponding quantizer level.

Figure 13.56 shows the process of generating the random binary sequence RBS that matched the 4-bit sinusoidal data. The first five rows of the RBS are shown. Cell B5 contains the seed of PRNG that generates 33 random integers in B6 to B38 to match the 33 sinusoidal values. Cell C5 contains the 33rd random number in column B (in B38) that is the seed for the random number in column C. Similarly, the 66th number in C38 is the seed for column D, and that in D38 the seed for column E. Each random number forms one random bit in columns G through J. The formulas shown in columns G and H are almost identical to those used in columns I and J except they refer to values in columns D and E respectively.

Figure 13.57 shows the encryption of the 4-bit sinusoidal data by performing the bit-by-bit Exclusive-OR function of the 4-bit data bits with those of the RBS. The data and RBS were copied and paste/special/values into their respective columns to make the Exclusive-OR operation simple to accomplish. The ExOR logical function was implemented with Excel logical functions. For example, K2 contains

```
=IF(OR(AND(A2,NOT(F2)),AND(NOT(A2),F2)),1,0)
```

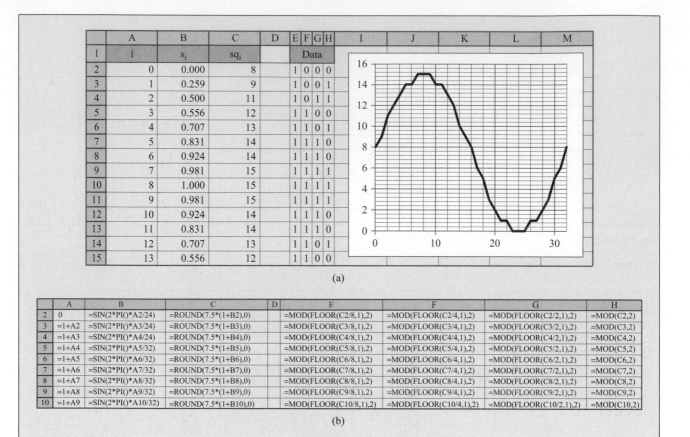

Figure 13.55

Worksheet that generates 4-bit data that encode a sinusoidal waveform: (a) worksheet and (b) formulas.

	A	B	C	D	E	F	G	H
2	0	=SIN(2*PI()*A2/24)	=ROUND(7.5*(1+B2),0)		=MOD(FLOOR(C2/8,1),2)	=MOD(FLOOR(C2/4,1),2)	=MOD(FLOOR(C2/2,1),2)	=MOD(C2,2)
3	=1+A2	=SIN(2*PI()*A3/24)	=ROUND(7.5*(1+B3),0)		=MOD(FLOOR(C3/8,1),2)	=MOD(FLOOR(C3/4,1),2)	=MOD(FLOOR(C3/2,1),2)	=MOD(C3,2)
4	=1+A3	=SIN(2*PI()*A4/24)	=ROUND(7.5*(1+B4),0)		=MOD(FLOOR(C4/8,1),2)	=MOD(FLOOR(C4/4,1),2)	=MOD(FLOOR(C4/2,1),2)	=MOD(C4,2)
5	=1+A4	=SIN(2*PI()*A5/32)	=ROUND(7.5*(1+B5),0)		=MOD(FLOOR(C5/8,1),2)	=MOD(FLOOR(C5/4,1),2)	=MOD(FLOOR(C5/2,1),2)	=MOD(C5,2)
6	=1+A5	=SIN(2*PI()*A6/32)	=ROUND(7.5*(1+B6),0)		=MOD(FLOOR(C6/8,1),2)	=MOD(FLOOR(C6/4,1),2)	=MOD(FLOOR(C6/2,1),2)	=MOD(C6,2)
7	=1+A6	=SIN(2*PI()*A7/32)	=ROUND(7.5*(1+B7),0)		=MOD(FLOOR(C7/8,1),2)	=MOD(FLOOR(C7/4,1),2)	=MOD(FLOOR(C7/2,1),2)	=MOD(C7,2)
8	=1+A7	=SIN(2*PI()*A8/32)	=ROUND(7.5*(1+B8),0)		=MOD(FLOOR(C8/8,1),2)	=MOD(FLOOR(C8/4,1),2)	=MOD(FLOOR(C8/2,1),2)	=MOD(C8,2)
9	=1+A8	=SIN(2*PI()*A9/32)	=ROUND(7.5*(1+B9),0)		=MOD(FLOOR(C9/8,1),2)	=MOD(FLOOR(C9/4,1),2)	=MOD(FLOOR(C9/2,1),2)	=MOD(C9,2)
10	=1+A9	=SIN(2*PI()*A10/32)	=ROUND(7.5*(1+B10),0)		=MOD(FLOOR(C10/8,1),2)	=MOD(FLOOR(C10/4,1),2)	=MOD(FLOOR(C10/2,1),2)	=MOD(C10,2)

(b)

Figure 13.56

Worksheet that generates pseudo-random binary sequence: (a) worksheet and (b) formulas.

Figure 13.56 (a):

	A	B	C	D	E	F	G	H	I	J
1	A=	B=	h=	seed=						
2	778456	45232	10000	13						
3										
4	i		Rand Int					RBS		
5	0	13	2944	1040	9776					
6	1	5160	9696	9472	1088		1	1	1	0
7	2	8192	4608	464	5360		1	0	0	1
8	3	6784	480	8816	9392		1	0	1	1
9	4	736	4112	3328	3984		0	0	0	0
10	5	8848	6304	6800	3936		1	1	1	0

(a)

Figure 13.56 (b):

	A	B	C	D	E	F	G	H	I	J
1	A=	B=		h=		seed=				
2	778456	45232		10000		13				
3										
4	i			Rand Int					RBS	
5	0	=D2		=B38		=C38		=D38		
6	=1+A5	=MOD($A$2*B5+$B$2,$C$2)		=MOD($A$2*C5+$B$2,$C$2)	=MOD($A$2*D5+	=MOD($A$2*E5+$	=IF(B6<$C$2/2,0,1)	=IF(C6<$C$2/2,0,1)	=IF(D	=IF(E
7	=1+A6	=MOD($A$2*B6+$B$2,$C$2)		=MOD($A$2*C6+$B$2,$C$2)	=MOD($A$2*D6+	=MOD($A$2*E6+$	=IF(B7<$C$2/2,0,1)	=IF(C7<$C$2/2,0,1)	=IF(D	=IF(E
8	=1+A7	=MOD($A$2*B7+$B$2,$C$2)		=MOD($A$2*C7+$B$2,$C$2)	=MOD($A$2*D7+	=MOD($A$2*E7+$	=IF(B8<$C$2/2,0,1)	=IF(C8<$C$2/2,0,1)	=IF(D	=IF(E
9	=1+A8	=MOD($A$2*B8+$B$2,$C$2)		=MOD($A$2*C8+$B$2,$C$2)	=MOD($A$2*D8+	=MOD($A$2*E8+$	=IF(B9<$C$2/2,0,1)	=IF(C9<$C$2/2,0,1)	=IF(D	=IF(E
10	=1+A9	=MOD($A$2*B9+$B$2,$C$2)		=MOD($A$2*C9+$B$2,$C$2)	=MOD($A$2*D9+	=MOD($A$2*E9+$	=IF(B10<$C$2/2,0,1)	=IF(C10<$C$2/2,0,1)	=IF(D	=IF(E

(b)

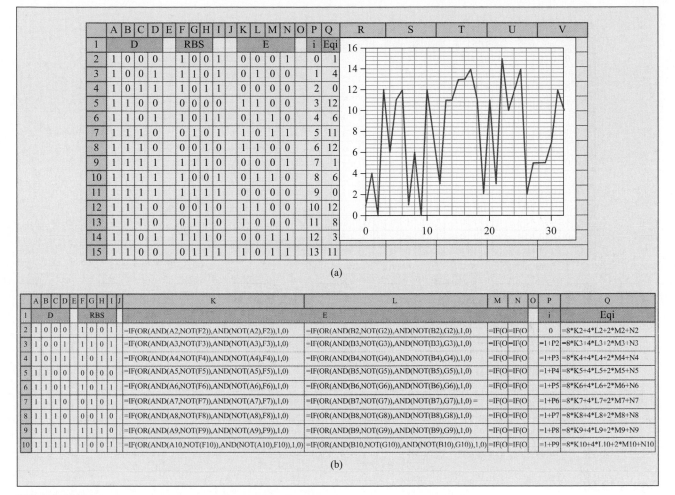

Figure 13.57

Worksheet that encrypts the data by performing the bit-by-bit Exclusive-OR of the data with the RBS: (a) worksheet and (b) formulas.

to form the encrypted bits in columns K through N. Column Q formed the quantizer levels corresponding to the encrypted bits and these are plotted in the chart to illustrate the random appearance of the encrypted sinusoidal waveform.

Figure 13.58 shows the decryption of the encrypted 4-bit sinusoidal data by performing the bit-by-bit Exclusive-OR function of the 4-bit encrypted data bits with those of the same RBS used in the encryption. The ExOR logical function was implemented with Excel logical functions to form the decrypted bits in Columns K through N. Column Q form the quantizer levels corresponding to the decrypted bits and these are plotted in the chart to show they agree with the original sinusoidal waveform.

Secure key transmission   EXAMPLE 13.51

Figure 13.59 shows the almost-secure transmission of the encryption key that is used as the seed in the PRNG. Because the successful modulus operations require the precision in the least significant digits, the process fails when Excel begins to represent numbers in scientific notation. The successful process results with values in D16 and F16 being equal.

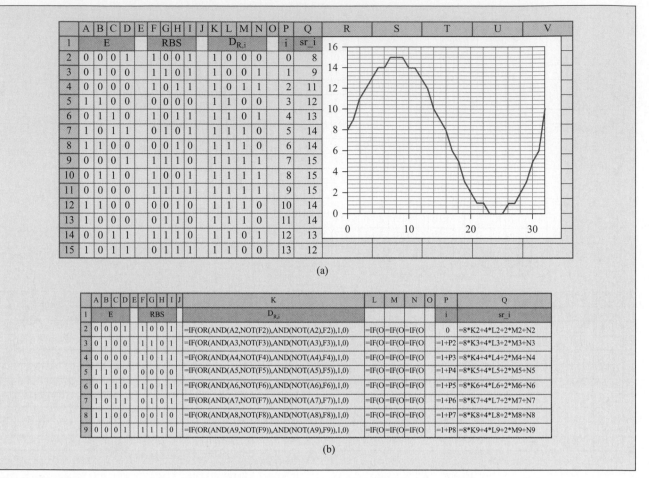

Figure 13.58

Worksheet that decrypts the data by performing the bit-by-bit Exclusive-OR of the data with the RBS: (a) worksheet and (b) formulas.

13.15.10 Channel Coding

EXAMPLE 13.52

Bit-level error correction

Figure 13.60 shows worksheet that corrects single errors when a data bit is repeated three times by forming a truth table that has the input section formed by logic values DA_r, DB_r and DC_r. A simple Excel formula determines the correct value D_r under the single error assumption.

EXAMPLE 13.53

Block data error correction

This example illustrates block error correction in the following steps:

STEP 1. Figure 13.61 generates a random data block having 4 rows and 4 columns. The data values change each time F9 is pressed. To form a constant data block that does not change when Excel recalculates the worksheet, the block is copied and paste/special/values are used to copy these values into the transmitted data worksheet.

	A	B	C	D	E	F
1	a	N		T		R
2	11	21				
3				x		y
4				9		11
5						
6				a^x		a^y
7				2357947691		285311670611
8						
9				X		Y
10				8		2
11						
12				Y^x		X^y
13				512		8589934592
14						
15				K_T		K_R
16				8		8

(a)

	A	B	C	D	E	F
1	a	N		T		R
2	11	21				
3				x		y
4				9		11
5						
6				a^x		a^y
7				=A2^D4		=A2^F4
8						
9				X		Y
10				=MOD(D7,B2)		=MOD(F7,B2)
11						
12				Y^x		X^y
13				=F10^D4		=D10^F4
14						
15				K_T		K_R
16				=MOD(F10^D4,B2)		=MOD(D10^F4,B2)

(b)

Figure 13.59

Worksheet that shows the almost-secure transmission of the encryption key: (a) worksheet and (b) formulas.

	A	B	C	D	E
1	DA_r	DA_r	DA_r		DT_r
2	0	0	0		0
3	0	0	1		0
4	0	1	0		0
5	0	1	1		1
6	1	0	0		0
7	1	0	1		1
8	1	1	0		1
9	1	1	1		1

(a)

	A	B	C	D	E
1	DA_r	DA_r	DA_r		DT_r
2	0	0	0		=IF(SUM(A2:C2)<2,0,1)
3	0	0	1		=IF(SUM(A3:C3)<2,0,1)
4	0	1	0		=IF(SUM(A4:C4)<2,0,1)
5	0	1	1		=IF(SUM(A5:C5)<2,0,1)
6	1	0	0		=IF(SUM(A6:C6)<2,0,1)
7	1	0	1		=IF(SUM(A7:C7)<2,0,1)
8	1	1	0		=IF(SUM(A8:C8)<2,0,1)
9	1	1	1		=IF(SUM(A9:C9)<2,0,1)

(a)   (b)

Figure 13.60

Worksheet that corrects single errors when data bit is repeated three times: (a) worksheet and (b) formulas.

Figure 13.61

Worksheet that generates a random 4 × 4 data block: (a) worksheet and (b) formulas.

	A	B	C	D
1	0	1	0	1
2	0	0	0	0
3	0	0	0	1
4	1	0	0	1

(a)

	A	B	C	D
1	=IF(RAND()<0.5,0,1)	=IF(RAND()<0.5,0,1)	=IF(RAND()<0.5,0,1)	=IF(RAND()<0.5,0,1)
2	=IF(RAND()<0.5,0,1)	=IF(RAND()<0.5,0,1)	=IF(RAND()<0.5,0,1)	=IF(RAND()<0.5,0,1)
3	=IF(RAND()<0.5,0,1)	=IF(RAND()<0.5,0,1)	=IF(RAND()<0.5,0,1)	=IF(RAND()<0.5,0,1)
4	=IF(RAND()<0.5,0,1)	=IF(RAND()<0.5,0,1)	=IF(RAND()<0.5,0,1)	=IF(RAND()<0.5,0,1)

(b)

STEP 2. Figure 13.62 shows unchanging data block that was formed by copy/paste/values into the data block data in B2:E5. Odd-parity of the columns PC was computed in the Row 6 below data. Odd-parity of the rows was computed PR in the column F next to the data.

STEP 3. Figure 13.63 shows the received data block formed by paste/special/values of the transmitted data block and by forming one error by switching one of the transmitted data bits. Odd-parity of the received data columns PC' was computed in the Row 7 below data. Odd-parity of the received data rows was computed PR' in the Column G.

Conditional statements compare the parity bits computed by the receiver with those transmitted along with the data block and indicate the row and column where differences occur. The cell that is at that row and column contains the error.

Row 8 and column H include conditional statements that display text indications of the column and row that contain the error.

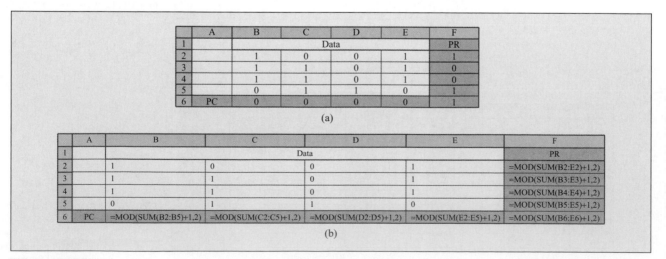

	A	B	C	D	E	F
1			Data			PR
2		1	0	0	1	1
3		1	1	0	1	0
4		1	1	0	1	0
5		0	1	1	0	1
6	PC	0	0	0	0	1

(a)

	A	B	C	D	E	F
1			Data			PR
2		1	0	0	1	=MOD(SUM(B2:E2)+1,2)
3		1	1	0	1	=MOD(SUM(B3:E3)+1,2)
4		1	1	0	1	=MOD(SUM(B4:E4)+1,2)
5		0	1	1	0	=MOD(SUM(B5:E5)+1,2)
6	PC	=MOD(SUM(B2:B5)+1,2)	=MOD(SUM(C2:C5)+1,2)	=MOD(SUM(D2:D5)+1,2)	=MOD(SUM(E2:E5)+1,2)	=MOD(SUM(B6:E6)+1,2)

(b)

Figure 13.62

Transmitted data worksheet has non-changing data and computes odd parity of data first in columns (PC) and then rows (PR): (a) worksheet and (b) formulas.

	A	B	C	D	E	F	G	H
1		Data				PC	PC'	
2		1	0	0	1	1	1	
3		1	1	1	1	0	1	<
4		1	1	0	1	0	0	
5		0	1	1	0	1	1	
6	PR	0	0	0	0	1		
7	PR'	0	0	1	0	1		
8				^				

(a)

	A	B	C	D	E	F	G	H
1		Data				PC	PC'	
2		1	0	0	1	1	=MOD(SUM(B2:E2)+1,2)	=IF[F2=G2," ","<"]
3		1	1	1	1	0	=MOD(SUM(B3:E3)+1,2)	=IF[F3=G3," ","<"]
4		1	1	0	1	0	=MOD(SUM(B4:E4)+1,2)	=IF[F4=G4," ","<"]
5		0	1	1	0	1	=MOD(SUM(B5:E5)+1,2)	=IT[F5=G5, ,<]
6	PR	0	0	0	0	1		
7	PR'	=MOD(SUM(B2:B5)+1,2)	=MOD(SUM(C2:C5)+1,2)	=MOD(SUM(D2:D5)+1,2)	=MOD(SUM(E2:E5)+1,2)	=MOD(SUM(F2:F5)+1,2)		
8		=IF(D6=B7," "," ")	=IF(C6=C7," ","^")	=IF(D6=D7," ","^")	=IF(E6=E7," ","^")	=IF(F6=F7," ","^")		

(b)

Figure 13.63

Received data worksheet has computes odd parity of received data first in columns (PC') and then rows (PR'). (a) Worksheet. (b) Formulas.

Reduction of channel capacity with range   EXAMPLE 13.54

This example computes and plots the reduction in the channel capacity with range due to the decrease in the detected signal energy dictated by the inverse square law. Let the signal energy at your phone equal $\mathcal{E}_s = 10^6 \sigma_N^2$, measured at $r_o = 0.5$ km from the antenna tower. The signal energy decays according to the inverse square law as

$$\mathcal{E}_s(\alpha r_o) = \frac{\mathcal{E}_s(r_o)}{\alpha^2}$$

The channel with bandwidth $B = 10^4$ Hz has the capacity at range $r = \alpha r_o$ computed as

$$C(\alpha r_o) = B\log_2\left(1 + \frac{\mathcal{E}_s(\alpha r_o)}{\sigma_N^2}\right) = B\log_2\left(1 + \frac{\mathcal{E}_s(r_o)}{\alpha^2 \sigma_N^2}\right) \text{ bps}$$

Figure 13.64 computes $\mathcal{E}_s(r)/\sigma_N^2$ and displays $C(r)$ as a function of range. The range values are increased by a factor of $\sqrt{2}$ to form an interesting set.

13.15.11 Data Networks

Data Packet Collisions   EXAMPLE 13.55

Figure 13.65 shows an Excel simulation of data packet collisions to compute the collision probability as a function of packet duration. The worksheet computes the average data throughput for a single user for non-colliding data packets.

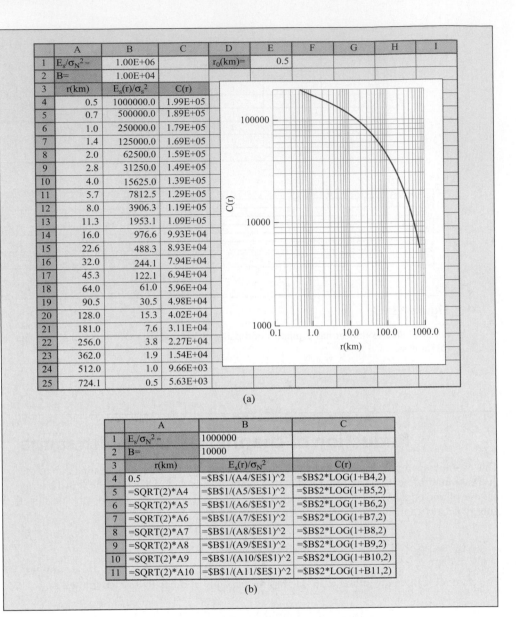

Figure 13.64

Worksheet that plots the capacity as a function of range: (a) worksheet and (b) formulas.

Data packets—P1 from user 1 and P2 from user 2—have the same duration that is specified in E2. P1 starts at a random time during the transmission interval determined in B1 and P2 starts at a random time determined in B3. The P1 and P2 packets are displayed in respective columns B and C by contiguous sets of 1's having duration specified in E2 that start at their respective start times. These are highlighted for visualization. The product of P1 and P2 is computed in column D, which will be greater than zero only when the packets overlap (the collision condition). A collision is detected in E3 by a summing D7:D27 and finding a result greater than zero.

VBA Macro *transmit* tallies the count of transmissions and collision to estimate the collision probability. It also computes the throughput in G9 by accumulating user 1's number of data bits (number of 1's in P1) in packets that do not collide. The average number of bits that are successfully transmitted is computed in G10 as the ratio of the data throughput divided by the number of transmissions.

```
Sub transmit()
If Range("E3").Value = 1 Then    ' First check if error
    Range("G7").Value = Range("G7").Value + 1 ' increment collision count
Else
    Range("G9").Value = Range("G9").Value + 10 * Range("E2").Value ' increment throughput
End If
Range("G6").Value = Range("G6").Value + 1 ' increment xmits and gen new P's
End Sub
-------
Sub reset()
Range("G6").Value = 0
Range("G7").Value = 0
Range("G9").Value = 0
End Sub
```

(a) VBA code

	A	B	C	D	E	F	G
1	P1 start=	0.70					
2	P1 end=	1.20		duration=	0.5		
3	P2 start=	0.30		collision?	1		
4	P2 end=	0.80					
5							
6	time (s)	P1	P2	P1 * P2		transmits=	58
7	0.0	0	0	0		collisions=	19
8	0.1	0	0	0		collision prob=	0.33
9	0.2	0	0	0		data throughput=	195
10	0.3	0	1	0		data/transmit=	3.36
11	0.4	0	1	0			
12	0.5	0	1	0			
13	0.6	0	1	0			
14	0.7	1	1	1.1			
15	0.8	1	0	0		Transmit	
16	0.9	1	0	0			
17	1.0	1	0	0			
18	1.1	1	0	0			
19	1.2	0	0	0			
20	1.3	0	0	0			
21	1.4	0	0	0			
22	1.5	0	0	0		RESET	
23	1.6	0	0	0			
24	1.7	0	0	0			
25	1.8	0	0	0			
26	1.9	0	0	0			
27	2.0	0	0	0			

(b) Worksheet

Figure 13.65

Worksheet and Macros that simulate data packet collisions to compute the collision probability as a function of packet duration: (a) worksheet and (b) formulas. (*Continued on next page*)

	A	B	C	D	E	F	G
1	P1 start=	=FLOOR(10*(2-$E$2)*RAND(),1)/10					
2	P1 end=	=B1+E2		duration=	0.5		
3	P2 start=	=FLOOR(10*(2-$E$2)*RAND(),1)/10		collision?	=IF(SUM(D7:D27)>0,1,0)		
4	P2 end=	=B3+E2					
5							
6	time (s)	P1	P2	P1 * P2		transmits=	58
7	0	=IF(AND(A7>=$B$1,A7<$B$2),1,0)	=IF(AND(A7>=$B$3,A7<$B$4),1,0)	=1.1*B7*C7		collisions=	19
8	=A7+0.1	=IF(AND(A8>=$B$1,A8<$B$2),1,0)	=IF(AND(A8>=$B$3,A8<$B$4),1,0)	=1.1*B8*C8		collision prob=	=IF(G6>0,G7/G6,0)
9	=A8+0.1	=IF(AND(A9>=$B$1,A9<$B$2),1,0)	=IF(AND(A9>=$B$3,A9<$B$4),1,0)	=1.1*B9*C9		data throughput=	195
10	=A9+0.1	=IF(AND(A10>=$B$1,A10<$B$2),1,0)	=IF(AND(A10>=$B$3,A10<$B$4),1,0)	=1.1*B10*C10		data/transmit=	=IF(G6>0,G9/G6,0)

(c) Formulas

Figure 13.65

(*Continued*)

13.15.12 Symbology

EXAMPLE 13.56   Self-clocking waveform

This example generates a self-clocking waveform that is typical of that produced by a credit card magnetic stripe. Let $T_p = 8$, that is, eight samples per "0" period in a credit card code waveform w_i. Figure 13.66 shows the waveform that encodes the 2-bit sequence 11 and starts and ends with two additional 0-valued bits, for a total of 6 bits (001100). The waveform states at a high level, making $w_0 = 1$.

Figure 13.66 shows the rows 1 to 15 of the worksheet, where the w_i values were entered manually. With $T_p = 8$ encoding each of the 6 data bits, the waveform w_i has 48 samples, for $0 \leq i \leq 47$.

A 0 bit maintains the same value over T_p samples. With $w_0 = 1$, the 0-bit value makes the first eight values $w_i = 1$ for $0 \leq i \leq 7$. A $1 \to 0$ transition occurs at $i = 8$ and the 0-bit value makes the next eight values $w_i = 0$ for $8 \leq i \leq 15$. A $0 \to 1$ transition occurs at $i = 16$ and the 1-bit value makes the next four values, $w_i = 1$ for $16 \leq i \leq 19$, and the next four values, $w_i = 0$ for $20 \leq i \leq 23$. The chart shows the values for w_i, for $0 \leq i \leq 47$.

Figure 13.66

Worksheet that generates a self-clocking waveform that encodes binary values 001100.

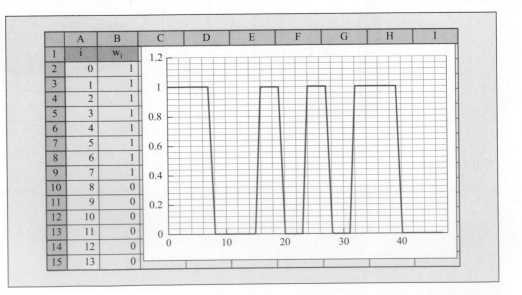

13.16 Summary

By the time you reach this point, you should have achieved the following goals;

1. A good grasp of Excel and VBA

2. Knowledge of the common approaches to analyze data

3. An ability to simulate an entire system and compute a measure of its capabilities

4. An understanding of how to present your results effectively

All three aspects are important. Probably the biggest disappointment I can have with work presented by a student occurs when (s)he has an insightful approach to data analysis, yet produces a final product that looks sloppy, disorganized, and leaves the impression that the student does not understand the problem. I prefer an approach that is well-organized and, while initially falling short of the solution, provides indications where the difficulties lie and the corrective actions that need to take place.

α—(Greek alpha) range factor or parameter in a PRNG.

β—(Greek beta) parameter in a PRNG.

δ—(Greek lower-case delta) histogram bin width.

Δ—(Greek capital delta) quantizer step size.

Δ_{PWM}—voltage step size produced by PWM waveform.

μ—(Greek mu) one millionth.

Ω—(Greek sigma) Ohm, unit of resistance.

σ—(Greek sigma) statistical standard deviation (SD).

σ^2—statistical variance.

σ_N—noise SD.

σ_N^2—noise variance.

θ—(Greek theta) angle with respect to x axis in complex plane.

A—Ampere (Amp), unit of electrical current.

ACK—acknowledge packet sent by destination in TCP/IP.

ADC—analog-to-digital converter.

ASCII—American Standards Committee for Information Interchange.

Asynchronous transmission—randomly-timed.

ATM—asynchronous transfer mode, or automatic teller machine.

b—bit (binary digit).

B—byte (8-bit data unit).

$\mathcal{B}$—bandwidth in channel capacity equation.

Bandwidth—interval of frequencies present in a waveform, or measure of frequency transmission quality of a channel.

BCD—binary coded decimal, 4-bit code words used integers 0-9.

Bit—binary digit, a 0 or 1.

Bps—bytes per second.

bps—bits per second.

Byte—8-bit data unit.

C—clear input of T-FF.

$\mathcal{C}$—channel capacity.

CAPCHA—machine-unreadable figure.

CD—compact disk used for storing data in a digital format.

CDMA—code division multiple access.

Character—transmission waveform enclosing a code word.

Code word—set of bits that represents a symbol.

Combinational logic—logic whose output depends only on the current input values.

Combinatorial logic—another term for combinational logic.

CPU—central processing unit in a computer.

CRC—cyclic redundancy check—Internet code word used for data packet error detection.

CSD—check sum digit, used for error correction.

$\mathcal{D}_S$—source data rate.

$\mathcal{D}_V$—video source data rate.

DAC—digital-to-analog converter.

Data packet—collection of code words transmitted as a unit.

dB—decibel, equal to ten times the base-10 logarithm of a power ratio.

Dibit—data unit formed by two bits.

$\mathcal{E}$—electrical energy.

$\mathcal{E}_{min}$—minimum signal energy for reliable data communication.

$\mathcal{E}_s$—signal energy.

EEPROM—electrically erasable programmable read-only memory, used to store digital data in thumb drives and smartphones.

Encryption—data coding that permits access by only intended recipients.

EOL—end-of-line character.

Ethernet—communication channel over dedicated cable.

ExOR—exclusive-OR logic gate or operation.

f_a—alias frequency, caused by sampling waveforms too coarsely in time.

FDMA—frequency division multiple access—form of orthogonal signals.

FET—field-effect transistor.

f_o—frequency of analog sinusoidal waveform.

f_s—sampling frequency.

giga-—prefix meaning one billion, or 10^9.

GHz—gigahertz, or 10^9 Hz.

GPS—global positioning system.

$\hat{\mathcal{H}}_f$—effective file entropy, number of bits that are needed to encode a file.

$\mathcal{H}_s$—source entropy in units of bits/symbol.

$\hat{\mathcal{H}}_s$—effective entropy computed using effective probabilities.

Hz—Hertz, or cycles per second.

i—index of time-domain samples.

I/O—input/output, or a connection that can both receive and transmit data.

IC—integrated circuit.

IR—infrared.

JPEG—Joint Photographers Experts Group.

jpg—image file using JPEG standard.

kB—kilobyte, or $2^{10} = 1,024$ bytes.

kbps—kilobits per second.

kilo-—prefix meaning one thousand.

kHz—kilohertz, or 10^3 Hz.

km—kilometer, or 10^3 m.

LAN—local-area network.

LCD—liquid crystal display.

LED—light-emitting diode.

LRC—longitudinal redundancy check—code word appended data packet for error correction.

LSB—least significant bit, usually the rightmost bit in a sequence.

LTE—long term evolution.

m—meter or number of symbols in a source vocabulary.

magnitude—absolute value of amplitude.

MB—Megabyte, or 2^{20} bytes.

Mbps—Megabits per second (a data rate).

MCU—microcontroller unit.

mega-—prefix meaning one million, as in 1 megaHertz (1 MHz).

MHz—megahertz, or 10^6 Hz.

micro-—prefix meaning one one-millionth, abbreviated μ, as in 1 microsecond (1 μs).

milli-—prefix meaning one one-thousandth, as in 1 millisecond (1 ms).

mm—millimeter, or 10^{-3} m.

MPEG—Motion Picture Experts Group.

ms—milliseconds.

MSB—most significant bit, usually the leftmost bit in a code word.

MUX—digital multiplexer.

mV—millivolt, or 10^{-3} V.

mW—milliwatt, or 10^{-3} W.

n—number of countable objects.

n_d—number of non-zero samples in TDMA window of size n_x.

n_t—number of trials in a simulation.

n_T—number of symbols produced by a source.

n_x—number of sample points in a time sequence.

N—random noise value.

N_i—sample of random noise.

n_{steps}—number of steps in a quantizer.

nano-—prefix meaning one one-thousand-millionth, or 10^{-9}, as in 1 nanosecond (1 ns).

ns—nanosecond, or 10^{-9} s.

Outcome—the result of performing an experiment.

$P[X]$—theoretical probability of symbol X.

$P_e[X]$—effective probability of symbol X determined from its relative frequency of occurrence.

Parity—forming an even or odd count of 1's in a data character included for error detection.

PDF—probability density function.

pico-—prefix meaning one one-million-millionth, abbreviated p, as in 1 picosecond (1 ps).

pixel—picture element.

PMF—probability mass function.

PRNG—pseudo-random number generator.

protocol—set of rules for data transfer.

pseudo-random numbers—random numbers generated by an algorithm.

PWM—pulse width modulation.

Q—flip-flop output.

$\overline{Q}$—complement of Q.

QR symbol—quick response symbol.

Quantizer—device that converts infinite-precision values into discrete (finite-precision) values.

$\mathcal{R}$—source transmission rate measured in symbols/second.

RAM—random access memory.

Random—unpredictable and obeying the laws of probability.

Redundant—computed from available data.

r_o—range at which a measurement is taken (calibration range).

ROM—read-only memory.

RMS—root-mean-square value (square root of the average of the squared values).

RS-FF—reset-set flip-flop.

RV—random variable.

s—second.

s_i—time signal value (non-random).

SEA—single error assumption.

Sequential logic—logic whose output depends on the current inputs and past outputs.

Signal—informational time waveform.

SNR—signal-to-noise ratio, typically expressed as signal energy divided by noise variance.

Start bit—first bit in asynchronous transmission character that indicates its beginning.

Stop bit—last bit or bits in asynchronous transmission that separate characters.

Symbol—informational unit generated by a source.

T_B—time interval representing a single bit in a data waveform.

T_C—data signal waveform duration.

TCP/IP—Transmission control protocol/Internet protocol.

TDMA—time division multiple access.

tera-—prefix meaning one thousand billion, or 10^{12}, as in teraherz 10^{12} Hz.

T-FF-—toggle flip-flop.

THz—(teraherz) 10^{12} Hz.

T-FF—toggle flip-flop.

T_m—time midpoint ($= T_p/2$) in self-clocking code.

TOF—time-of-flight in a sonar or radar ranging system.

T_{OFF}—time duration when PWM waveform is off.

T_{ON}—time duration when PWM waveform is on.

T_P—time period of a zero bit in self-clocking code.

T_{PWM}—time period of PWM waveform.

T_s—sampling period in ADC.

T_x—duration of pulse caused by laser passing over UPC black bar.

TTL—time-to-live counter in an Internet data packet.

UDP/IP—Universal datagram protocol/Internet protocol.

UPC—Universal product code.

USPS—United States Postal Service.

V—volt, unit of electrical potential.

V_{ave}—average of waveform produced by PWM.

VOIP—voice over Internet protocol.

V_{min}—minimum voltage in a quantizer.

V_{max}—maximum voltage in a quantizer or maximum voltage allowed for a signal.

V_S—supply voltage that powers a circuit.

W—watt, unit of power.

WAN—wide-area network.

Wi-Fi—wireless network.

WWW—World Wide Web.

X_{RMS}—value measured with RMS meter.